SESSILE ANIMALS OF THE SEA SHORE

SESSILE ANIMALS OF THE SEA SHORE

Vernon A. Harris

BSc, PhD (London)

Formerly Senior Lecturer in Zoology
Australian National University
Canberra

CHAPMAN AND HALL

LONDON · NEW YORK · TOKYO · MELBOURNE · MADRAS

UK	Chapman and Hall, 11 New Fetter Lane, London EC4P 4EE
USA	Chapman and Hall, 29 West 35th Street, New York NY10001
JAPAN	Chapman and Hall Japan, Thomson Publishing Japan, Hirakawacho Nemoto Building, 7F, 1-7-11 Hirakawa-cho, Chiyoda-ku, Tokyo 102
AUSTRALIA	Chapman and Hall Australia, Thomas Nelson Australia, 480 La Trobe Street, PO Box 4725, Melbourne 3000
INDIA	Chapman and Hall India, R. Sheshadri, 32 Second Main Road, CIT East, Madras 600 035

First edition 1990

Typeset in 10/12pt Sabon by EJS Chemical Composition, Bath, Avon
Printed in Great Britain by TJ Press (Padstow) Ltd, Padstow, Cornwall

ISBN 0 412 33760 6

British Library Cataloguing in Publication Data

Harris, Vernon A.
Sessile animals of the sea shore.
1. Seashore. Invertebrates
I. Title
592.0909′46

ISBN 0–412–33760–6

Library of Congress Cataloguing-in-Publication Data

Harris, Vernon A.
Sessile animals of the sea shore/Vernon A. Harris.
p. cm.
Includes bibliographical references.
ISBN 0–412–33760–6
1. Marine invertebrates. 2. Animal colonies.
3. Seashore biology. I. Title
QL121.H35 1990
592.092—dc20 89–23914
CIP

Contents

Preface

The intertidal zone offers an excellent opportunity to study the interaction between biological systems and the environment because, although conditions created by the rise and fall of the tide are potentially lethal to marine animals, many species have evolved structural and physiological specializations which enable them to avoid or withstand such adverse conditions.

Nowhere are these adaptations more clearly demonstrated than among the sessile animals of rocky shores. Many radically different animal groups have succeeded in colonizing the shore, and the various ways in which they have achieved this makes a fascinating study. Here the subject will be restricted to animals living permanently attached to the substratum on temperate rocky shores between tide marks or in the accessible sublittoral (the infralittoral fringe). The following account attempts to portray the intricate interplay of structural, physiological and behavioural adaptations that enable one animal to live where its congeners cannot.

After the physical and biotic conditions on the shore are outlined, each of the successful littoral groups is described in the order in which they appear, starting with those at the highest level and progressing to the sublittoral. Only such of their structures and physiological modifications as are necessary to understand how they are able to live on the shore are described; however, other aspects of more general biological interest are also discussed where relevant. Adaptations are the outcome of selective pressures from both the physical and the biotic aspects of the environment, and some thought is given to the nature of the forces that have brought about their evolution. Aspects of propagation, dispersal, colonization and interaction between species are discussed in relation to community composition and distribution of animals on the shore.

Comparison between living animals in their natural habitat in different parts of the world can be both instructive and stimulating. My own experiences in Britain, West Africa and Australia have not only broadened my perception and understanding of littoral species, but the subtleties that have arisen in comparing closely-related animals have thrown new light on many problems. The views expressed here are a personal interpretation of my observations and of the published work of other workers. My views may differ in some places from conventionally held ideas.

I have not considered it illogical or inappropriate to present a mixture of species from the Northern and Southern Hemispheres, for by so doing, the more familiar species are supplemented. The introduction of new examples focuses our attention on features or problems that might otherwise be overlooked through familiarity. Australian readers will find relevant information concerning animals familiar to them. The illustrations are nearly all based on my own collection of specimens and histological material and are designed to supplement and interpret the text.

This book would not have been completed without the help and encouragement of the publishers and the constructive criticisms of their editorial staff and their assessors. I am also indebted to my wife for her continual help and forbearance throughout the writing of this book.

V.A.H.
Canberra
1990

1

Sessile animals of the sea shore

1.1 THE SHORE COMMUNITY

The sea shore is the meeting-place of land and sea where tidal movements create a strip of land that is exposed as the tide recedes and covered as it rises. Below this there is a region where wave action profoundly influences life, and this region will be included in the following account of the sea shore.

In comparison with the relatively stable conditions for life in the open sea, this intertidal strip or littoral zone is an inhospitable place with wide and unpredictable fluctuations in environmental conditions; yet a surprisingly large number of animals, representing nearly all the phyla, can be found there.

The resident littoral community is overwhelmingly biased towards marine species adapted to withstand the force of waves and tolerate periodical exposure to the atmosphere (Table 1.1(a)). A few species of terrestrial origin live there (Table 1.1(b)), and some non-residents invade the tidal zone in search of food or for reproductive purposes at certain phases of the tide (Table 1.1(c)), but neither of the latter groups will be considered.

Littoral animals can be divided into two broad categories: free-living and sessile species. The former can move about to select optimal conditions, avoid dangerous environmental changes, escape enemies and search for food or suitable reproductive partners, but sessile animals are denied all of these abilities. They are permanently fixed in one place and rely entirely upon their capacity to withstand dangerous changes in the environment to survive.

The following account deals with sessile animals that can be found attached to rocks or other firm substrata while the tide is out, or in the shallow water below low tide level. The main groups are depicted in Fig. 1.1(a) and (b). A wide range of adaptations have evolved enabling these animals to live on the shore; these will be discussed together with points of general biological interest.

Table 1.1 Animals found on rocky shores

Phylum	*Free-living animals*	*Sessile animals*	*Chapter* *
	(a) *Common animals living on rocky shores*		
Protozoa	foraminiferans, flagellates, ciliates	*Zoothamnion*	*
Porifera		sponges	12
Cnidaria		sea anemones, corals	10
		hydroids	11
Platyhelminthes	turbellarian flatworms		*
Nemertini	boot-lace worms		*
Entoproct polyzoa		*Pedicellana*	*
Annelida	errant polychaetes, sedentary polychaetes		*
		fan worms,	7
		serpulid worms	7
Sipunculida	*Phascolosoma*		*
Crustacea	copepods, ostracods, tanaids, amphipods,		*
	isopods, crabs		*
		barnacles	5
Chelicerata	pycnogonids		*
Mollusca	chitons, gastropods, cephalopods		*
		mussels, oysters	6
Brachiopoda		lamp-shells	*
Bryozoa		sea mats	9
Phoronida			*
Echinodermata	sea stars, urchins, sea cucumbers		*
Chordata (Tunicata)		sea squirts	8
	(b) *Animals belonging to terrestrial groups*		*
	woodlice		
	millipedes		
	insects (bristle-tails, earwigs, caddis flies, beetles, kelp flies)		
	marine mites, spiders		
	pulmonate molluscs		
	(c) *Visitors to the shore*		*
	wading birds, gulls (feed whilst the tide is out)		
	sea snakes, turtles (lay eggs or give birth whilst tide is out)		
	various fish (whilst tide is in)		

* Denotes groups not dealt with in this book.

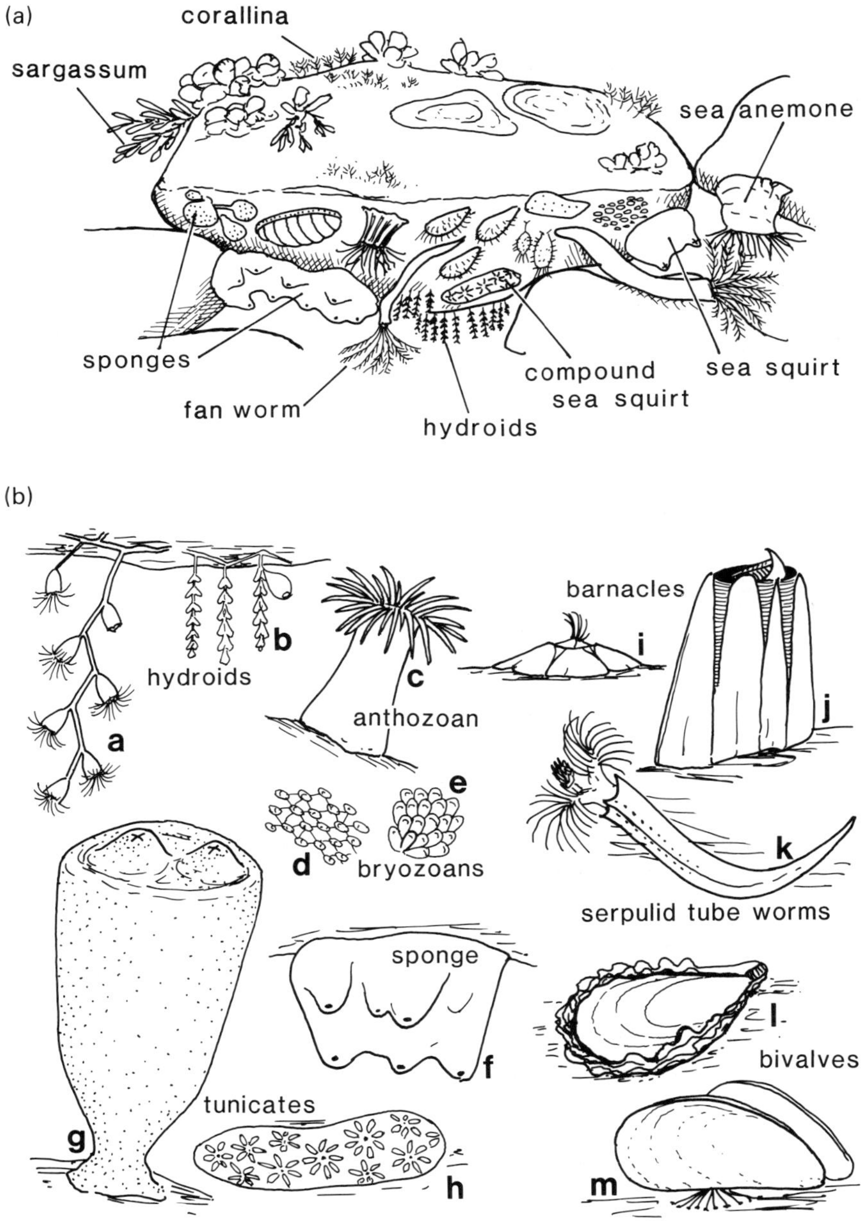

Fig. 1.1 (a) Typical plant and animal community on boulder rocks at very low tide level (New South Wales, Australia). Note inverted orientation of animals under rock. (b) Representative sessile animals from the shore: a, *Thyroscyphus*; b, *Amphisbetia*; c, *Actinia*; d, *Beania*; e, *Watersipora*; f, sponge; g, *Pyura*; h, *Botryllus*; i, *Chthamalus*; j, *Megabalanus*; k, *Galeolaria*; l, *Saccostrea*; m. *Mytilus*.

1.2 FEATURES OF THE SESSILE LIFE STYLE

The sessile mode of life must offer advantages which outweigh the disadvantages of immobility for so many different animals to adopt it.

In coastal waters, there is an abundant supply of food in the form of suspended particles (detritus and plankton) that can be exploited more efficiently by remaining still than by moving about. Because the food is already in a finely divided state, energy-efficient ciliary mechanisms can be used to collect it, pass it to the mouth and move it through the gut. In many animals the food particles are taken directly into the cells lining the gut and digested there. This means that there is no need for complex locomotory and sensory organs to capture food, masticatory organs to break it down into small particles, or elaborate digestive organs to digest the food; consequently body structure remains relatively simple. The main advantage of a sessile life style resides in its energy allocation. Mobile animals use a significant portion of the energy derived from their food in moving about to locate and capture that food. Moreover, locomotory organs require considerable energy expenditure for their construction and maintenance, even when they are not being used. In contrast, very little energy is expended by sessile suspension feeders on food collection and therefore, from the point of view of energy utilization, they must be considered highly efficient. This opens up a number of possibilities. A higher proportion of their energy budget could be channelled into food reserves, growth or reproductive effort. Alternatively, they could subsist on a less nutritious food source, or consume less food. The latter might be important to intertidal animals where the food supply is cut-off during low tide.

The absence of bulky muscle means that the body tissues of most sessile animals are relatively thin and usually in close contact with seawater, therefore diffusion alone is adequate to supply their oxygen demands. Even in very large sponges and sea squirts, the tissues are in close contact with the current that flows through them, so a circulatory system is not essential.

Many sessile animals have a branching, plant-like pattern of growth that tends to extend the animal equally in all directions. This branching growth is well suited to exploiting both space and a diffuse food source. Two processes play a part in this spreading growth: asexual reproduction and metamerism (the vegetative repetition of parts of the body such as feeding (trophic) or reproductive subunits). The former process will give rise to a clone of genetically identical individuals if they separate, or a colony if they remain in metabolic contact. In either case, there is a great increase in the number of trophic units. With dichotomous branching and replication of trophic units, there is a geometric increase in food-collecting capacity. Such growth allows rapid utilization of the available space and exploitation of a favourable food source without having to pass through the hazards of larval life, settlement, metamorphosis and juvenile growth. It allows the founding individual to increase its reproductive potential. Predators seldom kill the whole clone, colony or branched individual, and the parts that escape destruction rapidly undergo vegetative growth or asexual reproduction to rebuild the clone or colony.

The sessile way of life is not without its handicaps. Prominent among these is the problem of dispersal and the colonization of new areas. This function is taken over by the larval stage, which in the majority of cases is motile and planktonic.

Loss of mobility is a major disability under adverse environmental conditions, for an animal cannot move to a more favourable place, nor can it go in search of a better food source if the present one is inadequate. Far more serious is the fact that it is 'sitting prey' for even the slowest of predators, and a wide range of antipredator strategies have evolved that give some measure of protection.

Sessile animals cannot search for or select their mating partners. Except for the barnacles, they shed sperm into the sea, and fertilization becomes a random event. A further disadvantage is that animals cannot select a place to deposit their eggs which would be favourable to their future development; however, many animals overcome this by incubating the eggs internally.

In view of these disabilities and the harsh nature of the intertidal environment, it is remarkable that so many sessile species live there. Strong selective pressures must operate to bring about the evolution of modifications necessary to live in the littoral zone, and it is likely that these come from interactions with the other members of the shore community rather than from physical aspects of the environment.

2

The sea shore environment

The presence and abundance of an animal at a particular place is the outcome of a dynamic interaction among three forces: the animal's structure and physiological capacities, the physico-chemical (abiotic) environment, and all the other organisms that make up the local community (the biotic environment).

On the sea shore, tidal movements and waves are the primary factors that cause alternate submersion in seawater and exposure to air; they will be discussed first.

2.1 MOVEMENTS OF THE SEA

Tides

Tidal movement is the single feature that distinguishes the sea shore from all other habitats; it imparts a regular oscillation between the extremes of submersion and emersion.

The tidal cycle arises in a complex but predictable way as a direct result of the Earth's rotation and its movements relative to the Sun and Moon. The reason for two tides per day is not obvious until the forces generating them are considered.

Fig. 2.1 Tides. (a) Tide-generating forces. C, centre of rotation of Earth–Moon couple (vertical line, axis of rotation). A,B, diametrically opposite points on the Earth's surface. The gravitational acceleration of the Moon on point B(m′) is greater than on A(m″). The centrifugal force acting on B(cf′) is less than on A(cf″). Tide-generating forces (arrows below Earth) are the vectoral sum of gravitational and centrifugal forces at A and B (broad arrows). (b) Inequality of tides. (c),(d) Spring and neap tides respectively, see text for explanation.

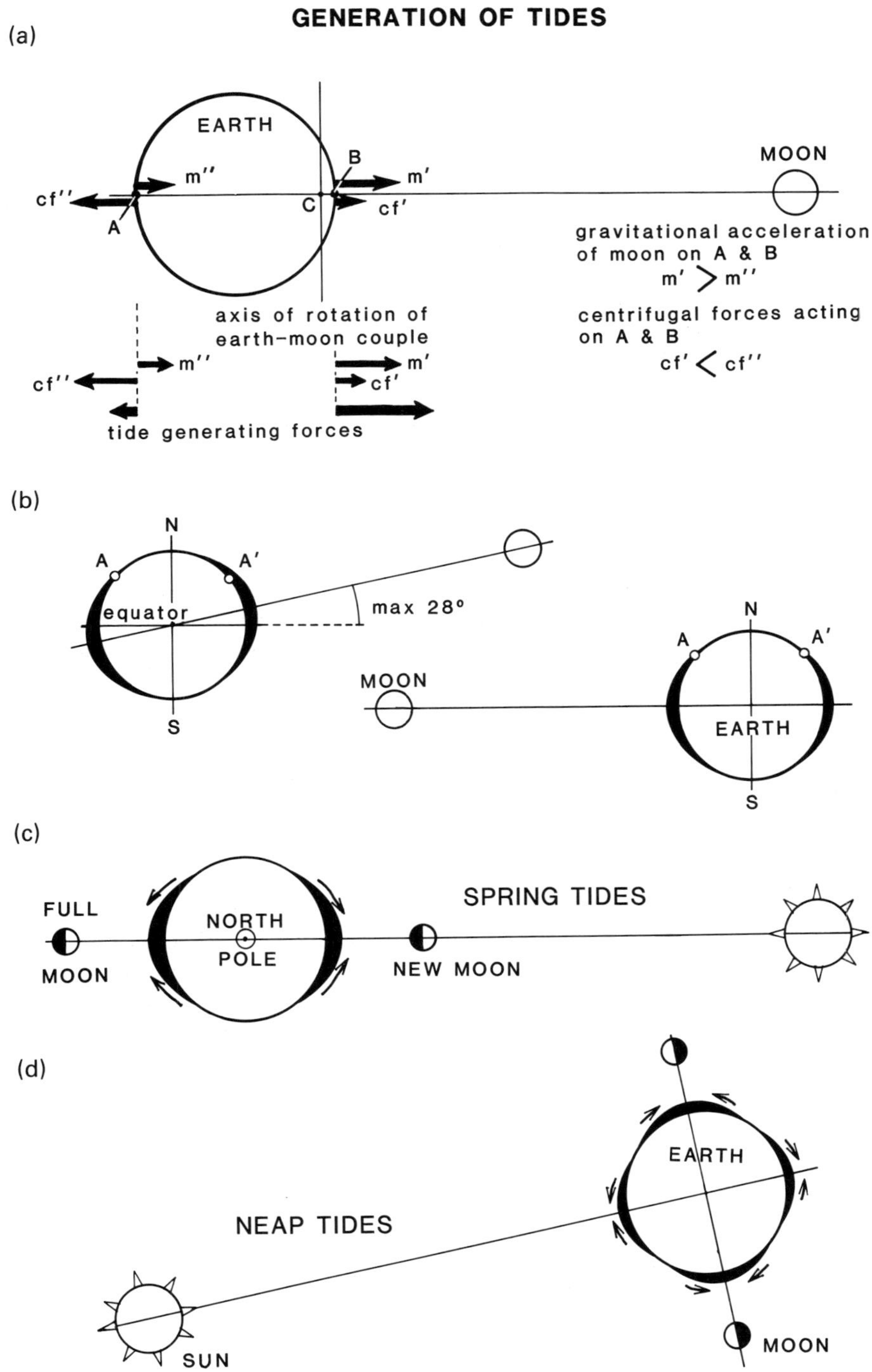
GENERATION OF TIDES
(a)
EARTH
MOON
m''
m'
cf''
cf'
A
B
C
gravitational acceleration of moon on A & B
m' > m''
axis of rotation of earth–moon couple
centrifugal forces acting on A & B
cf' < cf''
tide generating forces
(b)
N
S
A
A'
equator
max 28°
MOON
EARTH
(c)
FULL MOON
NORTH POLE
NEW MOON
SPRING TIDES
(d)
EARTH
NEAP TIDES
SUN
MOON

The Earth and Moon constitute a couple rotating about a common point (C) called the centre of rotation (Fig. 2.1(a)). Any point on the surface of the Earth (or Moon) will be subjected to two forces: a centrifugal force tending to displace it *away* from C, and the gravitational acceleration of the moon tending to displace it towards the moon. If two points A and B on opposite sides of the Earth are considered, both will be subjected to centrifugal and gravitational forces, but the magnitude and vectoral sum of these forces will be different because A lies further from the centre of rotation than B. The vectoral sum at both A and B will be directed perpendicularly away from the Earth's surface. This causes water to be displaced to form two high tides on diametrically opposite sides of the Earth (Fig. 2.1(a)(b)).

These two tides appear to pass round the Earth due to the Earth's axial rotation. The period of rotation of the Earth–Moon couple, however, is 29.53 d, thus there is an average retardation of the time of each tide by 24.5 min.

Exactly the same argument can be applied to the couple formed by the Earth and Sun, although here the dimensions are very different. The centre of rotation is closer to the Sun because of the latter's greater mass, but the tide-generating forces are approximately 46% of those produced by the Moon, owing to the vastly greater distance between Earth and Sun. Thus there are two pairs of tides with different periodicities; those caused by the Sun occur at intervals of 12 h, and those caused by the Moon at an average of 12.4 h. Consequently the two sets of tides slowly drift in and out of phase over a period of *c.* 14.7 d. When the tides caused by the Sun coincide with those caused by the Moon, they summate, and maximal high and low tides result – these are called spring tides. When the two sets of tides are out of phase, tidal oscillations are at a minimum; these are called neap tides. Thus spring and neap tides are related to the phase of the Moon as indicated in Figs 2.1(c) and (d) and 2.2.

The declination of the Moon with respect to the Earth's equator (Fig. 2.1(b)) gives rise to inequalities in the magnitude of the two tides each day at certain latitudes and at different times of the year. This is clearly seen in Fig. 2.2.

The time and magnitude of the tides at any particular place are considerably influenced by topography of the coast line, but two features of the local tides are of great biological importance: the length of time an organism is exposed to the air during periods of emersion, and the time of day

---

Fig. 2.2 Tide chart for Fort Denison, Sydney, showing predicted height of successive tides throughout the month of January and the corresponding phases of the moon (based on tidal predictions for 1987, supplied by Tide Laboratory, Flinders Institute of Atmospheric and Marine Science, South Australia).

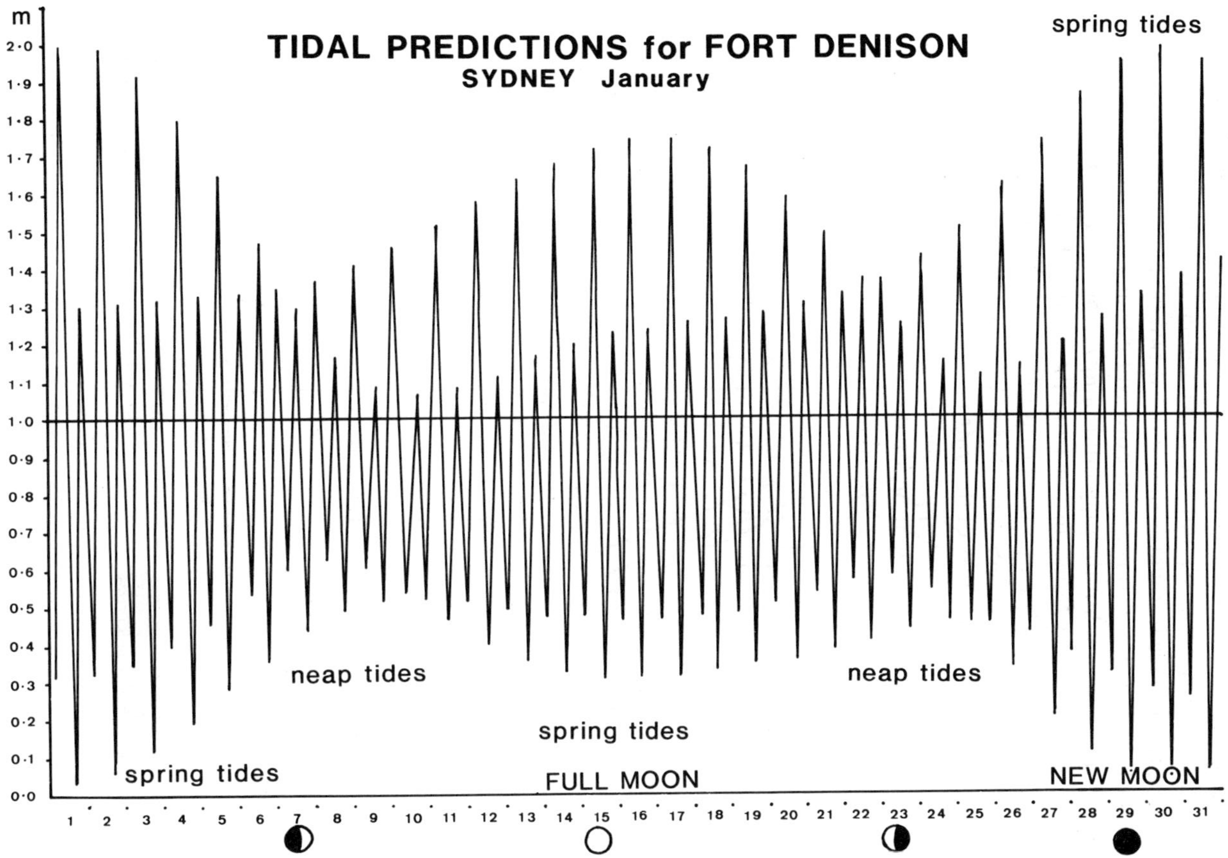
TIDAL PREDICTIONS for FORT DENISON
SYDNEY January
m
2·0
1·9
1·8
1·7
1·6
1·5
1·4
1·3
1·2
1·1
1·0
0·9
0·8
0·7
0·6
0·5
0·4
0·3
0·2
0·1
0·0
spring tides
neap tides
spring tides
neap tides
spring tides
FULL MOON
NEW MOON
1 2 3 4 5 6 7 8 9 10 11 12 13 14 15 16 17 18 19 20 21 22 23 24 25 26 27 28 29 30 31

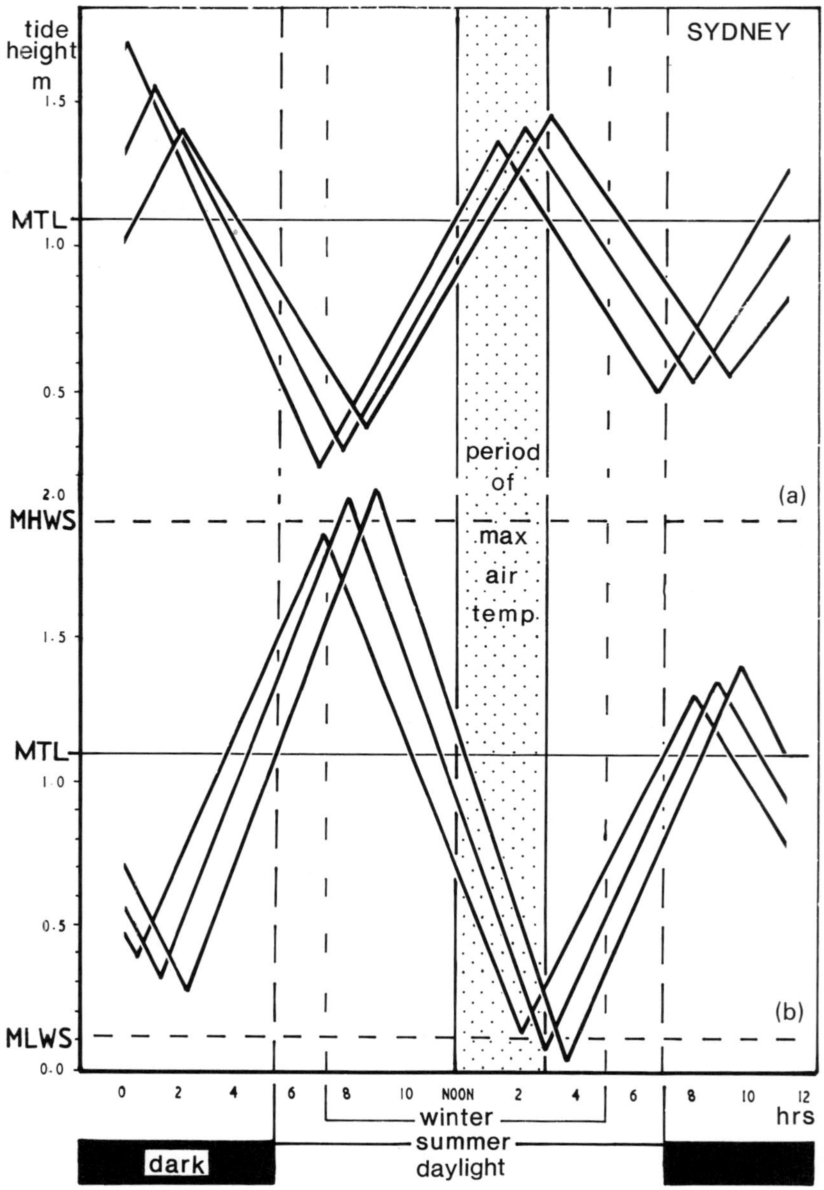

Fig. 2.3 Comparison of exposure to sun during daytime tidal cycles near Sydney, Australia. (a) During neap tides, organisms on the middle shore are immersed during the period of maximum insolation (stippled). (b) Maximum periods of exposure to the sun occur *c.* 2 d before maximum spring tides, when low tide occurs between 13.30 and 15.30. MTL, mean tide level; MHWS, mean high water of spring tides; MLWS, mean low water of spring tides.

at which this occurs. The course of three consecutive spring tides during midsummer on the coast at Sydney, New South Wales, has been plotted in Fig. 2.3(b), and the three neap tides occurring 7 d later in Fig. 2.3(a). Solar radiation on both days would have been intense during the middle of the day (10.00–15.00 local time), but the effect upon organisms living on the shore would be very different on the two occasions. During spring tides, the middle shore is exposed to the sun throughout the period of low tide, and both animals and algae suffer severe stress or mortality. On the other hand, the shore is covered by the tide during the middle of the day at neap tides, and the sun would have little or no harmful effect on organisms living there. At night there is no harmful radiation during low tide, and many more animals are active than during daytime low tides.

The situation on the upper shore is very different. During neap tides, organisms may not be submerged for several days in succession, whereas at least one period of submersion occurs each day during spring tides. Low water of spring tides (LWS) occurs at about the same time of day throughout the year for a particular locality. From Table 2.1 it will be seen that this occurs at about midday in some places, and there exposure to the summer sun is likely to be severe. Such localities provide good collecting sites.

Tide levels

It will be seen from Fig. 2.2 that high tide and low tide levels are by no means constant, thus average heights rather than actual values are a more useful concept. The position of any point on the shore can be referred to, either by its

Table 2.1 Predicted times of low water on days of full and new moon (spring tides) for some places in Britain* and Australia

Locality	*Time*
Aberdeen	07.00–10.00; 19.00–22.00
Dover	06.00–09.00; 18.00–21.00
Lowestoft	03.00–06.00; 15.00–18.00
Milford Haven	12.00–15.00; 24.00–03.00
Brisbane	03.00–04.30; 15.00–16.30
Sydney	01.30–03.30; 13.30–15.30
Melbourne	05.00–07.00; 17.00–19.00
Adelaide	10.00–12.30; 22.00–24.30

* Source: Lewis (1964).

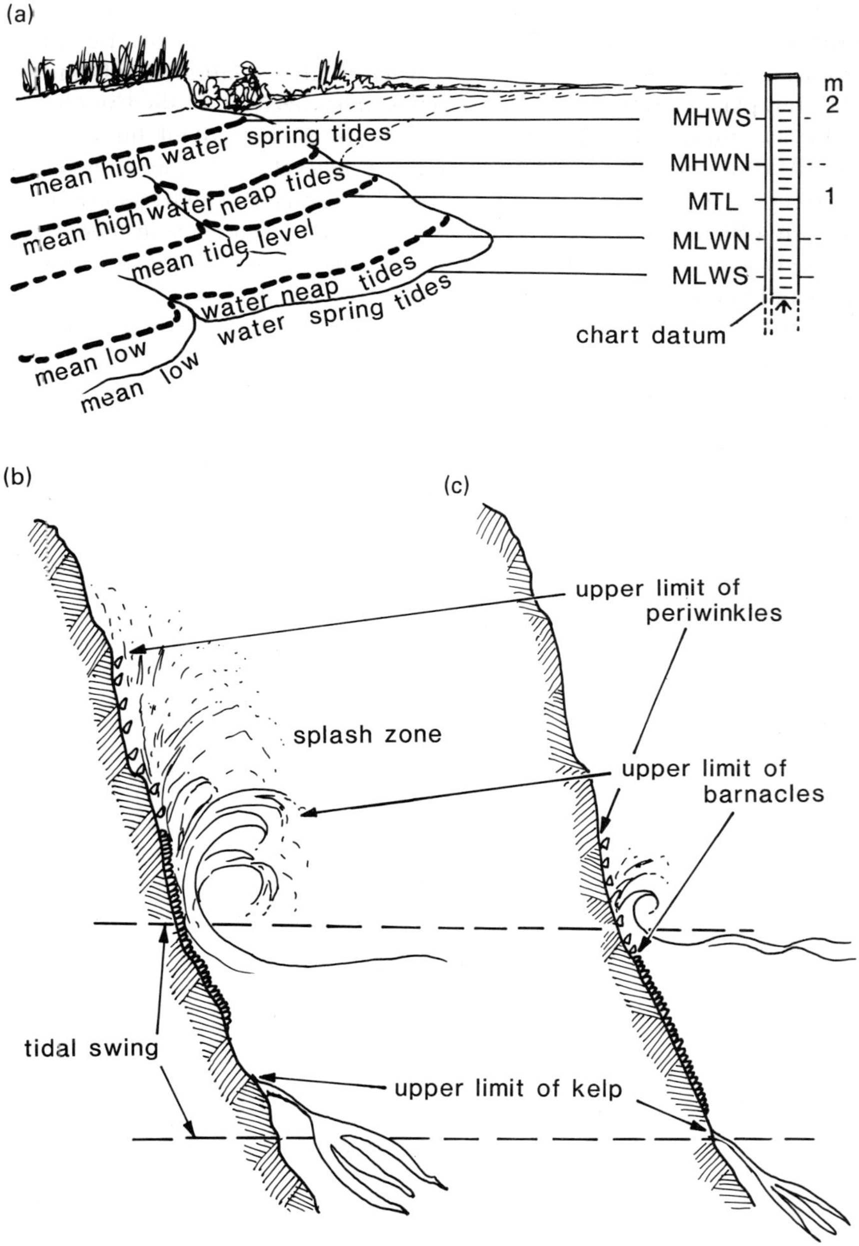

Fig. 2.4 (a) Tide levels (Sydney). (b) Exposed vertical rock face. (c) Sheltered vertical rock face.

height above Ordnance or Chart Datum, or by imaginary horizontal lines representing average high and low tides. The former, although precise, suffers from the disadvantage that the exact level of Chart Datum is only known for a few places (usually major ports) unless it has been established by survey.

By averaging the height of tidal oscillations, horizontal lines can be imagined on the shore representing mean tide level (MTL), mean high and low water of neap tides (MHWN and MLWN) and mean high and low water of spring tides (MHWS and MLWS) (Fig. 2.4(a)). Reference to these levels can only be made after they have been established by repeated observations at high and low tides for a particular location.

Mean tide levels give a good indication of the period of time a particular place on the shore will be submerged or exposed, but they do not reflect the sum total of environmental conditions at that point or its suitability for colonization. In general, the vertical zonation of animals and algae does not correlate well with tide levels or height above Chart Datum, and a more useful concept is a biological zonation of the shore (Table 2.2) based upon the known distribution of a few plant and animal indicator species.

Table 2.2 The biological zonation of rocky shores

Zones			*Biological indicators*
Maritime zone			Halophytes
	Littoral fringe	↑ Upper limit of marine species	Periwinkles, *Verrucaria*
Littoral zone		↑ Upper limit of barnacles	
	Eulittoral zone*		Barnacles
			Mussels and tube worms
Sublittoral zone	Infralittoral fringe†	↑ Upper limit of kelps	Kelp seaweeds
	↑ Extreme low water of spring tides) (ELWS)		

* Also called the middle shore.
† Also called the lower shore.

Zonation of the shore

Marine animals and plants are not spread at random over the shore. As the tide recedes, conspicuous bands appear on the rocks as a result of the colour of organisms dominating a particular level roughly parallel to the water line (Fig. 2.5(a)). An extensive account of zonation on British coasts has been given by Lewis (1964). There is a remarkable similarity between the types of animals and plants and their relative position on the shore in various parts of the world. This can be illustrated by comparison of similar rocky coasts in Pembrokeshire, England and New South Wales, Australia (Fig. 2.5(b)).

On both coasts, orange patches of lichens occur above the highest reach of the tides. Below this, there is a band where rocks are blackened by other lichens, particularly *Verrucaria*, and dotted with numerous small periwinkles (*Littorina neritoides, L. saxatilis, L.* (= *Melaraphe*) *unifasciata*, and *Nodilittorina pyramidalis*). This periwinkle zone or littoral fringe is widest on open shores where waves keep the rocks wet with spray, and it all but disappears on sheltered shores.

The periwinkles spread to lower levels on the shore, but a clear-cut boundary is usually present between the black lichens and a wide band of acorn barnacles. The barnacle or eulittoral zone is dominated by species characteristic of the region: *Semibalanus balanoides* and *Chthamalus stellatus* in West England and *Chamaesipho columna* in NSW. On sheltered shores, like the lochs of Scotland, the barnacle band is obscured by luxuriant growth of fucoid algae, particularly *Ascophyllum nodosum* and *Fucus vesiculosus*.

The lower limit of the barnacle zone is less clearly defined. Limpets are common and their browsing may keep the rocks clear of barnacles or seaweeds. On some shores, there is a conspicuous black band of mussels below the barnacles, and in NSW there is a white band of tube worms, *Galeolaria caespitosa*, at this level. This is the lower extent of the eulittoral zone. There is now another marked change due to the presence of large brown kelp seaweeds. The upper limit of the kelp marks the beginning of the sublittoral zone.

The upper part of the sublittoral zone is only occasionally exposed at extremely low spring tides; nevertheless, it is affected by the churning action of waves and must be considered as part of the shore. It is sometimes referred to as the infralittoral fringe (Stephenson and Stephenson, 1972). It is a fortunate circumstance that many of the animals belonging to the sublittoral community can be found in tide pools higher on the shore. Many of these are sessile animals and will be dealt with in the following account. Table 2.2 summarizes the main features of shore zonation based on plants and animals. No vertical scale is given because distribution of most species is related to intensity of wave action, not tide height. In the following chapters the

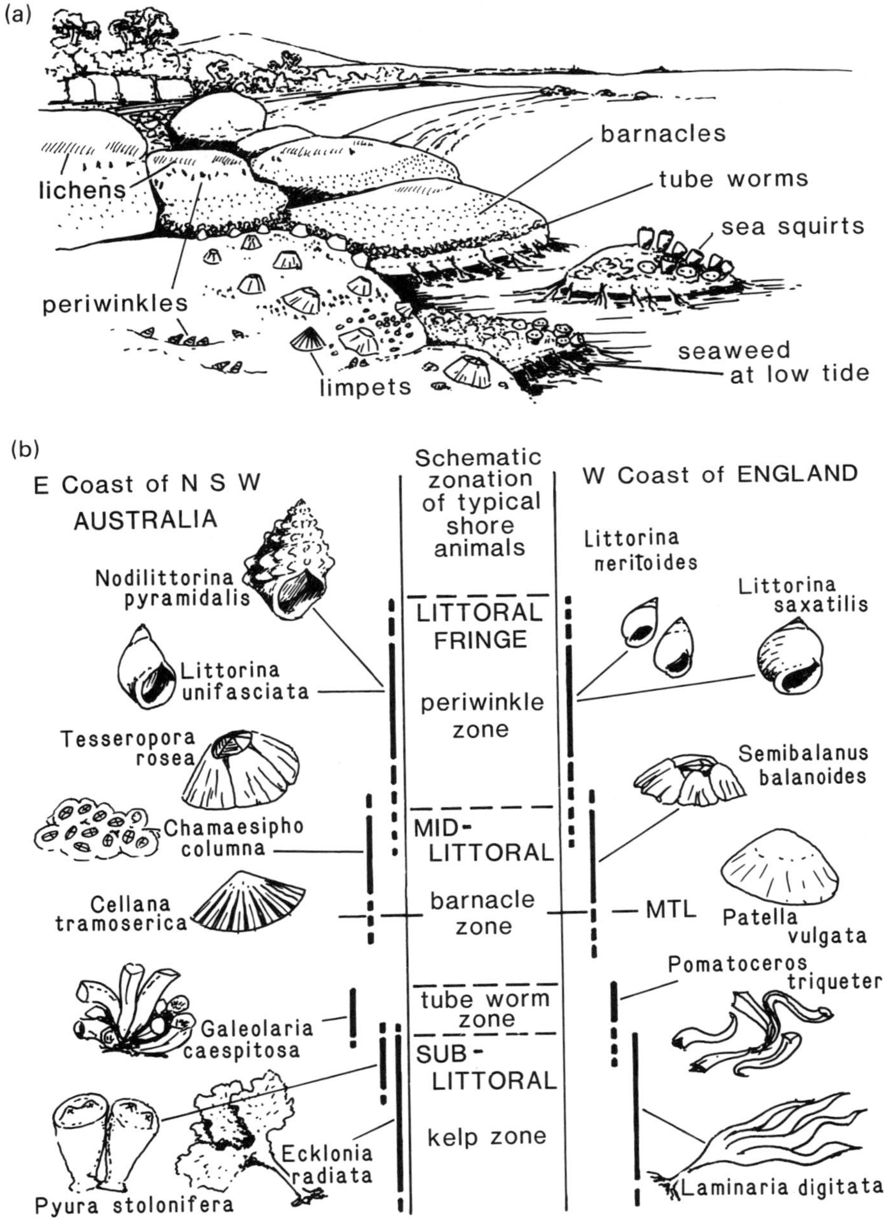

Fig. 2.5 Zonation of the shore. (a) Low tide on an Australian shore, showing horizontal banding. (b) Relative vertical distribution of typical shore animals from Sydney, New South Wales (tidal swing *c.* 2 m) and Dale, Pembrokeshire (tidal swing *c.* 6 m).

eulittoral zone will frequently be referred to as the middle shore, and the upper part of the sublittoral (the infralittoral fringe) as the lower shore.

In passing down the shore, there is a noticeable increase in the number of marine species encountered. The eulittoral zone is dominated by gastropods, barnacles, mussels or tube worms, but in the sublittoral these give way to a great assemblage of mobile and sessile animals (sponges, hydroids, sea anemones, polychaete worms, bryozoans, tunicates, molluscs, crustaceans, echinoderms and many others) that are poorly adapted to emersion.

Waves

Winds caused by variations of barometric pressure in the atmosphere influence the sea in two ways: by generating waves and by increasing the rate of evaporation.

Waves breaking on the shore play a major role in shaping the coastline. The size of a wave, and hence its energy, is determined by the velocity of the wind, the time it has been blowing and the unobstructed distance over which it blows (the fetch). Ocean swell can transfer prodigious energy, generated by storms at sea, over vast distances with little loss before it is expended against some distant coast, degrading some regions and building others.

For littoral plants and animals, particularly sessile species, this can be devastating. Their habitat is destroyed when rocks or boulders are turned upside-down or transported elsewhere. They may be crushed, torn apart, buried or exposed to solar radiation.

Waves are of far greater biological importance than tides, for their churning action mixes surface water with deeper water and circulates bottom sediments. Nutrients, essential to attached and planktonic algae for growth, are brought to surface waters; oxygen is distributed to deeper water much more rapidly than by diffusion; particles of organic matter suitable as food are distributed widely and toxic substances (like hydrogen sulphide produced by certain bacteria) are dispersed or denatured. Indeed, without the churning and circulation caused by waves, very few animals could survive in the sea.

On gently sloping shores, such as surf beaches and horizontal rocky reefs, most of a wave's energy is dissipated in friction with the substratum and finally ends in strong horizontal movement of the water back and forth. Pressure changes are small, but the churning action of the waves suspends fine organic particles and keeps the water well oxygenated. If sand is present this can have a strong scouring or abrasive effect and by destroying newly settled larvae may prevent species like barnacles from colonizing certain areas.

Heavy swell generated by storms at sea causes tremendous changes in hydrostatic pressure. When a wave breaks against a vertical cliff on an open coast, its velocity is brought abruptly to zero, causing sudden pressure changes. Harger (1970) found that mean wave impact ranged from

1.4×10^3 N m^{-2} to 9.0×10^5 N m^{-2} (i.e., from *c.* 0.01 to 9 atmospheres) on the coast of California. Presumably, pressures during storms could rise considerably above these values. These pressure changes are not the same as the slow, regular rise and fall in hydrostatic pressure during the tidal cycle. The high pressures are caused by the inertia of moving water and changes are very rapid, thus air trapped in crevices or empty shells will be compressed and then expanded with explosive violence. This is a major factor in storm damage to the shore and is responsible for stripping dead barnacles from the rock surface.

When waves break against a steep rocky shore, they wet the rocks many metres above sea level (Fig. 2.4(b)). This keeps the rocks moist for longer periods and allows both algae and marine animals to live higher on the shore than elsewhere. To the leeward of promontories or in sheltered bays and estuaries, the force of the waves is significantly diminished, and in land-locked fiords or lochs the effect of oceanic waves is reduced to zero. In such quiet waters, only locally-generated waves of low energy lap the shore, and animals do not extend as high on the rocks (Fig. 2.4(c)).

Rapidly moving water caused by waves also exerts a strong lateral force (drag) on any object that projects above the substratum. The direction from which this force comes is unpredictable and varies from moment to moment; nevertheless, shore animals are regularly exposed to such conditions and have evolved many adaptations to counteract them.

It is customary in discussing shore ecology to use the term 'exposure' to describe the severity of wave impact. A shore washed by unobstructed oceanic waves and receiving the full impact of swell is described as 'extremely exposed'. Where intervening land masses, wave reflection or interference reduce the energy of the waves before they strike, the shore can be described as 'moderately exposed' or 'sheltered'.

Exposure is difficult to measure, but it has two clearly visible effects upon the algae and animals that live on the shore. Firstly, on exposed, wave-beaten shores, both algae and animals extend much higher up the rocks than on moderately exposed or sheltered shores. The periwinkles and barnacles defining the upper limit of the littoral fringe and midlittoral zones (Table 2.2) are elevated far above extreme high water springs (EHWS). On the other hand, in very sheltered places the upper limit of the littoral fringe does not reach EHWS, thus there is a clear quantitative difference in vertical distribution with degree of exposure. Secondly, there is a qualitative difference between calm water and wave-exposed shores in the species that are present. Many species cannot tolerate rough conditions and are only found in sheltered situations, while other species cannot live under calm water conditions. This difference is very noticeable for seaweeds, barnacles and molluscs, but it is equally characteristic of the less conspicuous animals. Ballantine (1961) has shown that it is possible to define a biological exposure

(a)

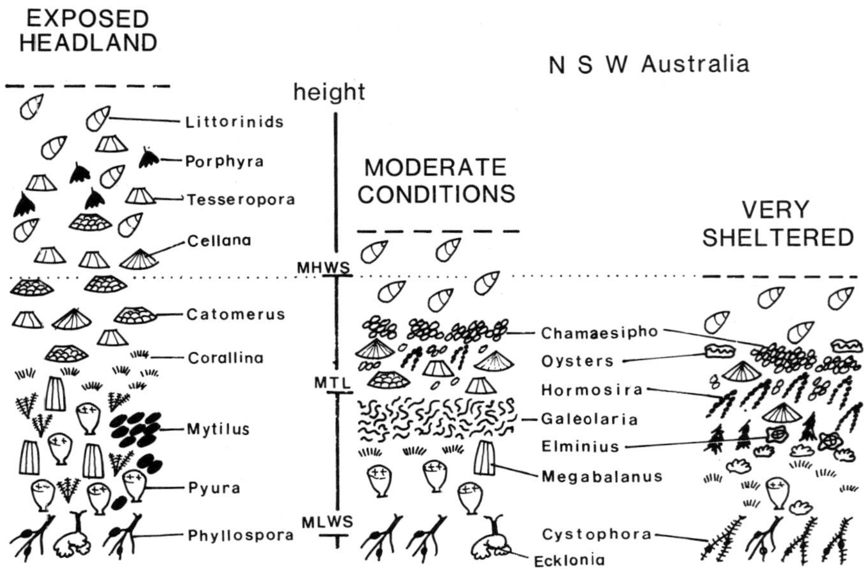

(b)

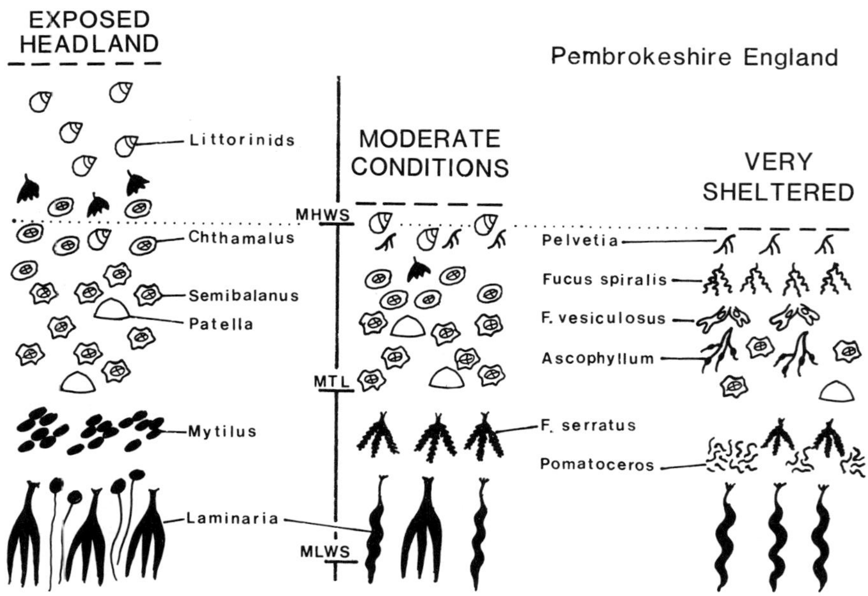

Fig. 2.6 Diagrammatic comparison of organisms on exposed and sheltered shores in (a) New South Wales, Australia and (b) Pembrokeshire, Great Britain.

scale for shores in the SW of England, based upon the relative abundance of certain species of animals and seaweed. A similar exposure scale of wider application to British coasts has been defined by Lewis (1964), and it would appear that the same could be done for other parts of the world. Examples of the broad faunal and floral difference between exposed, moderate and sheltered conditions are shown in Fig. 2.6, for NSW, Australia and SW England.

Finally, the energy of wave impact is responsible for grading loose material on the shore. On high-energy exposed shores, fine particles are carried away and only larger gravel or boulders remain. Abrasion and percussion may be severe. On quiet shores, sand, silt and even mud may be deposited and organisms are liable to be buried.

2.2 EXPOSURE TO THE ATMOSPHERE

The rise and fall of the tides subjects intertidal animals to two contrasting environments: air and seawater. Some of the physico-chemical properties of these two extremes are given in Table 2.3. During daytime periods of

Table 2.3 Comparison between seawater and the atmosphere: physical properties and chemical composition

Physical properties	
Density	SW 1024 kg m^{-3} at 20 °C and salinity (S) = 34‰*. Air 1.29 kg m^{-3} at standard temperature and pressure. The inertia of moving water is much greater than that of moving air. In air there is a loss of buoyancy; a rigid skeleton or high turgor pressure is needed to support body. Gills collapse in air and do not function. Cilia will not function if surface is dry.
Viscosity	Negligible for air.
Hydrostatic pressure	Pressure increases 1.013 × 10^4 N m^{-2} for each metre of depth (1 atmosphere per 10 m).
Diffusion	Oxygen diffuses *c.* 10^5 times faster in air. In water, respiratory surfaces need to be irrigated.
Thermal properties	Specific heat of SW = 3.9 × 10^3 J kg^{-1} °C^{-1}. Specific heat of air = 10 J kg^{-1} °C^{-1}. Air temperatures rise rapidly in the sun but water temperatures rise very slowly. Daily fluctuation of air temperatures is wide; sea temperatures remain constant. Water evaporating when tide is out exerts a cooling effect. Seawater (S = 34‰) freezes at −1.87 °C.

Table 2.3 Continued

Radiations	In air, white surfaces absorb less heat than black ones. Ultraviolet radiation is absorbed in the first few cm of water.
Chemical composition	
Water content	The amount of water vapour needed to saturate air is dependent on temperature. The drying power of air is proportional to the saturation deficit.
Ions	Na^+, K^+, Ca^{2+}, Mg^{2+} and Cl^-, the main ions in seawater, are absent in air. Salts present in seawater lower the freezing point and vapour pressure, and increase osmotic stress. Na^+, Ca^{2+} and K^+ are essential for life. Normal salinity of seawater is *c.* 34 g salts per litre (i.e. 34 parts per thousand, ‰). Salts reduce solubility of gases.
Gases	Air contains 210 l oxygen m^{-3}, but only a trace of carbon dioxide. Saturated seawater contains 6.44 l O_2 m^{-3} (at 10 °C and S = 34.3‰). Concentration of carbon dioxide as bicarbonate ion is variable but much higher than in air. It acts as a buffer and prevents fluctuation in hydrogen ion concentration.
Other dissolved mineral and organic matter	Present in seawater in very small amounts. Absent from air.

* ‰, parts per thousand.

emersion, animals will be exposed to solar radiation, temperature changes, humidity fluctuations and, during rain, reduced salinity.

Solar radiations

Solar radiation has components in the ultraviolet, visible and infra-red regions. Their intensity and duration depend upon latitude, season and time of day, and reach their maximum about midday.

The shortest wavelengths (below 290 nm) are filtered out in the upper atmosphere, but longer wavelengths of ultraviolet (290–390 nm) reach the earth's surface. They are absorbed by water in the upper few millimetres, so animals in the sea will not be affected. On land and during emersion, however, radiations pose a serious threat to unprotected animals. Many cases have been recorded where ultraviolet or intense sunlight has killed marine invertebrates, particularly larval stages (Segal, 1970).

Visible light is vital to the photosynthetic activities of plants. Algae, such as *Ulva, Cladophora* and *Porphyra*, living on the middle and upper shore are unaffected by strong sunlight, but the kelp, *Laminaria*, which is adapted to the low light intensities of the sublittoral zone, reduces its rate of photosynthesis in strong sunlight and during emersion. Long exposure causes destruction of photosynthetic pigments. Many seaweeds from the sublittoral zone are killed by 2 h exposure to direct sunlight out of the water. A good correlation has been found between sensitivity to light and the level at which seaweeds live (Hellebust, 1970).

Visible light (400–780 nm) is important to invertebrate animals in many ways. To some it is lethal, due to photosensitivity (Segal, 1970), but a more obvious effect is to inhibit activity. Many more animals are active on the shore during night-time low tides than during the day. Light may affect metabolic rate or retard growth. The barnacle, *Semibalanus balanoides*, growing in shade is often larger than nearby specimens in the sun. Wethey (1984b) found that this barnacle could survive higher on the shore when shielded from direct sunlight, although this might be a response to reducing the infra-red end of the spectrum. On the other hand, calcification and hence growth in corals takes place more rapidly during the hours of daylight (Goreau and Goreau, 1959). Reef-building corals and some sea anemones have symbiotic algae in their tissues which carry out photosynthesis in light. When coral is kept in complete darkness, the symbionts are gradually extruded from the tissues and lost.

Sunlight influences the behaviour of marine animals in a number of ways besides inhibition of activity. The direction of light is used by the larvae of many intertidal animals as a cue for orientation. The transition from daylight to darkness may entrain diurnal rhythms, and daylength (photoperiod) may determine the onset of breeding or the timing of events like spawning. There is evidence that daylength influences some intertidal species, for example, under conditions of constant light, *Semibalanus balanoides* does not come into breeding condition and the sea squirt, *Ciona intestinalis*, does not spawn; however, on return to alternating light and dark periods, these inhibitions are removed (Segal, 1970).

Although the intensity of moonlight is very low (*c.* 0.2 lux or 1×10^{-3} W m^{-2}, compared with sunlight in the order of 10^5 lux or 500 W m^{-2}) it profoundly influences the behaviour of many marine animals (Papi, 1960).

Temperature fluctuations

Radiations in the infra-red region (above 800 nm) cause heating when absorbed. In the sea this causes little or no change in temperature over a 24 h period because of the high specific heat of water and the vast volume of the sea. Seasonal changes in sea temperatures take place very slowly, and are influenced by oceanic currents as well as latitude. Water temperatures off the southern coast of Britain have an annual range of *c.* 6–18 °C, and a similar temperature range is found off Vancouver Island, British Columbia, but Halifax, Nova Scotia, although at a lower latitude, has a temperature range from *c.* 1 °C to 15 °C due to the cold Labrador current (Stephenson and Stephenson, 1972).

Air temperatures may have a daily swing that exceeds the annual fluctuations in the oceans. Temperatures on the shore in winter may fall 10 °C below sea temperatures, and in summer may rise 15 or 20 °C above them. Animals attempting to live on the shore must be physiologically adapted to tolerate these rapid changes in temperature.

Rock surfaces receiving direct insolation will become hotter than air temperature. Black basaltic rock heats up more rapidly and reaches a higher temperature than light-coloured coral limestone or chalk. Basalt reefs on the NSW coast of Australia frequently exceed 45 °C on summer days. The aspect of rocks influences surface temperatures; for example, north-facing rocks in Britain or south-facing rocks in Australia may not receive any direct sunlight and remain below or near air temperatures. At night rocks will lose their heat by radiation, although it is often the incoming tide that cools them.

Seawater (salinity 34‰) freezes at −1.87 °C. Shores in high latitudes, such as the Gulf of St Lawrence, are frozen during the winter and only those species that tolerate being frozen for long periods can survive there.

Humidity and evaporation

The amount of water vapour in the air is usually given as percentage relative humidity (%RH), but this has little biological meaning for it varies with temperature. A more useful concept is the drying power or saturation deficit (P_d) of air, for this gives an indication of the rate at which water will be lost from an animal's body by evaporation. The saturation deficit is proportional to the difference between the vapour pressure of water and the partial pressure of vapour in the air at a particular temperature:

$$P_d = P_s - P_o \tag{2.1}$$

where P_s is the vapour pressure of water at t °C and P_o is the partial pressure of vapour in the air at the same temperature. The greater the saturation deficit, the higher the rate at which water will be lost from the body. If P_o

exceeds P_s, vapour would condense on the body or on the water's surface. The saturation deficit can be calculated from the relative humidity by the following relationship:

$$P_d = P_s (1 - \%RH / 100) \quad (2.2)$$

The vapour pressure of water (P_s) is obtained from tables of vapour pressure. The vapour pressure of seawater is slightly less than that of pure water and varies with salinity. For seawater with a salinity of 34.5‰ (= chlorinity of 19‰), the vapour pressure is given by:

$$P_{s,SW} = P_s \times 0.9816 \quad (2.3)$$

and Table 2.4 shows values for the saturation deficit between fresh water, seawater and the air when the relative humidity is 50%.

Evaporation from an animal's body increases the concentration of ions in the body fluids, which in turn subjects the tissues to osmotic stress and withdraws water from them. Loss of water by evaporation during emersion is frequently referred to as desiccation, but death from osmotic stress is likely to ensue long before obvious signs of dehydration occur.

For survival from one tide to the next, the important factor is the size of the saturation deficit. After the tide retreats, the rocks will be wet, and at first saturation deficits remain very low. As air temperatures rise and the rock surface dries out, saturation deficits will increase. High deficiencies acting over a period of 3–4 h are unlikely to happen except during hot sunny days when low tides occur about midday. Above mean high water neaps

Table 2.4 Relationship between saturation deficit and temperature at a relative humidity of 50%

Temperature	*Pure water*		*Seawater* *	
°C	*Vapour pressure* ($N m^{-2}$)	*Saturation deficit at RH 50%* ($N m^{-2}$)	*Vapour pressure* ($N m^{-2}$)	*Saturation deficit at RH 50%* ($N m^{-2}$)
0	610	305	599	299
6	935	467	918	458
12	1402	707	1376	688
24	2982	1491	2927	1463
48	9581	4790	9405	4702

* Salinity 34.5‰.
Source of data: calculated from Equation (2.1).

(MHWN), their severity will increase because several days may elapse between successive submersions.

Some idea of the magnitude of the saturation deficit on an exposed shore can be obtained by calculating the values from published data. Southward

Table 2.5 Representative saturation deficits*

Plymouth, Autumn	RH = 93%	(Screen temp. 10 °C)
18 December 1953	Sea 12.2 °C	Shade 11.4 °C
	Barnacle body 12.4 °C	
	Saturation deficit between:	
	Sea and shade temp.	$140\ N\ m^{-2}$†
	Barnacle and shade‡	$160\ N\ m^{-2}$
Plymouth, Summer	RH = 49%	(Screen temp. 14.5 °C)
17 May 1954	Sea 12.2 °C	Shade 17 °C
In sun	Barnacle body 24.2 °C	
	Saturation deficit between:	
	Sea and shade temp.	$440\ N\ m^{-2}$
	Barnacle and shade‡	$2210\ N\ m^{-2}$
	Wet rock at shade temp. and shade	$950\ N\ m^{-2}$
Plymouth, Winter	RH = 57%	(Screen temp. 0 °C)
4 February 1954	Sea 6.3 °C	Shade 3.9 °C
Sunny	Barnacle body 10.5 °C	
	Saturation deficit between:	
	Sea and shade temp.	$480\ N\ m^{-2}$
	Barnacle and shade‡	$780\ N\ m^{-2}$
	Wet rock at shade temp. and shade	$330\ N\ m^{-2}$
Port Erin, Autumn	RH = 85%	(Screen temp. 6.5 °C)
22 November 1958	Sea 9.5 °C	Shade 7 °C
	Barnacle body 7 °C	
	Saturation deficit between:	
	Sea and shade temp.	$310\ N\ m^{-2}$
	Barnacle and shade‡	$130\ N\ m^{-2}$

For comparison with experimental atmospheres:

An atmosphere of 5% RH over 70% sulphuric acid is estimated to have a saturation deficit of *c.* $1800\ N\ m^{-1}$ at 17 °C, and $2800\ N\ m^{-2}$ at 24 °C.

At 17 °C, air of 95% RH has a saturation deficit of only $71\ N\ m^{-2}$.

* Source of temperature data: Southward (1958).

† $133\ N\ m^{-2} = 1$ mm Hg.

‡ Deficit between the seawater in the mantle cavity at the body temperature of the barnacle and the atmosphere at shade temperature.

(1958) measured sea, air and barnacle body temperatures during midday low tides at different seasons of the year at Plymouth and the Isle of Man. Table 2.5 shows the saturation deficits calculated from his data on the assumptions that the seawater had a chlorinity of 19‰ and air temperatures were the shade temperatures quoted. The body temperature of barnacles exposed to the sun may rise considerably above the surrounding air temperature. The values of saturation deficits calculated for wet rocks at shade temperatures represent a more realistic measure of the drying effect exposure would have on soft-bodied animals.

Air temperatures are no guide to sea or body temperatures for, even in winter, the latter may be several degrees above ambient if the barnacle receives direct insolation. Moreover, standard meteorological screen temperatures give no indication of actual conditions on the shore. Many of the saturation deficits are low, indicating that animals would not be under stress. The highest deficits occurred in sunshine during summer, and their drying effect would have been severe; however, barnacles could avoid them by closing their shell. These measurements of relative humidity and temperature are probably representative of the environmental conditions under which barnacles live; but they give little indication of the conditions in other microhabitats, such as under rocks or seaweed, in crevices or among pebbles, where most shore animals hide when the tide is out. In such places, temperatures may be closer to sea temperature and saturation deficits may be very low, so that animals living there are in no danger of serious water loss.

Salinity changes

Rain frequently falls on the shore whilst the tide is out, and bathes animals and seaweeds in fresh water. This may lead to osmotic stress if water is taken into the body by osmosis and dilutes the body fluids. In regions where the shore receives long periods of sunshine and little rain, evaporation from tide pools will cause them to become more saline. This will subject animals living there to osmotic stress and the danger of losing water from their body.

In estuaries where large volumes of fresh water enter the sea, salinity fluctuates with the tide, and animals living there will be subject to periodical osmotic stress.

3

The biotic environment

The seaweeds, micro-organisms and animals that live in the same habitat are just as much a part of the environment as the tides or temperature fluctuations. Any one member of this complex community may provide food, shelter or protection to any other member, or threaten its life by competition and predation.

3.1 SEAWEEDS AND MICRO-ORGANISMS

Seaweeds are the most important part of this biotic environment because they profoundly alter conditions of life for the animal component. As primary producers, they are the ultimate source of energy for nearly all other intertidal animals and micro-organisms, and the oxygen produced during photosynthesis replenishes coastal waters and tide pools.

Where green and brown seaweeds form a blanket over the middle shore, their presence softens the harshness of the physical environment during both submersion and emersion. Apart from a few small seaweeds like *Corallina* and *Amphiroa* or the wiry *Gelidium pusillum*, not many seaweeds can stand erect when out of water. The soft leathery thalli of nearly all the large brown algae collapse during emersion, trapping pockets of air and water. This blanket of weed will protect animals living underneath from direct solar radiation, desiccation and sudden changes of temperature on temperate shores, but in warmer latitudes and at levels where emersion may last for 6 h or more, intense insolation during low tide may cause photosynthesis to stop and oxygen levels to fall rapidly under the weed. Temperatures under the weed may rise, but not as high as those of nearby rock surfaces.

As the tide rises, seaweeds will be churned by the waves. Their flexibility and buoyancy result in surfaces being rubbed against one another, making colonization by animals difficult except for highly specialized species.

Silicularia bilabiata (Fig. 11.3) is a hydroid specialized to live on *Phyllospora comosa* in the turbulent surf region of Australasian shores, while *Fucus serratus*, living on the calm shores of Scottish lochs, is colonized by a variety of sessile animals such as *Spirorbis spirorbis* (Fig. 7.1(c)) and the bryozoans *Alcyonidium hirsutum* and *Flustrellidra hispida* (Fig. 9.1(d)). Support for the belief that it is the rubbing of seaweed surfaces that inhibits colonization is given by the stunted, bushy form of *Ecklonia radiata* which grows close inshore on New South Wales reefs. The broad, relatively stiff thalli bear numerous spines or projections which prevent the surfaces rubbing together. Old specimens are usually heavily covered with many hydroid species, such as *Amphisbetia minima, Phylactotheca armata* (Fig. 11.4(a) and (b)), and *Plumularia setacea* (Fig. 11.16(a)).

During high tide, seaweeds modify the environment in another way. The dense stands of floating thalli restrain the churning action of the waves so that at the surface of the rock, where the weeds are anchored by their holdfasts, the motion of the water is reduced to a gentle back-and-forth oscillation, ideally suited to suspension feeders and delicate hydroid polyps. Between holdfasts the rock surface may have been grazed clean by chitons and sea urchins, but on the holdfast and stipe of larger seaweeds a heavy incrustation develops of sponges, bryozoans, ascidians and the beautiful feather-like hydroids, *Aglaophenia plumosa** (Fig. 11.8(a)) and species of *Sertularia*. Numerous free-living animals, such as polychaetes, ophiuroids, sipunculids, crustaceans and bivalves, live in the crevices among the sessile animals on the holdfasts. By creating numerous new microhabitats, the seaweeds indirectly increase the diversity of the animal community.

Seaweeds are not eaten as such by sessile animals, but detritus derived largely from decaying algae is an important food source for suspension-feeding animals.

Detritus is the term used to describe the particulate matter to which dead animals, algae and animal faeces are degraded by autolysis and the activities of bacteria, marine moulds and protozoans living in the decomposing mass. The activity of marine bacteria is particularly important as they alone can hydrolyse the two most abundant carbohydrates, cellulose and chitin, converting them to glucose and amino acids. Newell (1965) found that the faeces of the intertidal snail, *Hydrobia ulvae*, when first voided had a nitrogen content of only 0.025%, but after 3 d in seawater it had risen to 1.75%. This increase was attributed to the synthesis of proteins by growing bacterial colonies using the cellulose in the faeces as an energy source.

A significant proportion of the organic substances released in the process of decomposition will pass directly into solution as dissolved organic matter. To this will be added organic molecules from the excreta of animals, as well as substances leached out of living seaweeds and phytoplankton.

* *A. plumosa* is Australian; *A. pluma* is a European species.

The concentration of dissolved organic matter is not great. It varies between 0.5 and 3 g C m^{-3} seawater (i.e., parts per million). However, this is an order of magnitude greater than the amount of particulate matter, including plankton and detritus, that is present in oceanic seawater (0.02–0.2 g C m^{-3}). It has been estimated that there are some 1.5×10^{12} tonnes of dissolved organic matter in the oceans (Raymont, 1963; Jørgensen, 1976).

Very little of this can be utilized by animals, for more than 90% is resistant to bacterial attack and is not available to living organisms. Of the remainder, a small percentage of free amino acids, glucose and other carbohydrates is available for any organisms that can extract them from the water. In oceanic water, amino acids are present at a concentration of 0.02–0.05 g m^{-3}. Inshore the concentration is higher and ranges from 0.02 to 0.2 g m^{-3} (North, 1975). Glycolic acid is the major carbohydrate that leaches from phytoplankton during photosynthesis (Shah and Fogg, 1973).

Dissolved organic matter can be utilized as a food source by micro-organisms, many planktonic organisms and some animals like the mussel, *Mytilis edulis* (Fig. 6.1(a)).

3.2 PLANKTON AND OTHER ANIMALS

An important part of the biotic environment is the multitude of planktonic organisms living in the water column which provide a source of food for many sessile animals.

Only planktonic organisms *c.* 1 mm or less will be suitable as food for the majority of suspension- and filter-feeding animals. The most important component in the size range 50–200 μm is the phytoplankton, consisting for the most part of diatoms and dinoflagellates together with other unicellular autotrophic organisms. Equally important are the very small diatoms and flagellates, known as the nanoplankton, that lie in the 5–50 μm size range. Some authors include bacteria in the nanoplankton. The abundance of phytoplankton may be much higher in coastal waters than in the open oceans, owing to nutrients being stirred up by the waves or delivered by rivers. Phytoplanktonic organisms carry out photosynthesis and are the primary producers of the sea.

Zooplankton consists of two components: animals that live permanently in the water column, such as calanoid copepods (the holoplankton), and the larvae of numerous benthic animals that only spend a short part of their life in surface waters (the meroplankton). Planktonic larvae may be extremely abundant in coastal waters at certain times of the year (depending on the breeding season) and constitute a major source of food for suspension and filter feeders.

Many animals living on the sea shore which have a hard shell or exoskeleton provide a firm substratum for the attachment of sessile species. Stebbing (1971) listed 43 species of sessile animals living epizoically on the sublittoral bryozoan, *Flustra foliacea*. Within the intertidal zone, mussels and oysters often have tube worms, bryozoans, hydroids and barnacles growing on them. Even the shells of animals that move about may be colonized; for example, the surf barnacle, *Tesseropora rosea* (Fig. 5.3(i) and (k)) frequently lives on the shell of a limpet, *Cellana tramocerica*. As barnacles are not present on the surrounding rock surface, it is likely that the browsing limpets destroy newly settled barnacles except for those on their shell.

Other animals may compete for food or space by growing larger or faster than adjacent species. Some produce chemical substances that inhibit growth of other species or prevent their larvae from settling in the vicinity. This and the subject of competition will be dealt with in later chapters.

SUMMARY

The rise and fall of the sea, caused by gravitational and centrifugal forces of the earth–moon couple, create the littoral zone which is subject to regular cycles of exposure to the atmosphere and submersion by the sea, approximately twice each day. Maximum tidal swing — spring tides — occurs around new and full moon; minimum tidal swing — neap tides — occurs around the first and third quadrature. Higher levels on the shore may not be submerged for several days during neap tides.

Winds and storms at sea generate waves and swell which have a more profound effect on shore conditions than the tides. Swell breaking on rocky shores gives rise to inertial pressures far greater than the changes in hydrostatic pressure caused by tidal movements. Waves stir up organic detritus (which provides food for suspension feeders) and sand (which exerts a strong abrasive effect); they oxygenate the water and disperse toxic substances (such as hydrogen sulphide). High energy waves produce strong shear forces and create a splash zone above high water mark which enables some animals to live higher on the shore than the tides would allow. Wind also enhances the drying and temperature-lowering effect of evaporation.

Animals exposed to the atmosphere may experience many potentially lethal effects, such as loss of water (desiccation), reduction in oxygen availability, reduced feeding time, osmotic stresses and problems with excretion of nitrogenous waste, and during the day; solar radiations which may cause photosensitization, reduction in photosynthesis, wide changes in temperature or death through UV light. Heat radiation during emersion may lead to freezing.

Seaweeds modify wave action, provide shelter from the sun and prevent water loss when the tide is out. Indirectly rotting seaweed provides food for suspension feeders in the form of detritus. Other food sources are planktonic organisms, dissolved organic matter and other members of the shore community such as amphipods, harpacticoid copepods that live on the surface of seaweeds.

Many different animals live in the littoral zone. Sessile animals feeding on plankton or detritus have an abundant supply of food which can be collected with very little expenditure of energy. They are disadvantaged by not being able to move away from harmful conditions or predators, but many counter strategies have evolved to overcome these disadvantages.

From the highest tide level down to the lowest region affected by the tides there is a wide range of environmental conditions of decreasing severity and increasing community complexity. Each species tends to occupy a discrete band or level on the shore. Barnacles dominate the middle shore with mussels and tube-living worms taking over towards the lower boundary. The lower shore is dominated by kelps and, in South Africa and Australia, large sea squirts.

4

Requirements for life on the shore

From the foregoing account of the littoral environment it is possible to predict a number of capacities an animal must possess before it can survive on the shore. Foremost among these is the ability to hold on to the substratum. Possession of a hard outer covering, to protect soft parts of the body from buffeting by waves and brief periods of exposure to the air, enables many animals to live in the sublittoral and on the lower shore, but for sessile animals to survive on the middle and upper shore, the evolution of physiological mechanisms enhancing their ability to withstand the stresses of respiratory interruption, water loss and wide fluctuations in temperature are essential. Other modifications might be desirable to overcome problems of turbulent currents, restricted feeding times, reproduction and interaction with other organisms.

4.1 ATTACHMENT TO THE SUBSTRATUM

Violent turbulence and currents caused by waves breaking on the shore demand that plants and animals have a firm attachment to the substratum, to prevent certain death from being washed up the shore and stranded above tide level, or the equally hazardous event of being washed out to sea. Free-living animals are able to cling to the rocks by means of claws, suckers and hooks, but these require energy expenditure either for attachment or release, and this would not be desirable for permanent attachment. Both sessile animals and seaweeds are attached to the substratum by some form of adhesive which requires no energy for maintenance beyond the initial production of the secretion.

There are many different ways in which animals have solved the problems of adhesion, but certain features are common to all. In most instances a solid or firm part of the animal's body is very closely applied to the substratum during growth, and the intervening space is filled with various organic secretions (mucopolysaccharides, glycoproteins or protein complexes). Protein materials are usually changed chemically or polymerized to form highly viscous or solid bonding material between base and substratum. The physico-chemical properties of the surfaces to be joined and the adhesive are important in determining the nature of this attachment.

The strength of an adhesive depends upon forces of attraction at the molecular level. If two surfaces come within *c.* 0.5 nanometres (1 nm = 10^{-6} mm), van der Waal's forces of attraction will start to act between the surface molecules. At *c.* 0.3 nm the somewhat stronger forces of hydrogen bonding come into play, and at 0.15 nm much-higher-energy chemical bonds (ionic and covalent bonds) may form. It is essential that the molecules of the adhesive are not easily pulled apart (i.e., they must be cohesive), and this is achieved by cross-bonding between the molecules. Bonding between molecules of the adhesive and the surfaces is possible only if all water between the surfaces (except, perhaps, a monomolecular layer) has been displaced. It is likely that one of the functions of mucopolysaccharide secretions is to remove the seawater, either by displacement or by absorption. It should be noted that even if all the water between the animal and substratum were to be removed, irregularities of the surfaces would prevent them coming close enough for molecular forces to act and hold them together. This is the function of the adhesive. It replaces the relatively wide gap between animal and substratum with two new surfaces of molecular dimensions, the substratum–adhesive and adhesive–animal interfaces, where bonding can take place (Fig. 5.6(d), zone a and b).

An ideal adhesive would be a polymer of high molecular weight and viscosity, with a limited number of polar or hydrophilic side chains that are capable of forming ionic or hydrogen bonds with the solid surfaces. Some internal cross-bonding would be desirable to increase its cohesion. It would be an advantage for the adhesive to be secreted as a monomer with low viscosity and low contact angle so that it would penetrate the space between the surfaces and displace water, and then harden by polymerization once in place.

Barnacles, bryozoans and sea anemones depend upon the high viscosity of the cement substance (mucopolysaccharides) for firm adhesion to the substratum, but adhesion of the byssus threads of mussels appears to depend upon hardening of the cement by a tanning process. The method of attachment of other sessile animals is not so well understood, but mucopolysaccharides are likely to be widely implicated as adhesive cements.

Strong attachment to the substratum will be of no avail unless the tissues of the body can withstand the strain imposed upon them by external drag forces. Therefore a requirement for intertidal animals is that their tissues have the necessary degree of cohesion. For many animals an exoskeleton will provide this strength. In soft-bodied sea anemones, a fibrous mesogloea anchored to the basal epithelium (Fig. 10.3(d)) gives the necessary cohesion.

Rapidly-flowing water exerts strong lateral pressure or drag on any object that projects above the substratum, thus shape is of considerable importance to littoral species. Animals that have a smooth surface and low profile will offer little resistance to water flow and will not be damaged by turbulent wave action. The sheet-like growth of many compound ascidians, bryozoans and sponges falls into this category. Water flows smoothly round the conical shell of barnacles or the smooth shell of mussels which offer little resistance to flow. There is a marked tendency for barnacles high on the shore to be smaller and have a lower conical profile than those living in the sublittoral zone; this might be interpreted as an adaptation to environmental conditions.

Flexibility of the body is an advantage to species living on seaweeds, where bending of the substratum occurs. It is also important to organisms such as hydroids that grow in height. Flexibility here is usually provided by joints and this subject will be considered later (see page 242).

Animals growing close together may give each other lateral support and, by reducing the surface area exposed to wave action, greatly reduce drag. This is seen in dense populations of barnacles and mussels (Seed, 1976). The large sea squirt, *Pyura stolonifera* (Fig. 1.1(b) g), living in turbulent surf water, often grows so close together that animals touch one another and form a new surface, against which the waves break, some 10 or 15 cm above the rock surface.

Tall tree-like or fan-shaped animals, such as sea fans, scleractinian corals and sponges found in deeper and calmer waters, are absent from the littoral zone, where they would be vulnerable to damage or death by wave action. A theoretical and largely speculative analysis of animal shape based on the form of coral reef species (Jackson, 1979) makes various predictions concerning life history strategies and body structure, but this analysis throws little light on littoral species beyond what might be inferred intuitively. Of far greater importance is the type of growth and its consequences.

Solitary animals like barnacles, mussels or sea anemones grow rapidly in size when young, but growth decreases with age and old animals grow slowly. They occupy a determinate area for attachment and usually depend upon their large size to overcome competition for space from surrounding animals or to escape predators. However, if they are killed or damaged by a predator or some environmental force, they can only be replaced by a new recruit. Organisms with vegetative or asexual reproductive capacities, on the other hand, continue growing at an increasing (geometric or exponential) rate. They tend to spread over the surrounding area, and when attacked by predators or damaged, they stand a good chance that some small part will remain unaffected, from which the organism can regenerate without the need for recolonization.

4.2 PROTECTION OF THE BODY

A protective outer covering of the body has been achieved in a number of ways, although none can be considered as a specific adaptation to shore life, for they are found in other marine animals not associated with the littoral zone. They can be grouped into three categories: (1) extraneous secretions, (2) exoskeletons bound or closely applied to the epidermis that has secreted them, and (3) the unique external covering of sea squirts. Details of these will be discussed later for each sessile group, but the salient features of the first two categories will be outlined here.

Cells in the epidermis of sea anemones secrete mucus. In *Oulactis muscosa*, often found in rock crevices adjacent to surf beaches in Australia, sand grains and shell fragments stick to the mucus, forming a sandy sheath round the upper part of the body which probably protects the anemone from abrasion. Similar attachment of sand has been noted for the North American Pacific Coast sea anemone, *Anthopleura elegantissima*. Hart and Crowe (1977) found that gravel attached to its column reduced evaporative water loss during emersion. Of animals with gravel attached, 68% survived 6 h exposure to winds of $8\,\mathrm{km\,h^{-1}}$ at 20 °C, but only 34% survived these conditions when the gravel was removed. Many tubicolous polychaete worms bind sand, shell fragments or faecal pellets with mucus to form tubes in which they live. An outstanding example is the massed tubes of sabellariid worms that may form spectacular rock-like mounds on the lower shore. Other external secretions result in the deposition of calcium carbonate to form a tube, as in serpulid worms, or a massive skeletal structure as in corals, both of which will be discussed in later chapters.

The second category of body covering embraces a wide range of structures secreted by the epidermis and forming a non-living exoskeleton. This may take the form of a single cuticular layer, as in hydroids, or a two-layered structure in the form of a shell. In both, various proteins are an essential component, although their properties are modified in various ways by sclerotization, incorporation of other organic compounds or mineralization.

Sclerotization involves hardening of the cuticular proteins. It has been studied extensively in insects, but is assumed to take place in other animals where hardening occurs. In insects this is brought about by a tanning process, in which quinones produced in the cuticle react with the free amino acid radicals of proteins, causing dehydration and extensive cross-bonding between adjacent protein molecules. The process is often accompanied by the formation of melanin pigments which give many arthropods their brown or black colour. The enzyme polyphenol oxidase is required to oxidize polyphenols, like *N*-acetyldopamine, to quinones. The presence of these compounds has been taken as evidence that quinone tanning occurs in other animals besides arthropods. Among the sessile animals discussed here, it occurs in the cuticle of barnacles, the periostracum of mussels (Waite, 1983b) and the outer cuticle of some ascidians and hydroids.

Anderson (1976) has proposed another form of sclerotization, in which proteins are bonded to the β carbon atom of *N*-acetyldopamine by their free amino groups. This has been called β-sclerotization and has been reported only for insects among arthropods. It does not result in darkening of the cuticle.

The two layers of a bilaminate exoskeleton differ in composition and functional role. The outer layer is usually very thin and forms a smooth, tough protective coat. The inner layer is usually much thicker and its properties depend upon various organic or mineral inclusions. Organic substances incorporated in the inner layers include chitin, high-polymer carbohydrates like cellulose, mucopolysaccharides and glycoproteins as well as pigments, lipids and even purine excretory products. Mineralization nearly always results from deposition of calcium carbonate.

Chitin is widespread in the animal kingdom and fungi; it occurs in the cuticular exoskeleton of crustaceans, bryozoans, hydroids and inarticulate brachiopods. Elsewhere it is found in the chaetae of polychaetes, the radula of gastropods, chaetognaths and various protozoa.

The chemical nature of purified chitin is a polymer of acetylglucosamine residues linked by β,1–4 glucosidic bonds to form large unbranched molecules. Hydrogen bonding between acetamido groups on adjacent molecules gives chitin its fibrous structure and great rigidity (Furth, 1980). However, chitin is nearly always found associated with proteins, and evidence from X-ray diffraction studies strongly suggests that the chitin molecules are bonded to the polypeptide chains of the protein to form a complex molecule. Chitin also combines with carotenoid pigments in the cuticle of many crustaceans.

Mineralization of the inner layer with calcium carbonate occurs in the hard shell of barnacles, mussels, articulate brachiopods and many bryozoans. Inarticulate brachiopods deposit calcium phosphate in their shell. Calcium carbonate may be laid down in either of its two crystallographic forms, calcite and aragonite, but some organic matrix is always present. In mussels this matrix is composed of protein (conchiolin). In some species of barnacles, appreciable amounts of chitin are present.

If the physical characteristics of a chitinous exoskeleton are compared with the mineralized shell of molluscs, several important points emerge.

The density of an organic exoskeleton is slightly greater than that of seawater, therefore it is well suited to planktonic barnacle larvae and only requires slight rigidity in hydroids to hold up tall branched forms in the water column. Its greater flexibility and elasticity make it ideally suited to small branched sessile forms, such as hydroids and bryozoans, living in turbulent waters. The size of an organic exoskeleton is fixed at the time of its secretion, and an animal cannot increase in size without shedding the old cuticle and producing a new one (as in arthropods) or producing new zooids of the same size (by budding in bryozoans and hydroids).

The tensile strength of mineralized exoskeletons depends to a large extent upon the amount of fibrous organic matrix present. When this is low, tensile strength is reduced and shells are vulnerable to fracture. The organic matrix probably increases the tensile strength by preventing the spread of fractures in the mineral component. The mineralized shells of barnacles and mussels offer another advantage: they allow continuous growth in size by addition at the edges. Calcareous exoskeletons are prone to solution, abrasion and boring by other organisms, thus continuous increase in thickness is an advantage. The main function of the organic cuticle will be to protect the inner layers from solution and abrasion.

4.3 PHYSIOLOGICAL ADJUSTMENTS FOR EMERSION

The first harmful effect likely to be experienced by an animal upon emersion is lack of oxygen. This might be due to the collapse of respiratory surfaces, or because water bringing oxygen can no longer be circulated through the body cavity (sea anemones and sea squirts) or the mantle cavity (barnacles and mussels).

Three major strategies are open to sessile animals for increasing their chances of survival during periods of emersion: structural or behavioural modifications that allow atmospheric oxygen to be used, reduction in energy demands, and change from aerobic to anaerobic metabolic pathways which effectively free an animal from oxygen dependence.

Some barnacles and mussels are able to utilize atmospheric oxygen with little structural modification. If part of the water in the mantle cavity is expelled and replaced with air, atmospheric oxygen can dissolve in the water to replenish the oxygen consumed by the animal. If the shell remains open, oxygen can diffuse from the surroundings into the mantle air as the animal removes oxygen from it. This would support a low rate of oxygen consumption and might be adequate for the first few hours of exposure. In the case of the mussel, it is possible that the moist surfaces of the mantle and gills can act as respiratory exchange surfaces. The process would be relatively inefficient, as the blood does not contain an oxygen-carrying pigment (haemocyanin) and deeper tissues are likely to suffer anoxia.

Aerial respiration would be unsuitable for long periods of exposure, because of the danger of dehydration and the buildup of metabolic wastes normally eliminated to the surrounding water. If saturation deficits between the mantle fluid and the surrounding atmosphere are high, water will evaporate from the mantle cavity, causing an increase in the concentration of the ions present. This, in turn, subjects the body to osmotic stress and loss of water. Under such conditions it would be necessary to close the shell tightly, but this would cut off the external oxygen supply, and oxygen tensions would drop to a level at which they no longer support aerobic metabolism.

Out of water, sessile intertidal species can no longer pump water or feed, and so they pass into a quiescent state with a low level of metabolism. The drop in metabolic rate can be considerable; in *Mytilus edulis* the energy demand dropped to 5% of the normal rate in water (De Zwaan and Wijsman, 1976). This strategy is beneficial over long periods of time because it conserves energy stores and at the same time lowers the rate at which metabolic wastes accumulate in the body.

By far the most important physiological adaptation to long periods out of water is the ability to switch from aerobic to anaerobic metabolism. The phenomenon of anaerobiosis in marine animals has been reviewed by De Zwaan (1977) and De Zwaan and Putzer (1985). In both barnacles and mussels, glycogen is the fuel degraded to yield energy-rich adenosine triphosphate (ATP) during anaerobiosis. The end product in barnacles is lactate, but mussels produce alanine and succinate initially, although after *c.* 24 h propionate accumulates (Gäde, 1983).

Wide salinity changes may cause barnacles and mussels to close their shell in order to avoid osmotic stress, but this inevitably leads to hypoxic conditions and a switch to anaerobic metabolism.

The value of a shell or cuticle to reduce the rate of evaporative water loss from the body has been mentioned earlier and, as already seen, seawater may be retained in the body or mantle cavity, and this could be another mechanism for preventing desiccation during emersion. However, the most important factor determining rate of water loss is the saturation deficit between the animal and the surrounding air, and so behavioural adaptations that enable microhabitats with very low deficits to be selected (rock crevices, under boulders or seaweed, places where runoff from tide pools keeps the rocks wet) will favour colonization of the shore.

An essential requirement for life on the middle shore (i.e., above MLWN tide levels) is a broad tolerance to wide fluctuations in air and rock temperatures, especially in positions exposed to the sun during emersion. Each species has both an upper and a lower limit, beyond which any further change in temperature cannot be tolerated and death ensues. Where tolerance ranges are known, they are generally wider for species living high on the shore than for those living in the sublittoral zone. The extensive literature on temperature tolerance, and the effects of temperature on all aspects of an animal's life, has been reviewed by Kinne (1970). Species extending over a wide latitudinal range may have genetically distinct physiological races whose temperature tolerances are closely tuned to local environmental extremes. This has been shown to be the case for many animals (Bullock, 1955).

When an animal living at one temperature is transferred to another temperature within its normal range, its metabolic rate is altered. Over a period of time there is a tendency for this rate to return gradually to its former level. This ability to compensate for relatively small movements in environmental temperatures is an important part of the process called acclimatization, whereby animals readjust their metabolic rate to slow, seasonal changes in environmental temperatures. Many other adjustments to seasonal changes occur; for example, *Mytilus galloprovincialis* can only survive anoxia for *c.* 2.5 d in summer, but during November to December it can survive 5 d (Zs-Nagy and Ermini, 1972).

The ability of an animal to live at a particular environmental temperature will depend upon the temperature-dependent characteristics of enzymes involved in metabolic processes, the subject has been reviewed by Somero and Hochachka (1976). The evolution of enzymes with slightly different temperature characteristics must have played an important part in enabling marine animals and algae to colonize the middle and upper shore. Somero (1978) has suggested that the mechanism responsible for adaptive changes in enzyme characteristics may depend upon molecular modifications at non-active sites. The latter could play an important role in influencing the activity of catalytic sites on the enzyme molecule, and consequently adjust metabolic rates to compensate for environmental temperature changes.

Mobile animals can avoid the dangers of high temperatures by behavioural means, such as hiding in crevices and under seaweed, or by migrating down the shore, but sessile animals cannot do this, although their larvae may have the ability to find sheltered sites for settlement. A light-coloured shell will reflect some of the infra-red radiation and keep an animal at a lower temperature than nearby black objects. Evaporative cooling is theoretically possible if sufficient water is available in the mantle cavity, but it has not been conclusively demonstrated in any barnacle or bivalve; if continued, it could lead to serious problems of desiccation.

Low temperatures may prove equally lethal to marine animals. The greatest danger will be experienced by species living on the middle and upper shore in high latitudes, where air temperatures can fall far below the freezing point of seawater (*c.* −1.8 °C). On temperate shores the sea seldom freezes, due to the very high heat capacity of water, but there is a marked tendency for barnacles and mussels normally found on the middle shore to live in the sublittoral or lower shore at high latitudes. Here they are protected from freezing by the sea. The body fluids of most marine invertebrates are iso-osmotic with the surrounding seawater and will not freeze unless temperatures fall below −1.8 °C.

Freezing of an animal's body is potentially lethal for several reasons. The formation of ice crystals would rupture cell membranes, denature macromolecules, concentrate solutes, and remove the aqueous medium in which all metabolic reactions take place. Many marine vertebrates and some invertebrates contain glycerol or other cryoprotective substances in their cells or body fluids which prevent freezing, but similar substances are not found in barnacles or mussels. In the latter, extracellular water may freeze, causing increase in the osmotic concentration of solutes. Some water within the cells may be bound to macromolecules and not free to crystallize (De Vries, 1974).

Intertidal sessile animals can feed only when covered by the tide, and so the time available for feeding is related to height on the shore. Any structural modification that increases efficiency of food capture, or physiological mechanism that improves utilization of the food captured, will be of benefit to intertidal species. Biological rhythms, which can anticipate periods favourable to feeding or reproduction, have been observed in many marine animals and may be of significance to sessile littoral species.

Barnacles, oysters and mussels have evolved sufficient of the desirable characteristics outlined above to enable them to live on the middle and upper shore. They must be regarded as the most successful of marine animals to invade the littoral zone. A few tube-living worms and sea squirts succeed in colonizing the lower shore, but the vast majority of species from the lower shore and sublittoral regions show few adaptations to emersion. They are best regarded as marine species adapted to turbulent wave action that can survive very occasional short periods of emersion.

The following account sets out to show the diverse ways in which the various groups of sessile animals have overcome the problems of living in the littoral zone. It starts with the barnacles and bivalves as the most highly adapted groups, proceeds to the less-well-equipped groups of tube-dwelling worms and sea squirts, and finishes with the bryozoans, sea anemones, hydroids and sponges.

SUMMARY

Successful colonization of the middle shore requires adaptations to six aspects of a marine animal's life.

1. Adhesion to the substratum must be greater than the drag forces experienced by an animal. Adhesion is achieved by secretion of a cement substance between the organism and the surface of the substratum. Cements are mucopolysaccharides or other protein substances which may be hardened. Cohesion, or tissue strength, is necessary to prevent drag forces tearing the body from the base. A streamlined shape or very low profile can significantly reduce drag. Firm attachment must not interfere with orientation to currents or restrict growth. A few species (mussels and some sea anemones) are able to move their position on the shore.

2. Softer tissues of the body must be protected from abrasion, being torn by

shear forces or attacked by predators and micro-organisms whilst submerged, and from water loss and harmful radiations when emersed. Protection is usually provided by an outer cuticle composed of protein material which may be sclerotized by tanning or combined with chitin. An inner calcareous layer may be present forming a shell. In sea anemones, where extensive changes in the size and shape of the body prohibit a cuticle, a mucus layer protects the body.

3. Sessile animals cannot feed during periods of emersion or turbulent wave action, thus food intake must be maximized during submersion. Some animals can utilize dissolved organic matter, others derive nutrients from symbiotic organisms in their tissues. In modular organisms multiplication of trophic units increases the area of food collection.

4. For life on the middle shore some modification of reproduction is required. Most sessile species on the shore are hermaphrodite and fertilization is external. Self-fertilization is avoided by incompatibility of gametes, or maturation of eggs and sperm at different times. Onset of reproduction and release of larvae may be phased by environmental factors (temperature, photoperiod, presence of certain food organisms) or initiated by pheromones. Behavioural adaptations of the larval stage to select suitable settlement sites and tolerate conditions on the shore is essential.

5. Physiological and metabolic modifications are necessary for life on the middle shore, but are not so important on the lower shore where emersion is infrequent and of short duration. Gills do not work efficiently out of water. This leads to oxygen shortage which may be overcome by reducing metabolism, utilizing atmospheric oxygen, or switching to anaerobic pathways. Modification of nitrogen metabolism may be necessary as ammonia cannot be excreted during emersion. Desiccation, exposure to lowered salinity and freezing all affect ionic and osmotic balance, and require mechanisms to counteract their effect.

6. Sessile animals must have evolved tolerance to high temperatures, freezing or fluctuating salinity in order to live on the shore in certain latitudes.

5

Acorn barnacles (*Crustacea: Cirripedia*)

Whatever criterion we use to measure success, whether it is height on the shore, the number of species, the density of individuals in a population or their geographical range, several species of balanomorph or acorn barnacles must be ranked among the most successful colonizers of the marine littoral zone. In temperate latitudes they dominate the middle and upper shore, and they are conspicuous sessile animals on rocky shores round the world. This success is all the more remarkable in view of the relatively recent evolutionary radiation of the Balanomorpha during the Miocene and Pliocene periods; indeed, Darwin (1854) suggested that the Recent epoch might be called the 'Age of Barnacles'.

Much of their success in colonizing a variety of microhabitats must be attributed to their crustacean origin and the way typical crustacean features (such as cuticular exoskeleton, biramous jointed limbs, haemocoelic blood system and planktonic nauplius larva) have been modified for a sessile mode of life.

Adaptations fall into two categories, those fitting the animal for sessile life, and those allowing colonization of the littoral zone. Profound modification of the limbs related to filter feeding, shortening of the body, and the loss of eyes and antennae fall into the first category. The loss of a stalk, the adoption of a low conical profile, and locking of the plates forming the shell all reflect adaptation to the influence of waves, while other tendencies, such as reduction in size, fusion of the shell, development of contractile fibres to anchor the shell to the substratum and numerous behavioural modifications are associated with improving their ability to live on the shore. Physiological modifications which enable them to survive out of water can only be interpreted as adaptations to tidal conditions and intermittent emersion.

To understand the nature and extent of these modifications, and to attempt to answer some of the problems they raise, it is necessary to outline the main features of balanomorph organization.

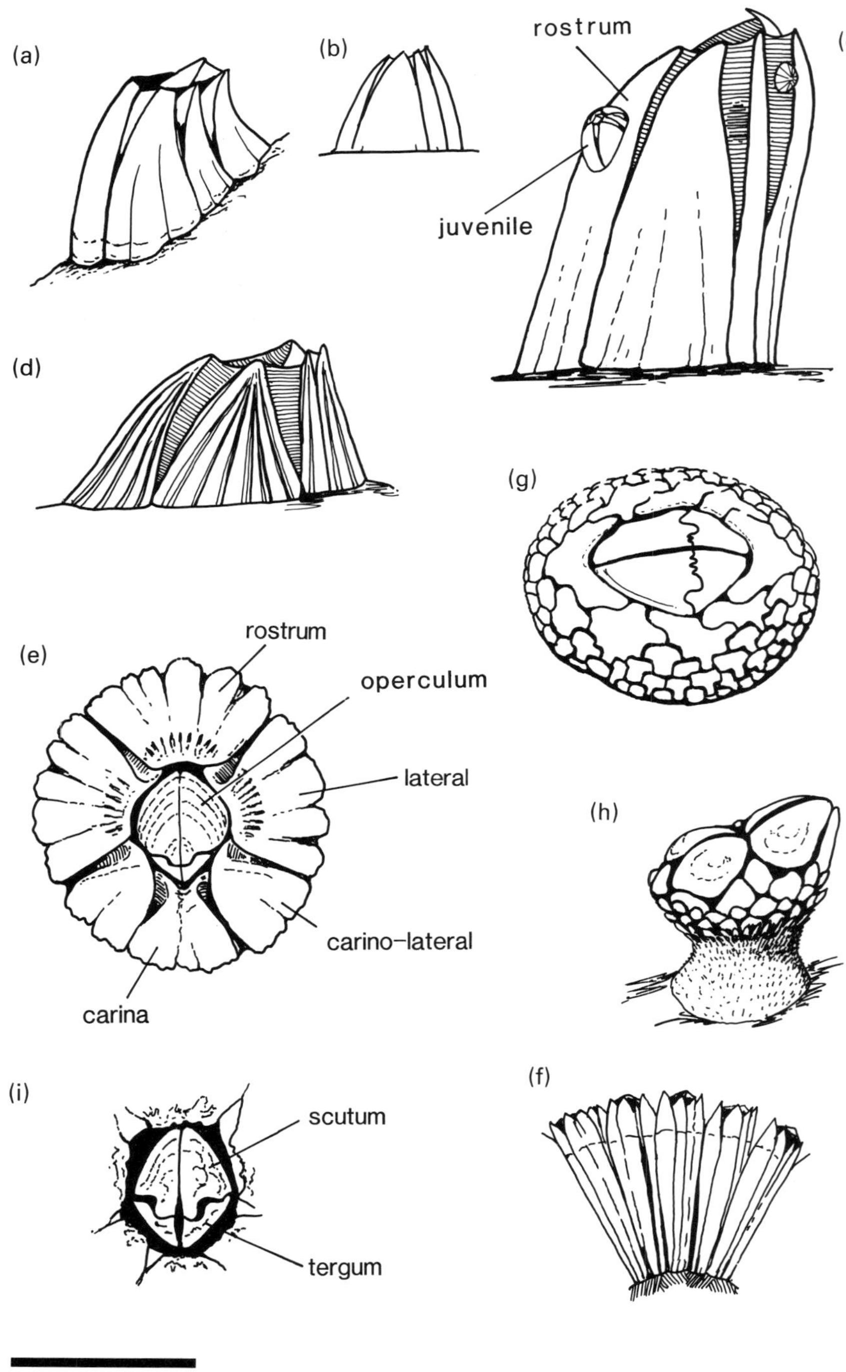
(a)
(b)
rostrum
(c)
juvenile
(d)
(g)
(e)
rostrum
operculum
lateral
(h)
carino–lateral
carina
(i)
scutum
tergum
(f)

True barnacles belong to the order Thoracica* and are divided into three groups: the more primitive stalked or goose barnacles (Lepadomorpha), curiously deformed asymmetrical barnacles (Verrucomorpha), and the acorn or surf barnacles (Balanomorpha) which are the main concern of this chapter.

5.1 STRUCTURE OF ACORN BARNACLES

The body is enclosed in a conical shell formed by four or six (sometimes more) calcareous wall plates or parietes, which are attached to the substratum without a peduncle or stalk (Fig. 5.1). The shell remains open at the apex, but the interior can be shut off from the environment by an operculum of four plates (Fig. 5.1(e), (g) and (i)) that lies in the crater-like sheath (Fig. 5.2). A slit-like aperture between the four plates can be opened to allow the limbs to be thrust out for feeding. The internal arrangement of the body can be seen in Fig. 5.2, where the shell is indicated by black and tissues of the body by stippling.

The body and gills are suspended in a large mantle cavity, filled with seawater, which is in communication with the outside world through the aperture. The body is short and consists of a prosoma, containing the digestive organs and testes, and a posterior thorax bearing six pairs of limbs called cirri. There is no distinct head or abdomen. The body is orientated in an upside-down posture with the anterior or head end attached to the anterior pair of valves (see below). This is a functional necessity for it allows the cirri (which no longer act as locomotory organs) to project up into the surrounding water as a feeding net. The body and cirri can be drawn back into the mantle cavity and the operculum closed for protection.

The general structure of a barnacle is well illustrated by *Megabalanus* (= *Balanus*) *nigrescens*, an impressive species up to 5 cm tall that is common on the lower shore of the East Coast of Australia. A striking feature of living specimens is a sudden display of the beautiful iridescent blue and brown mantle lips each time the aperture is opened for the cirri to protrude. The closely related *M. tintinnabulum* from African, E. Indian and Pacific coasts,

* See Appendix for classification of barnacles.

Fig. 5.1 Barnacles of the shore. (a) *Balanus crenatus*. (b,c) *Megabalanus nigrescens*, young specimen and adult with juvenile attached. (d) *Balanus amphitrite*. (e) *Semibalanus balanoides*. (f) *Semibalanus balanoides* growing close together. (g) *Catomerus polymerus*. (h) *Pollicipes polymerus*, natural size. (i) Eroded operculum of *Semibalanus balanoides*. Scale bar: (a) = 10 mm, (b) = 20 mm, (d) and (e) = 6 mm, (g) = 8 mm, (h) = 15 mm.

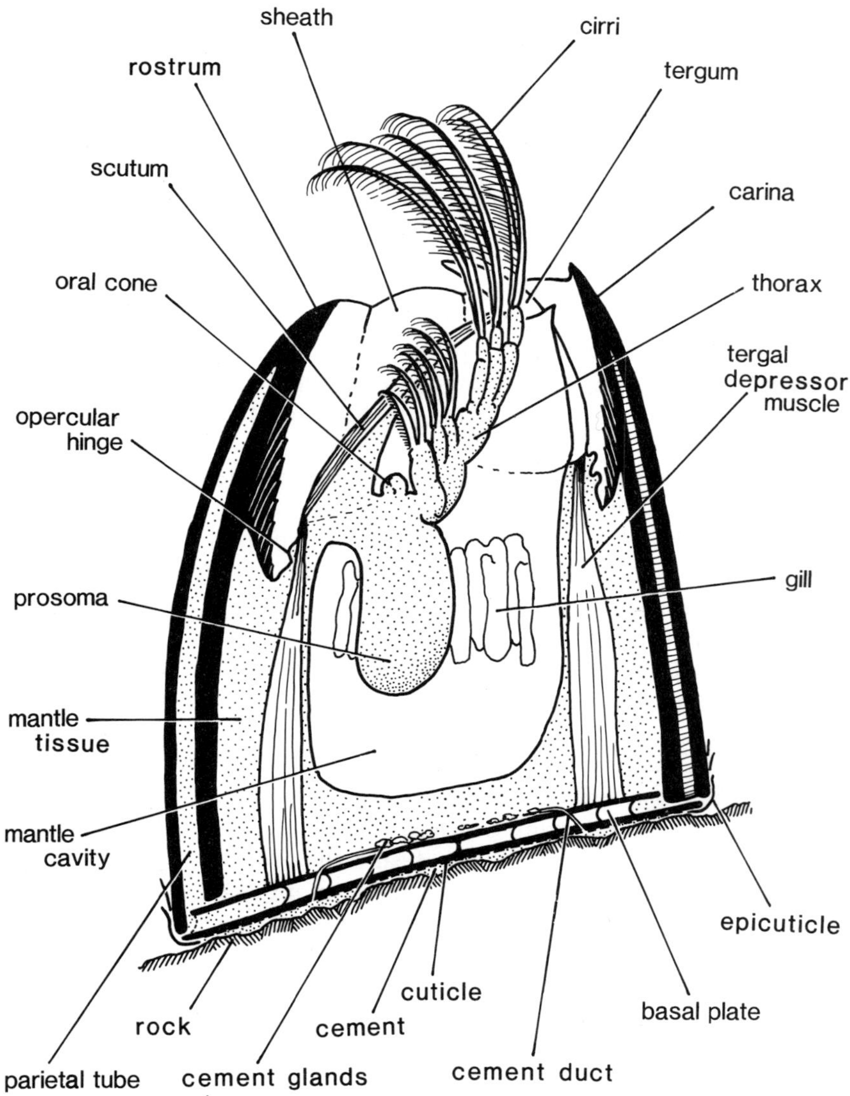

Fig. 5.2 Megabalanus nigrescens. Diagrammatic vertical section to show inverted posture of body and the mantle.

and *M. californicus* from California and Mexico, both show very similar structures.

The shell of adult *M. nigrescens* is cylindrical and only tapers slightly to the apex. The eroded shell of old specimens appears purple or indigo when wet, but the conical shell of young animals may be pink (Fig. 5.1(b) and (c)). It is not difficult to make out the six wall plates that interlock to form the shell. One of these marks the anatomical anterior of the shell and is called the rostrum, while the diametrically opposite plate, the carina, is posterior (Fig. 5.1(e)). The side walls are formed by the paired lateral and carino-lateral plates. Fig. 5.3(b) shows the way in which expansions of the wall plates, called radii and alae, overlap to give rigidity to the upper part of the shell. Below the sheath, the plates abut squarely with a toothed margin (Fig. 5.4(e) and (f)). The base of the shell or basis is a single plate attached to the rock surface by a layer of cement substance (Fig. 5.4(a) and (c)). From Figs 5.2 and 5.4(a) and (b), it will be seen that the wall plates and basis are not solid; they have numerous tubes or canals passing through them which are filled with mantle tissue and haemocoelic sinuses. These tubes, called parietal tubes, run vertically in the wall plates and radially in the basis (Fig. 5.4(a) and (c)).

The pyramid-shaped operculum lies deep in the crater or sheath formed by the upper part of the shell (Fig. 5.2). It is formed by two valves, each consisting of a posterior tergum (immediately recognized by a sharp fang-like process) and a deeply ridged scutum (Fig. 5.3(a)). Each scutum movably articulates with the tergum on the same side, and all four plates are attached to the base of the sheath by a flexible ring of cuticle called the opercular hinge (Fig. 5.2). This allows the whole operculum to be raised or lowered, tilted back and forth or rocked from side to side by the depressor muscles.

The mouthparts comprise paired mandibles, maxillules and maxillae, and are carried on a small protuberance on the prosoma called the oral cone (Figs 5.2, 5.5(e) to (g)). Immediately behind them are six pairs of shorter biramous limbs, the cirri. The three anterior pairs, called maxillipeds, are responsible for capturing fine food particles and passing food to the mouth, while the three posterior pairs of cirri form the main feeding net (Fig. 5.5(a) to (d)). The anus opens between the last pair of limbs close to the origin of the long tubular penis.

The tissue lining the inside of the shell is the mantle. It contains muscles, haemocoelic blood spaces, connective tissue fibres, ovaries and cement glands. Two sac-like outgrowths at the side are the gills. The whole of the mantle including the gills is covered with typical arthropod cuticle continuous with that of the body and opercular lips. From Fig. 5.2 it will be seen that the mantle cavity is open to the surrounding sea, and water can be passed over the gills when the valves of the operculum open.

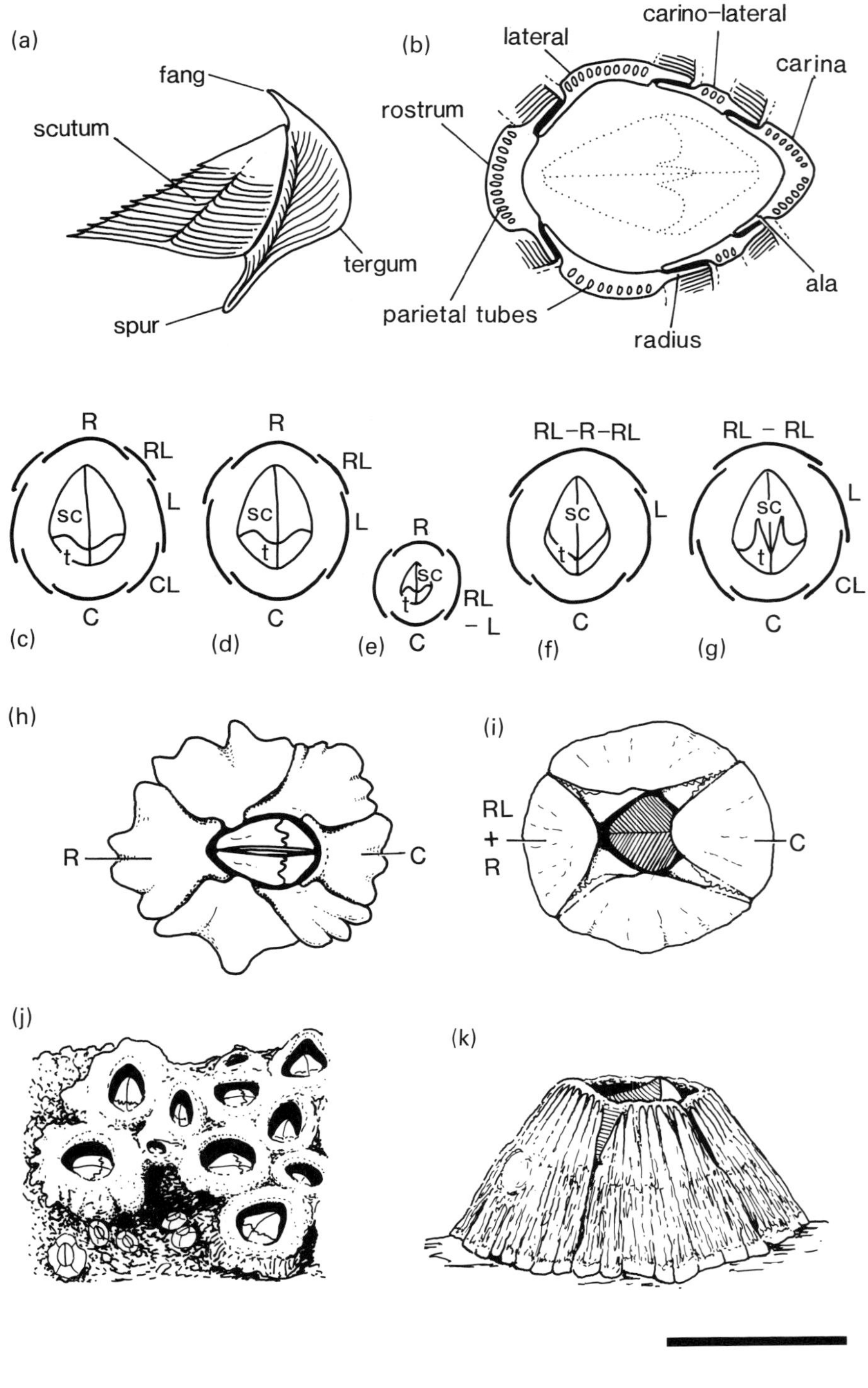
(a)
fang
scutum
tergum
spur
(b)
lateral
carino-lateral
carina
rostrum
ala
parietal tubes
radius
R
RL
L
sc
t
CL
C
(c)
(d)
(e)
RL
- L
RL-R-RL
(f)
RL - RL
(g)
(h)
(i)
RL
+
R
(j)
(k)

5.2 REPRESENTATIVE INTERTIDAL BARNACLES

Although *Megabalanus nigrescens* provides a good example of barnacle structure, little is known of its general biology and physiology; for this we have to turn to other species. Nearly all our knowledge of barnacle ecology, reproduction and physiology comes from studies on half a dozen or so species from the Northern Hemisphere. As frequent reference will be made to them in the following discussion, they are briefly mentioned here with points of difference from *M. nigrescens*.

The following barnacles are representative of the three superfamilies (Balanoidea, Balanomorphoidea and Chthamaloidea) into which the Balanomorpha is divided.

The genus *Balanus* resembles *Megabalanus* in structure. *Balanus balanus* and *B. crenatus* (Fig. 5.1(a)) are widely distributed in N. Atlantic and N. Pacific coasts where they occur on the lower shore and into deeper water. *B. glandula* is common on Pacific coasts of N. America and is found on the middle shore. *B. improvisus* is tolerant of lowered salinities and is found in estuaries in Britain, the East Coast of North America and the Baltic Sea. *Semibalanus* (= *Balanus*) *balanoides* (Fig. 5.1(e)) is extremely common on the cold temperate and subarctic shores of Britain, Europe, North America and the Pacific Coast of Alaska and British Columbia, where it is accompanied by the related *Semibalanus* (= *Balanus*) *cariosus*. It has attracted considerable interest, both from its economic importance and from its marked tolerance of harsh intertidal conditions. Studies on this species have made major contributions to our understanding of barnacle biology; a detailed account of its structure has been given by Stubbings (1975). *Semibalanus balanoides* is a gregarious species occurring in population densities up to *c.* 30 000 m^{-2} in a band between LWN and HWN, but on wave-washed shores it extends considerably higher. The shape of an isolated specimen is shown in Fig. 5.1(e)). It grows to a diameter of *c.* 1.5 cm and a height of 0.7 cm, a size and shape that offer far less frictional resistance to flowing water than the cylindrical shell of *Megabalanus nigrescens*. When very close settlement occurs (i.e., in the order of 50 000 m^{-2}), neighbours resist lateral growth, and the shell takes on a tubular form with a height many times the basal diameter (Fig. 5.1(f)). There is no fang to the tergum, but the

---

Fig. 5.3 (a) Scutum and tergum of *Megabalanus*. (b) Horizontal section through sheath to show interlocking plates. (c) to (g) Arrangement of wall plates: (c) Catophragmidae; (d) and (e) Chthamalidae (e, *Chamaesipho columna*); (f) Tetraclitidae; (g) Balanidae. (h) *Chthamalus antennatus*. (i) *Tesseropora rosea*. (j) *Chamaesipho columna*. (k) Eroded specimen of *Tesseropora rosea*. Wall plates in (c)–(g): C, carina; CL, carino-lateral; L, lateral; R, rostrum; RL, rostro-lateral; sc, scutum; t, tergum. Scale bar: (h) = 6 mm, (i), (j) and (k) = 8 mm.

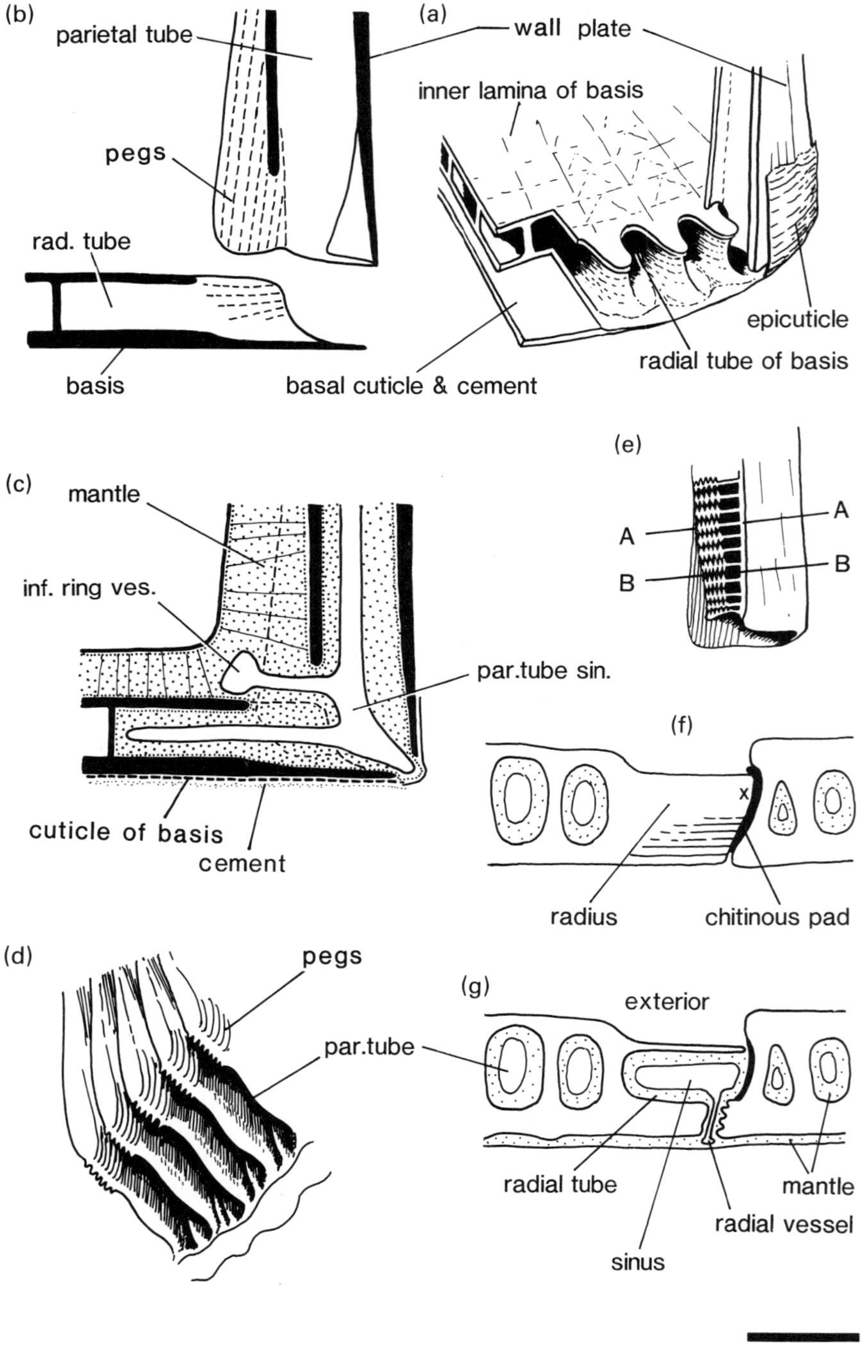
(b)
parietal tube
pegs
rad. tube
basis
(a)
wall plate
inner lamina of basis
epicuticle
radial tube of basis
basal cuticle & cement
(e)
A
A
B
B
(c)
mantle
inf. ring ves.
par.tube sin.
cuticle of basis
cement
(f)
x
radius
chitinous pad
(d)
pegs
par.tube
(g)
exterior
radial tube
sinus
radial vessel
mantle

wall plates overlap in the typical balanoid manner. Erosion may expose the parietal tubes on older specimens and change the appearance of the opercular valves (Fig. 5.1(i)).

The most important difference between *S. balanoides* and *M. nigrescens* is the absence of a calcareous basis. The basis is a chitinous disc that is cemented to the substratum. There are no pegs that interlock with the base, and the shell is anchored to the basis by attachment fibres which will be discussed later. It is evident from the high position it occupies on the shore that *S. balanoides* is physiologically well equipped to stand long periods of emersion and, unlike *M. nigrescens*, is capable of aerial respiration.

The small Australian barnacle *Elminius modestus*, introduced into Europe and now common on sheltered shores in the south of the British Isles, is yet another member of the Balanoidea, but differs from the foregoing species in lacking parietal tubes in the wall plates and in reduction of the number of plates to four.

All the foregoing species belong to the superfamily Balanoidea. The following examples illustrate the range of structure in other superfamilies.

The pink surf barnacle of East Australia, *Tesseropora rosea* (Fig. 5.3(i) and (k)), abundant well above mid-tide level on rocks exposed to extremely heavy wave action, and the closely related *Tetraclita squamosa*, found on the coast of California and widely throughout the Pacific Ocean, are both members of the Balanomorphoidea. Young specimens are white, but older individuals with badly eroded shells appear pink because of the exposed parietal tubes, which are partly filled with a hard pink calcareous deposit. There is an extremely thin calcareous basis, but it lacks radial tubes and does not interlock with the wall plates. The parietal tubes are very wide at their base and have internal buttresses to which the attachment fibres are fixed (Fig. 5.9(a)). The structure of the shell is not easily seen on old specimens, but in young animals four wall plates can be identified.

Fig. 5.4 Structure of wall plates and basis of *Megabalanus nigrescens.* (a) Cut-away diagram showing interlocking plate and basis. (b) Vertical section through wall and basis (separated). (c) Same as (b), showing mantle tissue and haemal sinuses. (d) Base of wall plate viewed from below. (e) End view of lower part of a wall plate where it abuts the neighbouring plate. A series of horizontal radial tubes (shown black) can be seen in the radius. (f),(g) Horizontal sections through the radius. (f) Section A–A passes through the thin lamella between adjacent radial tubes. (g) Section B–B passes through a radial tube and shows its connection with the blood system. The growing point is marked x (see text). Inf. ring ves., inferior ring vessel; Par. tube. sin., sinus in parietal tube.
Scale bar: (b), (c), (f), (g) = 0.3 mm.

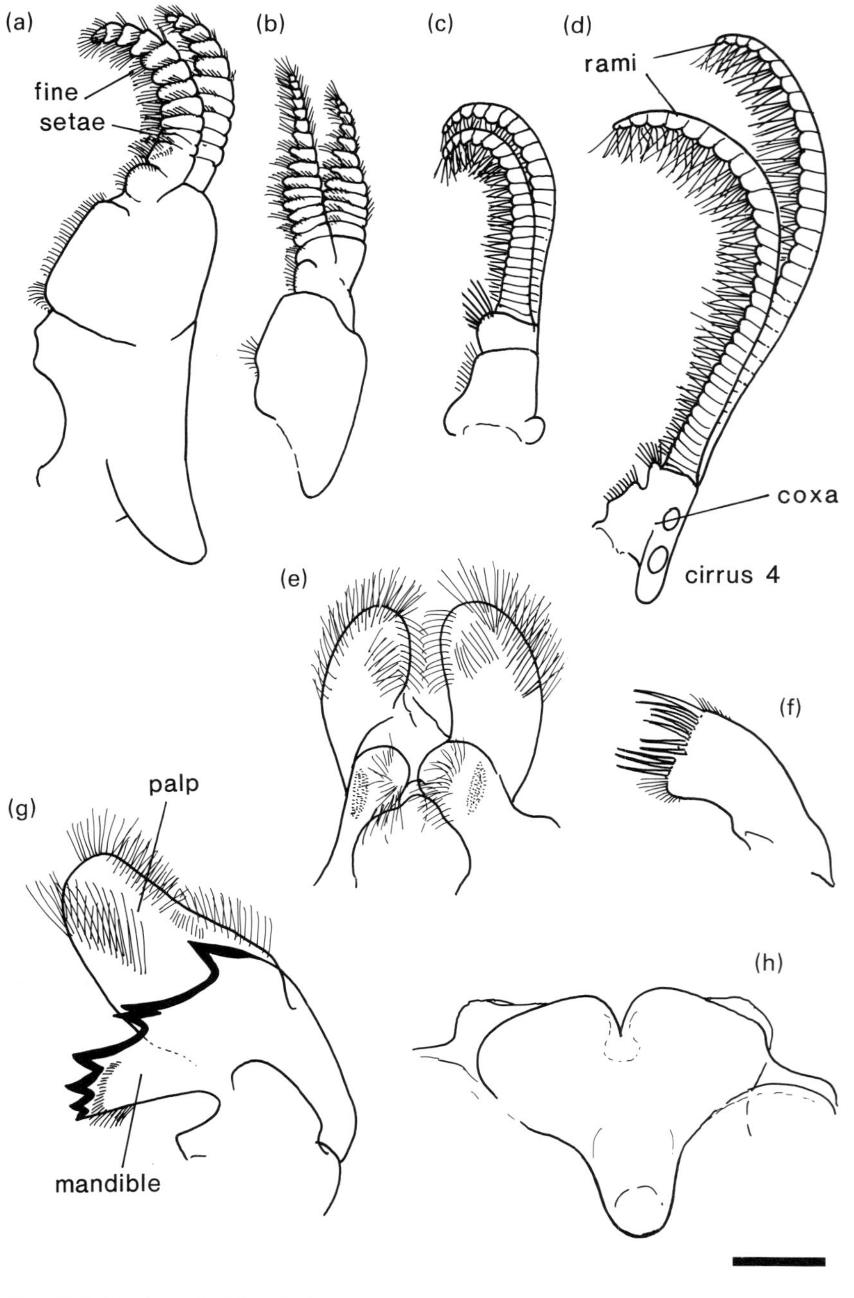

Fig. 5.5 Limbs of *Megabalanus nigrescens*. (a) First maxilliped. (b) Second maxilliped. (c) Third maxilliped. (d) Fourth thoracic limb. (e) Second maxilla. (f) First maxilla. (g) Mandible. (h) Labrum (upper lip).
Scale bar: (a)–(c) = 1mm; (d) = 2mm; (e)–(h) = 0.75 mm.

The highest level to which barnacles extend is usually occupied by various species of *Chthamalus*. This is clearly seen when *Semibalanus balanoides* is present on the same shore; for example, on the west coast of England, where the latter species dominates the shore up to about MHWN, *C. stellatus* extends above it to about MHWS or considerably above this on rocks exposed to heavy wave action. Similarly on Vancouver Island, British Columbia, *C. dalli* occupies the higher levels above *Balanus glandula*, and on the NSW coast of Australia, *C. antennatus* (Fig. 5.3(h)) occupies a level above the honeycomb barnacle, *Chamaesipho columna* (Fig. 5.3(j)).

Species of *Chthamalus* have six wall plates, but the alae and radii overlap in a different way to those of *Balanus*: the rostrum is overlapped by the adjacent rostro-lateral plates (Fig. 5.3(d)). The basis is chitinous and the wall plates are solid (without parietal tubes). *Chamaesipho columna* is also a member of the Chthamaloidea. It is found on Australian and New Zealand shores in astonishing numbers (population densities may range from 50 000 to more than 100 000 m^{-2}) and takes the place of *Semibalanus balanoides* as a zone indicator between MTL and MHWN. The four plates that form the shell can be seen clearly on young specimens (bottom of Fig. 5.3(j)), but the shells of adults are fused together to form the characteristic honeycomb appearance of this species.

The arrangements of the wall plates reflect the relationships among the various species of barnacles. Darwin (1854) considered that balanomorph barnacles evolved from stalked forms with numerous plates protecting the peduncle (similar to the present-day *Pollicipes polymerus*, (Fig. 5.1(h)), found on the Pacific Coast of North America, possibly through some intermediate form similar to the scaly surf barnacle, *Catomerus polymerus*, found among *Tesseropora rosea* on the East coast of Australia (Fig. 5.1(g)).

Newman and Ross (1976) suggested that each of the three groups (Chthamaloidea, Balanomorphoidea and Balanoidea) had a separate origin among lepadomorph ancestors, i.e., the Balanomorpha is a polyphyletic taxon.

The primitive *Catomerus polymerus* is placed in the Chthamaloidea. It has eight plates surrounding the operculum, plus four or five rings of small basal plates forming the wall of the shell. The inner ring of plates have been called rostrum, rostro-lateral, lateral, carino-lateral and carina respectively (Fig. 5.3(c)). In all three groups, there has been a tendency for plates to fuse or be lost. The rings of small scale-like plates (if they were present) must have been the first to disappear. In *Chthamalus* with six plates, the carino-laterals are missing (Fig. 5.3(d)). In *Chamaesipho* the rostro-laterals appear to have fused with the laterals to give four plates (Fig. 5.3(e)). The four-plate condition of *Tesseropora* has been arrived at by fusion of the rostro-laterals and rostrum (Fig. 5.3(f)). A different interpretation has been advanced for the Balanoidea. In species with six plates, the rostrum appears to have been lost, and the anterior plate to have been formed by the fusion of the rostro-laterals

(Fig. 5.3(g)). In this interpretation, the anterior plate is not homologous in the three groups, but to simplify terminology in the following discussion, the anterior plate will be called the rostrum irrespective of its derivation.

5.3 STRUCTURAL ADAPTATIONS FOR SHORE LIFE

A hard protective shell and firm attachment to the rocks must be considered the most important of the many features displayed by barnacles.

The conical shape of the balanomorph shell, its strength and its very large basal area for attachment may well have evolved as a response to the selective pressures imposed by turbulent wave action. At first it might appear a simple step to derive a rigid shell from a series of plates or scales like those seen in *Catomerus*, but the matter is far from simple. Serious problems in growth confront the development of a rigid conical shell and firm attachment. Insight into these problems and an understanding of the many solutions that have evolved can be gained by studying shell architecture.

Attachment

Only the basis is actually cemented to the rock surface. The wall plates could not be attached directly to the substratum, for this would prevent growth in height and diameter, which occurs at the basal perimeter.

Cement glands, located in the mantle tissue lining the basis, secrete a proteinaceous substance (Walker, 1972) that passes through the basis in fine ducts and is deposited near the periphery between the basis and the substratum (Fig. 5.6(e)). When it hardens it has considerable adhesive strength. Yule and Walker (1984a) found that cohesive failure of the cement of *Semibalanus balanoides* occurred at $9.3 \times 10^5\ \mathrm{N\ m^{-2}}$. New ducts are formed at each growth episode (Fig. 5.9(d)), and secretion probably occurs at the time of lateral expansion (described later). A detailed account of the histological structure of the cement glands and development of the duct system in *Balanus crenatus* has been given by Bocquet-Védrine (1970). When an attempt is made to dislodge large barnacles like *Megabalanus nigrescens* or *Tesseropora rosea*, the basis is usually left behind, testifying to the strength of adhesion and independence of the shell.

There are two ways in which the shell is anchored to the basis: by interlocking pegs or by attachment fibres. In *Megabalanus nigrescens* the internal basal face of the wall plates is thickened to form vertical ridges or pegs, with fine striations which interlock with the open (peripheral) ends of the radial tubes in the basis (Fig. 5.4(a), (b) and (d)). The pegs do not fuse with the basis, nor does any part of a wall plate actually touch the basis; a thin layer of mantle tissue always intervenes (Fig. 5.4(c)). It is extremely difficult

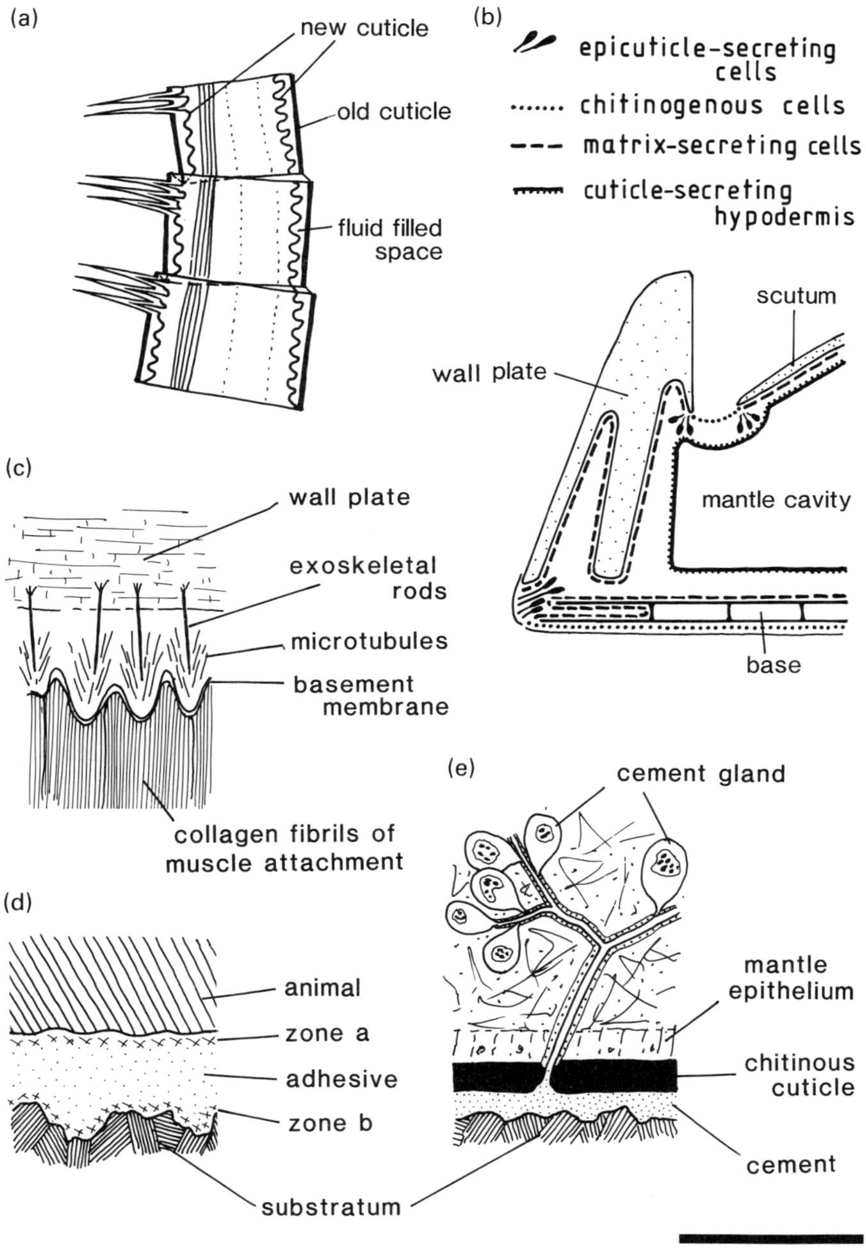

Fig. 5.6 Segments of cirrus in a pharate barnacle. (b) Regions of mantle secreting cuticle (see text). (c) Attachment of muscles to shell (based upon Bubel, 1975). Adhesion (see text). (e) Method of adhesion in *Semibalanus balanoides*.
Scale bar: (c) = 5 μm, (e) = 0.02 mm.

to separate the wall plates from the basis without damaging the pegs, but on specimens that have been treated with alkali to remove the mantle tissue, some movement can be felt between the pegs and their sockets, and by application of pressure they can be pulled out. Interlocking of the wall plates and the basis gives great rigidity to the barnacle shell and firmly anchors it to the rock, but it poses a difficult problem to explain how growth in height and girth take place.

In species such as *Tesseropora rosea*, *Semibalanus balanoides* and *Chthamalus stellatus*, which lack a calcareous basis and interlocking pegs, the shell is firmly anchored to the chitinous basis by numerous attachment fibres (Figs 5.8, 5.9(a)). These fibres are *c.* 0.5 mm long in *T. rosea* and lie along the basal extremity of the wall plates and their buttresses (Fig. 5.9(a) and (b)). Under polarized light a short middle segment is strongly birefringent, indicating that it consists of muscle fibres. Contraction of these would pull the shell down firmly onto the basis. Their contracted length is 0.1 mm, which suggests that the shell cannot be lifted more than 0.1–0.2 mm.

Strength of the shell

The tendency towards reduction in the number of wall plates by fusion (see page 51) would lead to greater rigidity and ability to resist high wave pressures, but the distribution of species on the shore does not lend much support to this idea. *Catomerus polymerus*, a multi-plate species, lives alongside *Tesseropora rosea* on rocks which receive exceedingly high wave impact, while another four-plate species, *Elminius modestus*, lives in sheltered waters where there is little wave action.

A closer correlation can be made between shell shape and severity of wave action. Many sublittoral species tend to have a cylindrical form, while species living on the middle and upper shore have a low conical shell which offers negligible frictional drag to water flowing past. An extreme example is shown by *Tetraclitella purpurascens* (with a height-to-diameter ratio of *c.* 1 : 8), which lives in narrow crevices between rocks or under boulders, where velocity of flow and inertial pressures of waves will be high.

Fig. 5.7 (a) a–e: Development of tubes in parietal plates (based on Costlow, 1956; see text). (b) to (d) Stages in moult cycle and formation of new opercular hinge (modified from Boquet-Védrine, 1965; see text for explanation). (e) Rostral plate of very young *Tetraclitella purpurascens*, showing growth lines corresponding to moult cycles and epicuticular bristles. chit. sec. c., cells secreting chitinous cuticle; epicut. r., epicuticular ridges; epicut. sec. c., epicuticle-secreting cells; lam. sh., lamellae of sheath; mant., mantle; mant. hyp., mantle hypodermis; matx sec. c., cells secreting matrix; new lam. scut., new lamina scutum; op. h., opercular hinge; par. pl., wall plate; scut., scutum.

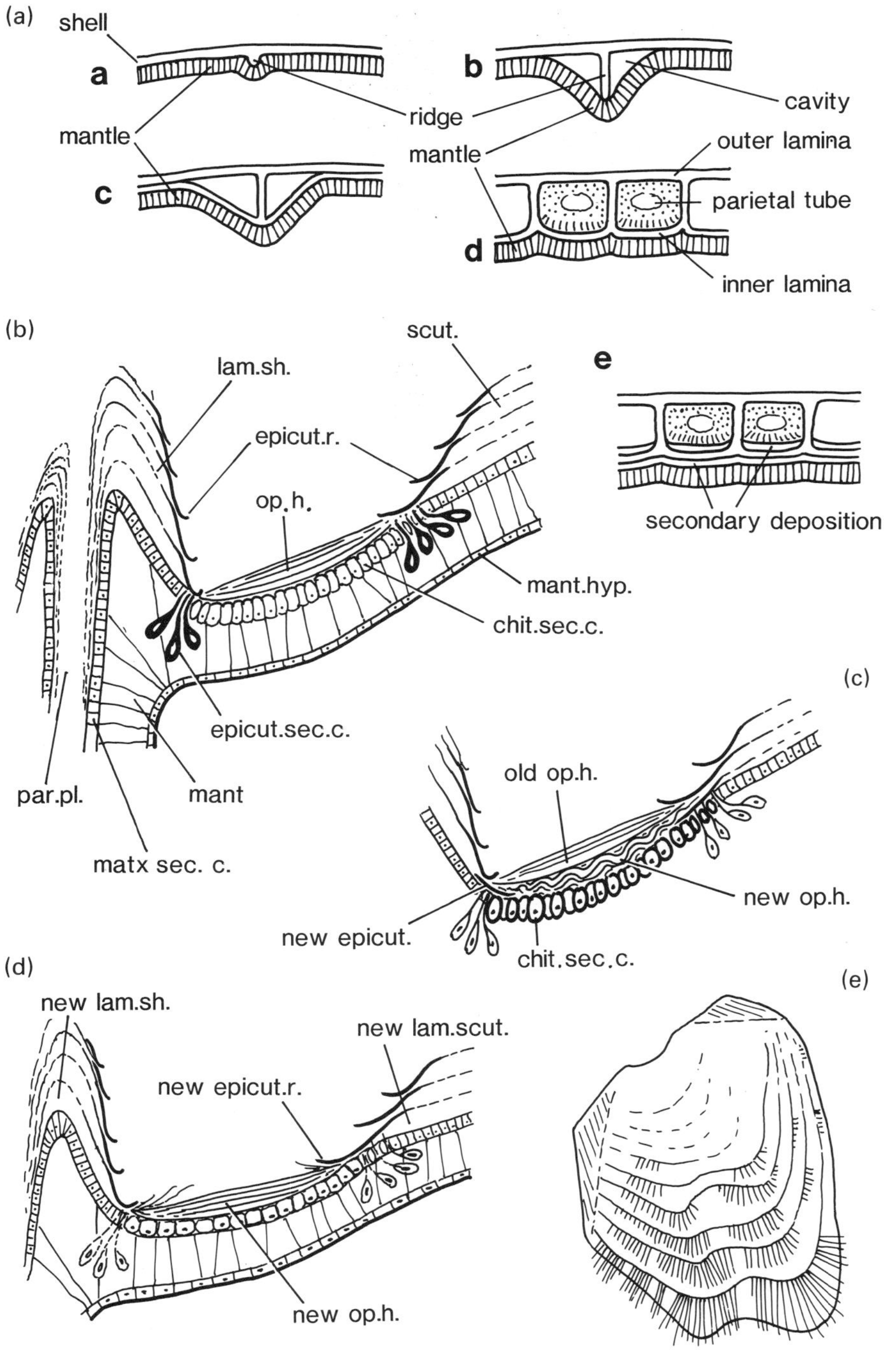
(a)
shell
a
b
c
d
e
mantle
ridge
cavity
mantle
outer lamina
parietal tube
inner lamina
secondary deposition
(b)
scut.
lam.sh.
epicut.r.
op.h.
mant.hyp.
chit.sec.c.
epicut.sec.c.
par.pl.
mant
matx sec. c.
(c)
old op.h.
new op.h.
new epicut.
chit.sec.c.
(d)
new lam.sh.
new lam.scut.
new epicut.r.
new op.h.
(e)

The wall plates of *Elminius modestus* are solid and can only be increased in thickness by deposition of shell on the inside (mantle) surface. In *Tesseropora rosea* and *Megabalanus nigrescens*, where there are vertical parietal tubes in the wall plates, the plates can be strengthened by secondary deposition of shell material within the parietal tubes as well as the inside surface. Vertical ridges are present on the outside of the shell in species like *Balanus amphitrite*, which strengthen the shell. Parietal tubes strengthen the shell in another way. Wall plates can be imagined as two sheets or laminae held apart by the solid partitions between adjacent tubes (Fig. 5.7(a)e). The partitions function in the same way as the web of an 'I' beam, resisting forces of compression (i.e., bending) perpendicular to the surface. In *Tesseropora rosea*, where the tubes are wide basally, secondary partitions or buttresses are formed, giving even greater strength against compression. In the development of parietal tubes we see a fine example of structural economy; the tubes give maximum strength and rigidity for a minimum amount of shell material.

5.4 GROWTH OF BARNACLES

A conical shell like that of a limpet can increase in both height and basal diameter by addition round the perimeter. A barnacle grows in a similar way by addition to the base of the wall plates, but a moment's thought will show that this cannot be the only process involved, for the aperture through which the cirri are protruded must increase in size as the limbs and body grow larger. Clearly growth at the base must be accompanied by changes in shape of the apical opening; this can only occur by erosion at the apex of the wall plates, or by lateral growth at the sutures between adjacent plates (Fig. 5.4(f) and (g)).

Some erosion seems to occur in all littoral species, partly owing to the fragile apex of the plates laid down by the very young barnacle, but also caused by other agencies such as abrasion, rasping by herbivorous gastropods and chemical solution by seawater or the digestive juices of sea stars which can blanket barnacles with their everted stomach.

Species like *Balanus amphitrite* show a pronounced lateral growth of the radii and also (Fig. 5.3(b)) near the apex of the shell, particularly in the younger stages (Fig. 5.1(d)), which rapidly increases the diameter of the sheath opening at each moult cycle by pushing the plates apart laterally. Differential growth in height and diameter may cause a gradual change in the shape of the shell with age; for example, young *Megabalanus nigrescens* up to 1 cm high are conical (Fig. 5.1(c)), but older specimens (more than 2 cm high) are cylindrical (Fig. 5.1(b)).

Growth of the barnacle shell is complex. Besides involving changes in size and shape of both wall plates and opercular valves, it is intimately connected with the moult cycle. Different species have evolved mechanisms unique to themselves, thus it is seldom possible to generalize about barnacle growth.

Moulting

Barnacles moult frequently. The cuticle lining the entire mantle cavity is lost, together with the cuticle of the body and cirri, but no part of the shell is affected. Crisp and Patel (1960) found that at 12 °C, juveniles of *Semibalanus balanoides* moulted at intervals of 6 d, whereas the interval for adults was *c.* 12 d. A marked seasonal variation in moulting frequency was noted. Towards the end of autumn, following sperm transfer and egg laying, the barnacles did not moult for 6–8 weeks. Moulting then resumed at a much reduced frequency until mid-spring, when the larvae were released and moulting reverted to its normal pattern. During the winter months the lining of the mantle cavity was not shed. This appears to be a direct adaptation to incubating the eggs and larvae. Moulting rhythm was not affected by tidal periodicity, starvation or emersion for several days, although it could not take place unless the barnacle was submerged. Injection of the moulting hormone, crustecdysone, into *S. balanoides* increased the frequency of moulting (Tighe-Ford and Vaile, 1972).

The ghost-like exuvium of a barnacle is shed in one piece showing every detail of structure and setation. It consists of the covering to the body, appendages and penis, together with the lining of the foregut and hindgut, the gills and the entire lining of the mantle cavity, but no part of the shell is ever shed in this manner, even though it is a product of the hypodermis. Indeed, loss of the shell would leave the animal unprotected and this might prove fatal.

Profound changes have occurred in the moulting process during the evolution of cirripedes, and balanomorph barnacles deviate from the typical crustacean moulting pattern in three important ways: the shell and opercular valves are never shed, even though they are part of the exoskeleton, the moult cycle does not affect all parts of the shell in the same way, and there are peculiarities with mineralization of the shell. These features call for some discussion.

Some parts of the shell, particularly the epicuticle, may be lost by abrasion, but this has nothing to do with moulting. It is clear that the shell escapes the normal process of ecdysis, although how it does so is not easy to answer. The hypodermal cells of crustaceans synthesize and secrete all the components of the cuticle (Stevenson, 1985). This takes place in a definite sequence or series of steps shown in Table 5.1.

A unique feature of the barnacle moult cycle is that the various steps outlined in Table 5.1 are separated in location and time sequence. Costlow

Table 5.1 Steps in the moult cycle of crustaceans

Step	*Process*	*End product*
1	mitosis and cell division	DNA synthesis
2	epicuticle secretion	proteins (and lipids?)
3	moulting fluid secretion	enzymes (proteinases and chitinase)
4	procuticle secretion	chitin and proteins
5	secretion of polyphenols and polyphenol oxidase	tanning of procuticle
6	endocuticle secretion	chitin and proteins
7	mineralization	deposition of calcium carbonate

(1956), studying *Balanus improvisus*, distinguished three cell types in the mantle adjacent to the shell which were specialized to secrete epicuticle, chitinous cuticle and organic matrix respectively. Each cell type is confined to specific areas of the mantle (Table 5.2; Fig. 5.6(b)).

A consequence of this segregation of cell functions is that some parts of the shell do not possess any epicuticle, while other parts may be without calcified matrix. Bubel (1975) investigated the fine structure and secretory activities of these cells in the mantle hypodermis of *Elminius modestus*. The cuticle produced by normal hypodermal cells lining the mantle cavity has a very thin layer of cuticulin, and a thicker laminated region of fibres consisting of a chitin–protein complex which may be tanned. This structure departs from the usual three layers (epicuticle, exocuticle and endocuticle) typical of

Table 5.2 Distribution of hypodermal cells in barnacles

Cell type	*Location*
Epicuticle-secreting	Basal extremity of wall plates
	Junction of sheath and opercular hinge
	Around periphery of opercular valves
	Between radii and alae, and the sutures between wall plates
Chitinogenous	Periphery of basis
	Between radii and alae
	Underlying the opercular hinge and opercular valves
Matrix-secreting	Inner (mantle) surface of wall plates and basis
	Lining parietal and horizontal tubes of wall plates, radial tubes of basis
	Lower edge of sheath
	Suture between wall plates

decapod crustaceans. The opercular hinge resembles the mantle cuticle except that the fibrous layers are much thicker and remain flexible. Epicuticle on the outside of the shell is secreted by specialized cells and appears to be a tanned protein.

The production of epicuticle and chitinous cuticle occurs at the same time that new cuticle is formed over the body and mantle, just prior to the moult. The secretory activity of matrix cells, however, appears to be continuous. The organic matrix is laid down as lamellae parallel to the surface, and gives rise to compartments in which crystallization of calcium carbonate takes place. By suitable staining of histological sections of the shell of *Semibalanus balanoides*, it can be shown that considerable quantities of chitin are present in the calcified matrix besides protein. Muscles are anchored to the basis and opercular valves by special cells containing clusters of microtubules surrounding exoskeletal rods that are embedded in the matrix of the shell or valves (Fig. 5.6(c)) (Bubel, 1975).

The percentage of mineral matter as calcium carbonate is very high (95% for *Lepas*), although the mechanism for its deposition is not known (Neville, 1975). There is no indication that calcium ions are removed from the shell during proecdysis; indeed, to do so would weaken the shell and defeat the very purpose of its existence. Deposition of matrix and calcium carbonate appears to continue throughout the inter-moult period.

Growth of the operculum

The operculum undergoes a modified moulting process at the same time as the body and mantle, but the outcome is different. The position of the epicuticular, chitinous and matrix-secreting cells of the opercular region is shown in Fig. 5.7(b). In *Elminius modestus*, a ring of epicuticle is laid down at the periphery of each opercular valve and at the base of the sheath (Bocquet-Védrine, 1965; Bubel, 1975). A new hinge is secreted by the chitinogenous cells beneath the old one. Later matrix is laid down and calcified under each valve as well as the wall of the sheath (Fig. 5.7(c)). The old hinge is lost during or after ecdysis of the body, but the old layers of epicuticle and calcified matrix on the valves and sheath remain, forming the characteristic ridges or growth bands (Fig. 5.7(b) and (d)).

Thus at each moult a new hinge is formed, the sheath grows in depth, and the valves increase in diameter as well as thickness. In theory the epicuticular ridges should provide a record of all the moults that have occurred, although in practice erosion at the apex soon spoils the record.

Growth of the wall plates

The external surface of the shell of *Elminius modestus* and very young *Tetraclitella purpurascens* is covered with epicuticle that is arranged in

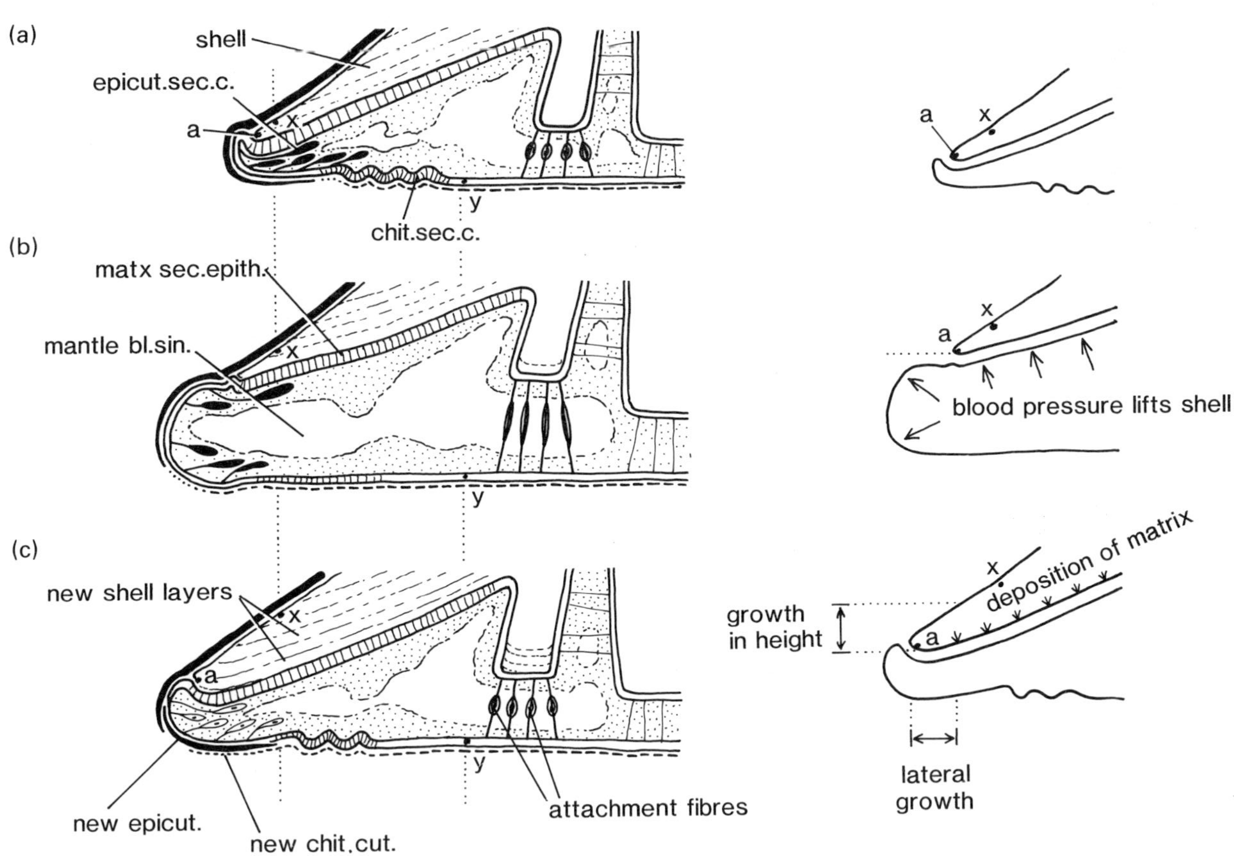
(a)
shell
epicut.sec.c.
a
x
y
chit.sec.c.
(b)
matx sec.epith.
mantle bl.sin.
x
y
(c)
new shell layers
x
a
y
new epicut.
new chit.cut.
attachment fibres
a
x
blood pressure lifts shell
deposition of matrix
growth
in height
lateral
growth

concentric rings or ridges bordered with bristles (Fig. 5.7(e)). Similar rings are found on other barnacles near the base and represent successive growth episodes.

Bourget and Crisp (1975) found that in *Semibalanus balanoides*, a new ring of epicuticle with setae is formed 6–12 h before ecdysis of the body cuticle, thus demonstrating that their formation is related to the moult cycle. During proecdysis, a new ring of epicuticle is laid down under the old one by epicuticle-secreting cells at the basal perimeter of the mantle (Fig. 5.8), and the basis is increased by a ring of chitinous cuticle at its periphery. Describing the process in *Elminius modestus*, Bocquet-Védrine (1965) showed that the old epicuticle split round its perimeter, exposing the new epicuticle. She attributed this to growth at the base of the wall plates.

Matrix-secreting cells of the mantle under the base of the wall plates lay down organic matrix which is then mineralized by deposition of calcium carbonate. This increases the height of the wall plates. Costlow (1956) found that the calcareous part of the wall plate in *Balanus improvisus* grew continuously and growth was not related to moulting. Similarly, Bourget and Crisp (1975) found that matrix was deposited and calcified in distinct bands at the base of the wall plates throughout the period between moults in *Semibalanus balanoides*. Deposition only occurred when the barnacles were submerged; thus, under natural conditions on the shore, two bands appear each day. However, under experimental conditions of continuous submersion, banding still appeared, which strongly suggests that shell growth follows an endogenous semidiurnal rhythm phased by tidal exposure.

To stretch the new cuticle and expand the basis, the shell must be lifted from the basis. This would require a vertical and a horizontal force. Bourget and Crisp suggested that pressure in the body fluids might lift the shell and at the same time push the new epicuticle laterally to form the convex rings, but did not explain the process.

High pressures can be generated in the mantle cavity when the operculum is tightly closed by contraction of the scutal and tergal depressor muscles. This pressure will be transmitted throughout the fluid of the body and hence to the shell wall, which will be lifted by it (Fig. 5.9(e)). The upward movement of the shell will be slight because of the attachment fibres anchoring it to the basis, but it will be sufficient to allow lateral expansion of the epicuticle and increase in diameter of the basis.

Besides growth in height, the wall plates increase in width by growth along the edge of the alae and radii (Fig. 5.3(b)). A chitinous pad is secreted between the ala and radius, and matrix is deposited along their edges (Fig. 5.4(f)).

Fig. 5.8 Growth at base of wall plates and mechanism of lifting wall (see text for explanation). Abbreviations as in Fig. 5.7; bl. sin., blood sinus.

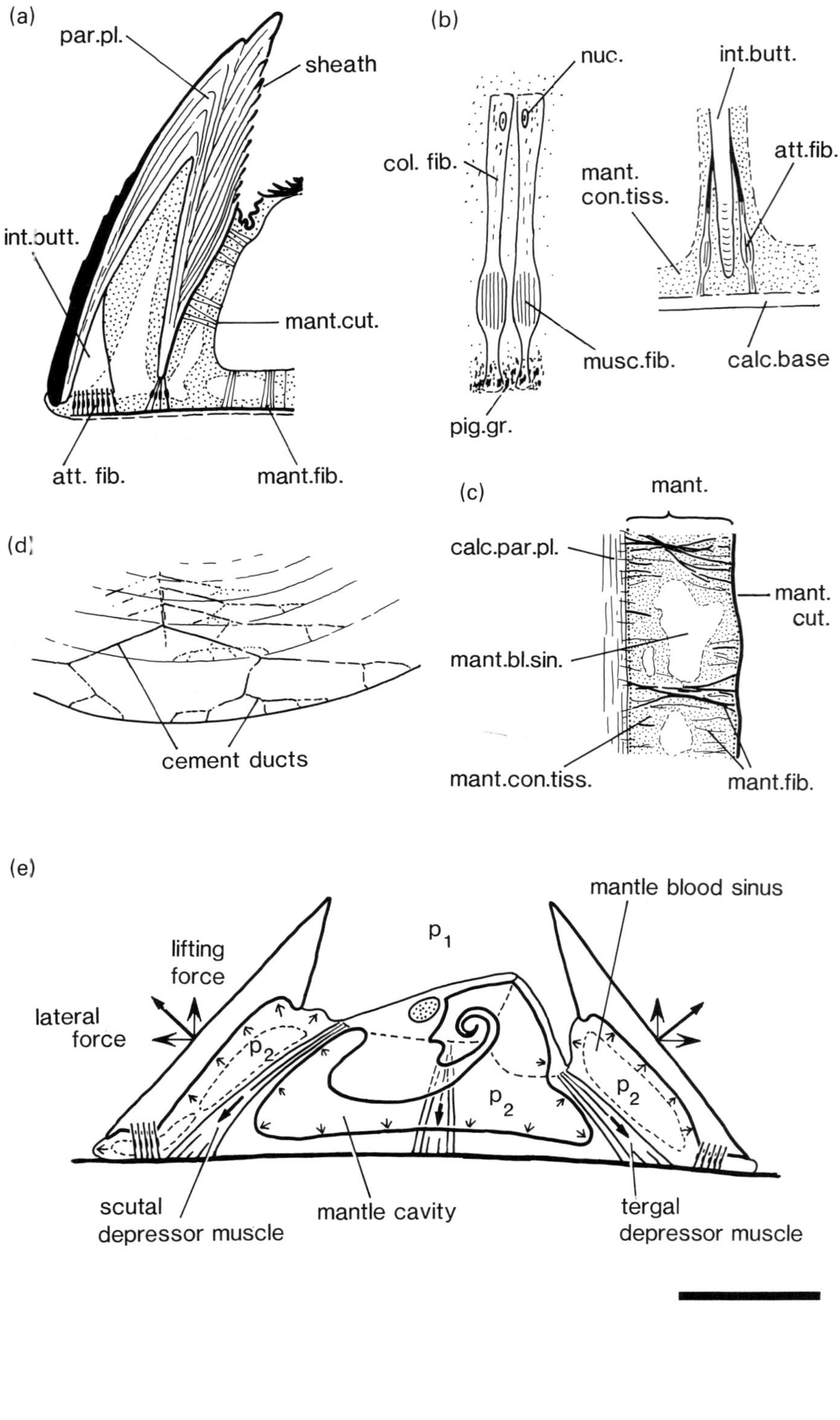
(a)
par.pl.
sheath
int.butt.
mant.cut.
att. fib.
mant.fib.
(b)
nuc.
int.butt.
col. fib.
att.fib.
mant.
con.tiss.
musc.fib.
calc.base
pig.gr.
(c)
mant.
calc.par.pl.
mant.
cut.
mant.bl.sin.
mant.con.tiss.
mant.fib.
(d)
cement ducts
(e)
mantle blood sinus
p1
lifting
force
lateral
force
p2
p2
p2
scutal
depressor muscle
mantle cavity
tergal
depressor muscle

Growth of the wall plates is more complex in *Megabalanus nigrescens* and *Balanus improvisus* both of which possess parietal tubes. An account of how these tubes develop in the young barnacle has been given by Costlow (1956). The first wall secreted by the mantle of a newly settled larva is a single calcareous lamina covered externally with epicuticle (Fig. 5.7(a) a). Longitudinal ridges appear on the inside which lift the mantle away from the wall (Fig. 5.7(a) b). The mantle now secretes a second or inner lamina, leaving spaces which become the parietal tubes (Fig. 5.7(a) c,d). Mantle grows up into the tubes, and in the course of time lays down more calcareous material on the inside of the inner lamina and within the parietal tubes, so that the walls increase in strength. Horizontal septa are laid down in some species (for example, *Semibalanus balanoides* and *Balanus improvisus*), with the result that the parietal tubes are cut off at intervals.

The shell of *Megabalanus nigrescens* grows in height in a manner similar to that described for *Elminius modestus*, except that the way in which the shell is lifted is not so easily explained, for the wall plates interlock with the basis. It is evident that very high pressures will be required to raise the pegs from their slots in the basis. It is possible that the parietal and radial tubes act as hydraulic rams which lift the shell. This topic will be discussed later.

At the top of the shell in the region of the sheath, the radii and alae have a broad area of overlap which is separated by a thin film of epicuticle. Lower down, alae and radii are poorly developed or absent, and adjacent plates abut squarely. In this region there is a wedge of chitinous cuticle separating the plates (Fig. 5.4(e), (f) and (g)) and a series of interlocking denticulations. In *Megabalanus nigrescens*, the radii consist of an outer and inner lamina, with horizontal septae and tubes just like the parietal tubes but arranged horizontally. Each tube contains mantle and a blood sinus that is connected with the blood system of the mantle lining the shell. The structure and arrangement of these horizontal tubes in the radii suggests that they may function as hydraulic rams to force the wall plates apart whilst new matrix is laid down along the edge of the plate.

Fig. 5.9 (a) *Tesseropora* vertical section, showing mantle and attachment fibres. (b) Isolated attachment fibres and tangential section of buttress to show arrangement of fibres. (c) Mantle fibres. (d) Cement ducts in base of *Megabalanus*, showing growth rings. (e) Distribution of pressure to lift wall plates. p_1, external pressure, p_2, increased internal pressure within body (mantle cavity and sinuses) caused by contraction of depressor muscles when operculum is closed. Scale bar: (b) = 0.05 mm, (c) = 0.1 mm. att. fib., attachment fibres; calc. base., thin calcareous base; calc. par. pl., calcified matrix of wall plate; col. fib., collagenous fibres; int. butt., internal buttresses; mant., mantle; mant. bl. sin., mantle blood sinuses; mant. con. tiss., mantle connective tissue; mant. cut., mantle cuticle; mant. fib., mantle fibres; musc. fib., muscle fibres; nuc. nucleus; par. pl., wall plate; pig gr., pigment granules.

It is reasonable to conclude that the shell passes through major growth episodes (during which epicuticle and chitinous cuticle are laid down) that are synchronous with the moult cycle and are probably regulated by the same hormonal system controlling exuviation of the body. Increase in the linear dimensions of the plates due to calcification is not linked to moulting, and follows a tidal rhythm.

There can be little doubt that these modifications of the moult cycle, together with the unique mode of shell growth, have been among the most important factors leading to the extensive adaptive radiation of the Balanomorpha.

5.5 MOVEMENTS OF BARNACLES

Once the barnacle larva has attached itself to the substratum and secreted a shell, further locomotion is impossible. Nevertheless, a wide range of movements of the limbs, body and operculum concerned with protection, respiration, feeding and reproduction can be made. The blood system plays a vital role in all these movements and will be considered first.

The vascular system

The blood system of interconnected haemocoelic sinuses found in barnacles is typical of crustaceans, but it has three unique features: there is no heart, blood pressures are unusually high, and it plays a major role in movements of the body. In *Megabalanus*, the blood sinuses form definite channels resembling blood vessels, but lacking the endothelial lining of true blood vessels, in which a one-way circulation occurs. The blood system of *Megabalanus californicus* has been described by Burnett (1977), and *M. nigrescens* appears similar.

A valve situated between the rostral sinus and basal vessel, and another between the scutal vessel and scutal sinus, compel blood to flow from the rostral sinus into the basal vessel and all the sinuses of the mantle (Fig. 5.10(a) and (b)). On its return, blood passes through the gills to the scutal sinuses, and from there, either to the viscera or the epineural sinus where it is directed to the cirri. Blood returning from the cirri is collected into a large thoracic sinus which eventually leads back to the rostral sinus (Burnett, 1977).

The absence of a heart, noted by Darwin (1854), raises the question as to how the blood is forced through the circulatory system. General movements of the body will compress some sinuses and drive blood into other parts of the body; the valves will ensure that this movement is unidirectional. In the majority of balanoid barnacles that have a regular rhythm of feeding movements, the rostral sinus will be compressed every time the body and cirri

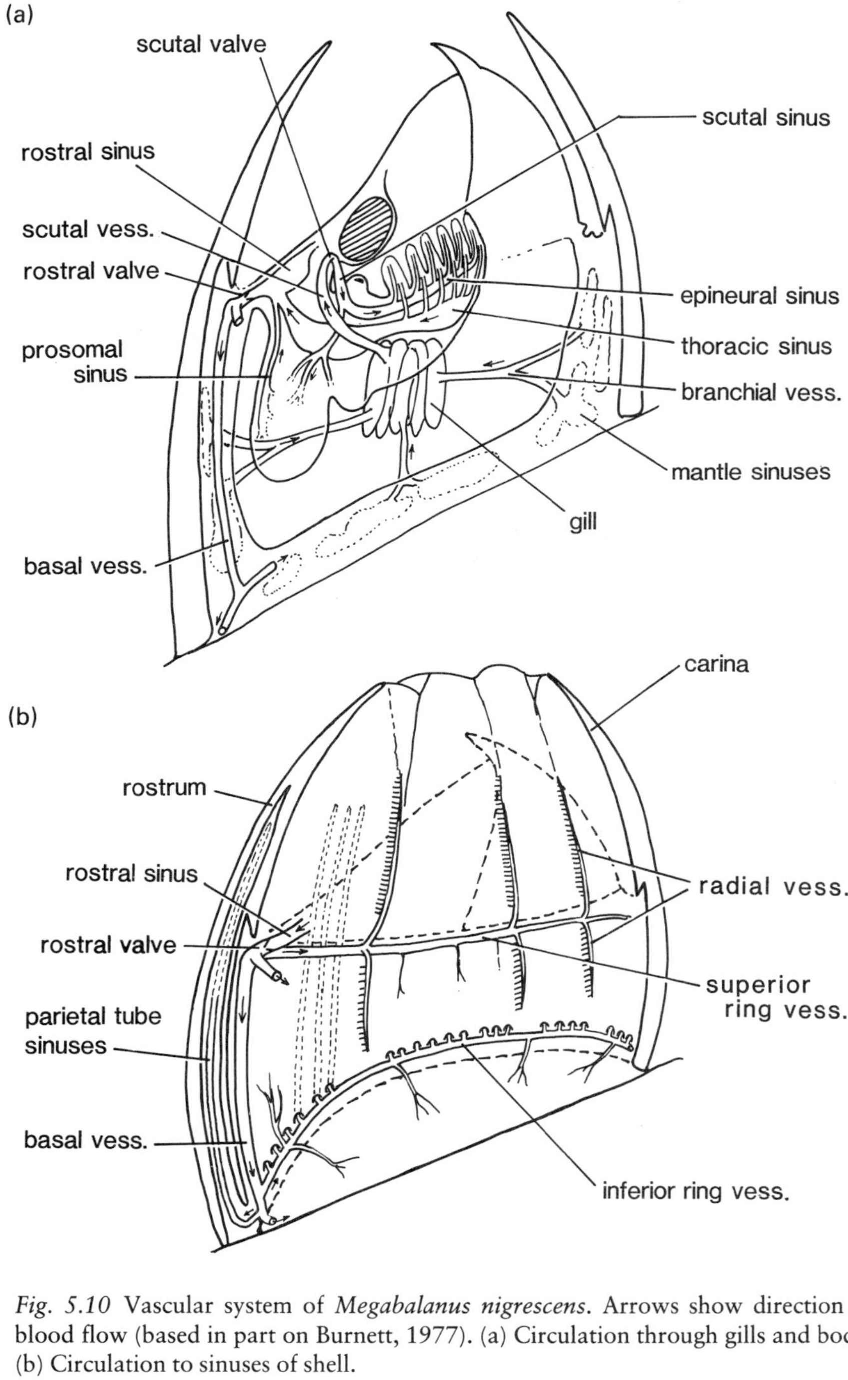

Fig. 5.10 Vascular system of *Megabalanus nigrescens*. Arrows show direction of blood flow (based in part on Burnett, 1977). (a) Circulation through gills and body. (b) Circulation to sinuses of shell.

extend, and the rise and fall of blood pressure would automatically cause circulation of the blood. This cannot be the only mechanism, for slight pulsations can be seen in the opercular hinge of *Megabalanus nigrescens* when the operculum is closed and the barnacle appears inactive. Moveover, a regular and persistent rhythm of blood pressure changes has been recorded by Fyhn *et al.* (1973) in the stalked barnacle, *Pollicipes polymerus*, which does not make rhythmic feeding movements.

The blood pressures recorded by Fyhn *et al.* were remarkably high for an invertebrate animal. They ranged from 1.47 to 3.43×10^3 N m^{-2} at 7–12 °C, and on contraction of the stalk, pressures went as high as 4.9×10^3 N m^{-2}. These are considerably higher values than those recorded from larger crustaceans (1.73×10^3 N m^{-2} for *Homarus*, 1.06×10^3 N m^{-2} for *Cancer* – Prosser and Brown, 1961), and strongly suggest that they are caused by contraction of muscles within the body rather than by a muscular pumping organ.

Amazing pressures were recorded by Tait and Emmons (1925) in *Megabalanus californicus*. Normal pressures ranged between 2.66 and 4.66×10^4 N m^{-2}, with a maximum recorded pressure of 5.73×10^4 N m^{-2} (normal systolic pressure in man is *c.* 1.6×10^4 N m^{-2}). It is unlikely that these represent actual blood pressures, for the way in which the measurements were carried out suggests strongly that they were pressure changes within the mantle cavity. When the opercular valves are closed tightly, contraction of the depressor muscles will exert pressure on the fluid within the mantle cavity to give abnormally high pressures. Such pressures will automatically be transmitted to the blood system, but they cannot be regarded as normal for the vascular system. The fact that it is possible for a barnacle to generate pressures of this magnitude throws light on the functional significance of a calcareous basis and interlocking pegs of the wall plates. Such a design would be well equipped to withstand internal pressures of half an atmosphere* or more without tearing the shell from its base.

The suggestion has already been made that the blood sinuses in the tubes of the wall plates, basis and radii may form part of a hydraulic system that could lift the wall plates or push them apart during growth episodes. It is now evident that the pressures generated within the blood system or mantle cavity are more than adequate to push the plates apart, but it is unlikely that this is done by contraction of the same muscles that move the body. Much higher pressures could be produced by the depressor muscles of the operculum, as previously explained (Fig. 5.9(e)).

One further important property of the mantle has to be pointed out. There must be some mechanism that prevents all the blood passing to a region of low resistance (for example the gills) when the blood pressure rises. Over-

*1 atm. = 101 325 Pa = 101 325 N m^{-2}

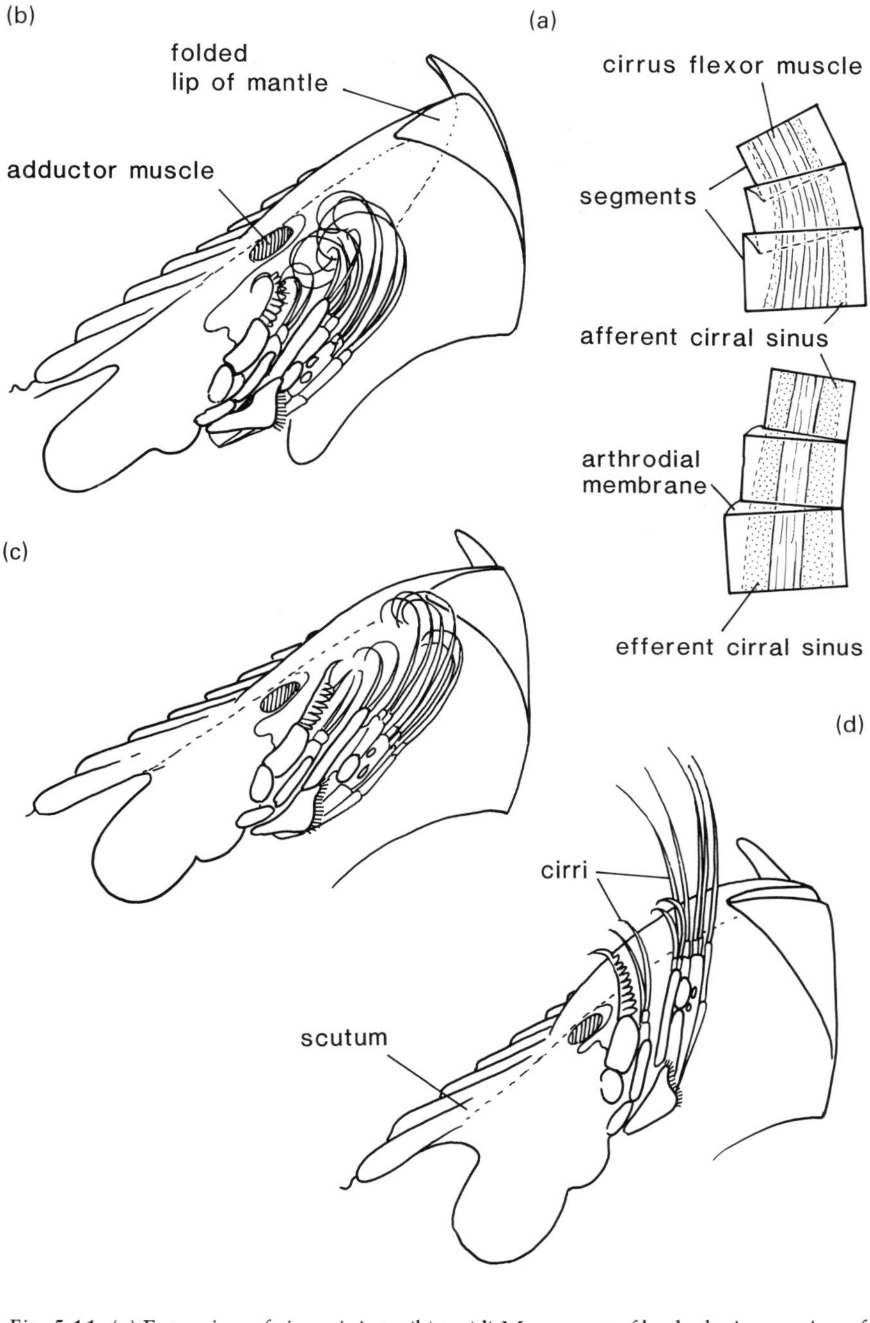

Fig. 5.11 (a) Extension of cirrus joints. (b) to (d) Movement of body during casting of cirri in *Megabalanus*.

inflation of any part of the mantle or gills is prevented by numerous elastic fibres (Stubbings, 1975), which traverse all regions of the mantle lining the shell and the gills (Fig. 5.9(a) and (c)). They probably allow some distension, but will maintain the pressure and prevent rupture or over-dilation of any of the sinuses within the mantle. The mantle is also anchored to the walls and basis by other fibres which fan out and are embedded in the matrix. This is essential to stop the powerful depressor muscles stripping the mantle away from the basis when they contract (Bubel, 1975).

Role of the blood in movement

It may be assumed that the blood carries out vital functions (distribution of nutrients, hormones and oxygen) as in other crustaceans, but in barnacles it takes on a new hydraulic role. It will transmit the forces generated by contraction of muscles in one part of the body to other parts of the body. This is clearly illustrated by extension of the cirri.

Limb movements

The two rami of each cirrus consist of many segments joined by a flexible arthrodial membrane and hinged at their posterior border (Fig. 5.11(a)). A single muscle, the cirrus flexor muscle, runs through each segment and rolls-up the ramus into a tight coil when it contracts. Uncoiling or extension is brought about by increased blood pressure, which forces blood into the afferent cirral sinus. The increase in volume causes the arthrodial membrane to bulge between the segments, pushing each one back and so uncoiling the ramus (Fig. 5.11(a)).

A similar mechanism is responsible for the extension of the penis (Fig. 5.14(a) and (e)). This organ can be extended to three times its resting length by blood passing into the longitudinal sinus when blood pressure is increased. This elongation is made possible by numerous annulations in the cuticle. Directional movements of the penis when it is extended are brought about by longitudinally arranged muscle fibres and two larger muscles at the base. Simultaneous contraction of these muscles will retract the penis if blood is allowed to leave the sinus.

Fig. 5.12 (a) Dorsal aspect of thorax, showing rod-like tergal sclerites. (b) to (d) Mechanism of body extension (see text). (e) and (f) Raising of operculum in *Semibalanus balanoides*. att. fib., attachment fibres; bas., basis; mant. cav., mantle cavity; mant. musc., muscles of mantle; mant. sin., mantle sinus; scut. dep. musc., scutal retractor muscle; terg. dep. musc., tergal retractor muscle; heavy lines indicate muscles in contraction.

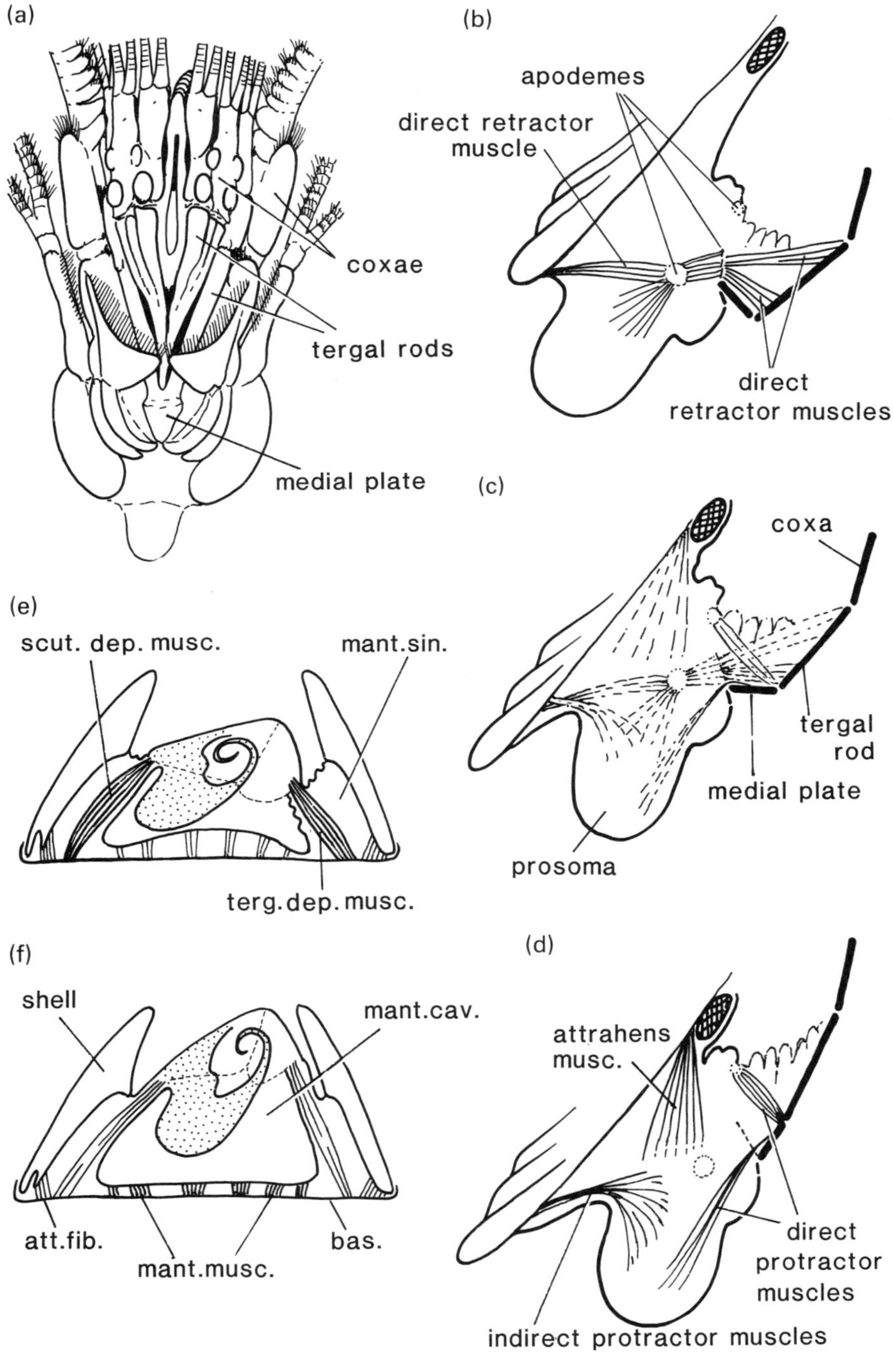
(a)
coxae
tergal rods
medial plate
(b)
apodemes
direct retractor muscle
direct retractor muscles
(c)
coxa
tergal rod
medial plate
prosoma
(e)
scut. dep. musc.
mant.sin.
terg.dep.musc.
(f)
shell
mant.cav.
att.fib.
mant.musc.
bas.
(d)
attrahens musc.
direct protractor muscles
indirect protractor muscles

Body movements

The body can be swung upwards by direct muscular contraction of the attrahens muscle (Fig. 5.12(d)). This may be accompanied by extension of the posterior thorax and uncurling of the cirri into the casting position (Fig. 5.11(b), (c) and (d)). From Fig. 5.12(a), it will be seen that the terga of the segments to which the last three pairs of limbs belong form long rods that run forwards as chevrons to articulate with a medial sclerite. In the retracted position (Fig. 5.12(b)), the joint between this medial plate and the tergal rods is deeply folded. The thorax is elongated by contraction of the direct trunk protractor muscles, which unfold this joint and push the tergal rods backwards (Fig. 5.12(c) and (d)). Extension is assisted by contraction of the attrahens and the anterior component of the longitudinal muscles. Their contraction also forces blood from the prosoma posteriorly into the thorax and cirri, causing the latter to uncurl. Retraction of the thorax and lowering of the whole body is brought about by a relaxation of the attrahens muscle and contraction of other components of the longitudinal muscles in the prosoma and thorax (Fig. 5.12(c) and (b)). In *Semibalanus balanoides*, muscles in the basal region of the mantle displace blood into the tergal sinus when they contract (Gutmann, 1960) (Fig. 5.12(e) and (f)). This lifts the operculum together with the body in the sheath. It will be seen that the opercular depressor muscles directly antagonize this hydraulic lifting mechanism.

Opercular movement

The aperture to the mantle cavity is closed by contraction of the scutum adductor muscle, which draws the scuta together (Fig. 5.13(a) a), but the way in which the aperture is opened is less clearly understood, for there are no direct muscles for this purpose. When slight lateral pressure is applied to the scuta on a barnacle that has been removed from its shell and has had the adductor muscle cut, the aperture opens. In the living animal, a similar lateral pressure could arise in a number of ways: from the narrowing walls of the sheath as the operculum is elevated, from the opercular hinge in the fully stretched position, or by contraction of the lateral depressor muscles when the operculum is raised, provided in each case that the adductor muscle relaxes. Gutmann (1960) considered that the attrahens muscle pushes the body against the scuta when it contracts, and this, together with the lateral depressors, pull the scuta apart (Fig. 5.13(a) a,b). On an isolated operculum of *Megabalanus nigrescens*, a more positive opening of the aperture is obtained when lateral pressure is applied to the tergal spur (Fig. 5.3(a)). In the living animal, lateral pressure might arise from increase in blood pressure in the tergal sinus. It is not unlikely that the body itself helps to push the valves

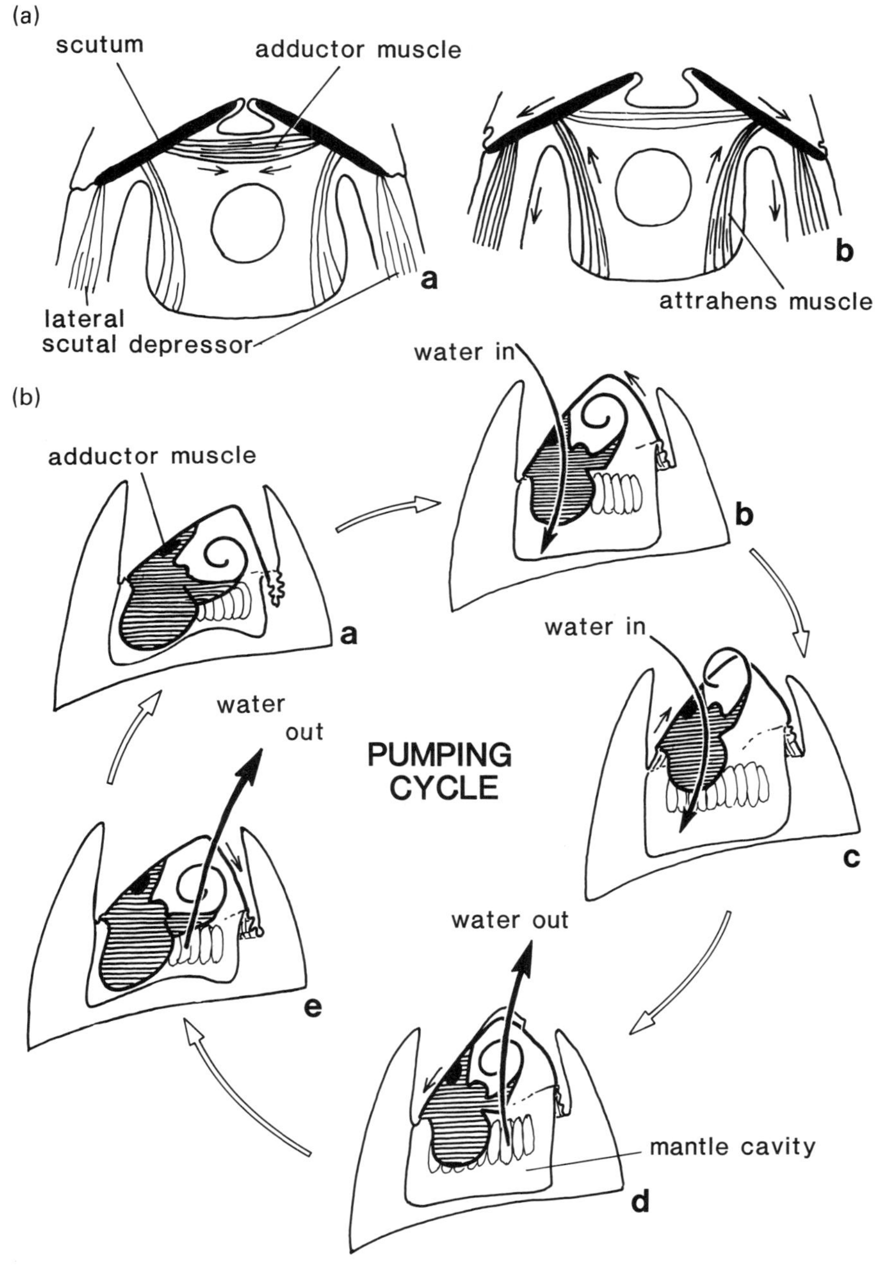

Fig. 5.13 (a) a,b: Method of opening aperture (see text). (b) Pumping cycle of *Megabalanus*, showing water currents through mantle cavity. Small arrows indicate movements of operculum; large arrows show water movements.

apart as it is lifted up into the casting position, and inflation of sinuses in the lips of the mantle surrounding the aperture cannot be ruled out as another mechanism, but there is little experimental evidence to indicate which of these possibilities is responsible for opening the opercular valves.

Three muscles, one scutal depressor and two lateral depressors originating from the basis, insert on the scuta, and a bunch of muscle fibres, the tergal depressors, attaches to the tergum (Figs 5.12(e) and (f), 5.13(a)). Contraction of these muscles pulls the whole operculum down into the sheath, or rocks the operculum from side to side or backwards and forwards.

Functional significance of opercular and body movements

The primary function of the operculum is protection. If any part of the operculum is touched, the aperture is closed and the terga is pulled down towards the carina so that the scuta come to lie in a horizontal position. This reaction may be accompanied by a jet of water from the mantle, which would eject any organism trying to gain access to the mantle cavity. When a finger is placed over the top of a large *Megabalanus nigrescens*, it is not unusual for the barnacle to deliver a sudden upward jab with the fang-like projections of the terga which can be disconcerting and painful. This is clearly a defensive reaction. By touching the opercular hinge with, say, the point of a pencil, the barnacle can be induced to rock the operculum from side to side and crush the implement against the sheath by powerful grinding movements of the scutum. This does not prevent larvae of bryozoans, hydroids or tube worms from settling on the scuta of old specimens and growing there, but the crushing movements would provide good protection against carnivorous prosobranchs (such as *Thais orbita* or *Cabestana spengleri*, common in the same zone as *M. nigrescens*) that might try to insert their proboscis in an attempt to bore through the relatively soft opercular hinge.

Movements of the body cannot be observed directly because of the shell, and only the extension of the cirri can be seen. However, Anderson (1981) was able to observe the movements of the body of *Balanus perforatus* through a hole in the base.

The body can make two types of movement: a rocking back and forth in the sagittal plane, and twisting of the thorax. The backward swing and extension of the body is accompanied by raising of the operculum (Fig. 5.13(b)); this draws water into the mantle cavity. The posterior cirri may be uncurled and spread as a net during the upward swing if the barnacle is feeding, but this does not occur during respiratory pumping. The reverse swing, in which the body is lowered and moved forwards, may be accompanied by sinking of the operculum and rotation towards the carina. This displaces water from the mantle cavity. The intermittent flow of water through the mantle cavity, caused by this pumping cycle, replenishes the oxygen supply to the gills and incubated eggs or larvae.

The other movement of the body is a twisting of the thorax that enables the cirral fan to be orientated in almost any direction (relative to the anterior–posterior axis of the shell) to face into the water flow.

The principal function of body movements is feeding. Feeding activities can be grouped into three categories: passive filtration, casting or normal beating, and microfeeding. Passive filtration is the method of feeding used by stalked barnacles and primitive balanomorphs, such as *Catomerus polymerus*. The cirral fan is extended slowly and held open for a relatively long period whilst water flows past. In still water, the cirri are folded away and the operculum closed (Anderson, 1983). Similar behaviour is seen in *Tesseropora rosea*, *Chthamalus stellatus* and other high-level barnacles that make use of the backwash after a wave has broken over a rock.

Other species rhythmically extend and withdraw their cirri while swinging the body as in pumping. Withdrawal occurs almost immediately after extension. This has been called normal beating (Crisp and Southward, 1961). Its frequency is characteristic of each species and varies with environmental conditions (see page 81). Each cast of the net is capable of capturing planktonic organisms, but it also creates a slow current of water over the shell which draws food towards the barnacle, thus it is particularly favourable to species living in relatively still water or sheltered places, such as *Elminius modestus*. Casting is used by *Semibalanus balanoides*, *Chthamalus stellatus* and many other species.

The cirri do not act as a passive net. Anderson and Anderson (1985) noted that a small item of prey might be captured by a single cirrus which curled downwards from the fan, although larger planktonic organisms were captured by rolling-up the whole fan.

Some of the larger species, such as *Balanus perforatus* and *Megabalanus nigrescens*, carry out regular pumping activity without full extension of the cirral net. Considerable volumes of seawater are passed through the mantle cavity and are filtered by the first and second maxillipeds, which bear fine setae capable of filtering very small planktonic organisms and detritus from the water current (Fig. 5.5(a) and (b)). This food is brushed off by the maxillae and passed to the mouth. Barnes (1959) has shown that *Balanus nubilus* is capable of collecting food particles of diameter 2 μm or less, and has called the process microfeeding. *Semibalanus balanoides* is also capable of microfeeding as well as normal beating.

5.6 REPRODUCTIVE SPECIALIZATIONS FOR SHORE LIFE

All balanomorph barnacles are hermaphrodite and usually capable of producing eggs and sperm at the same time. However, young *Balanus hameri* in their first season are functional males without ovaries (Walker, 1980). A similar phenomenon is found in *Megabalanus nigrescens*, where large specimens may have several young barnacles attached to the shell (Fig. 5.1(b)). Some of these have well-developed testes and penis and are presumed to be capable of inseminating the large barnacle. The high energetic demands of egg production could be met more effectively by large individuals; this might give a selective advantage to small animals functioning as males.

Exchange of sperm between adjacent individuals, leading to cross-fertilization, appears to be the rule among acorn barnacles, although Barnes and Crisp (1956) found that isolated individuals of *Chthamalus stellatus* frequently contained viable eggs or developing embryos in the mantle cavity. As barnacles do not shed sperm freely into the sea, they concluded that self-fertilization must have taken place (although the possibility of parthenogenesis does not seem to have been ruled out). Inbreeding resulting from self-fertilization is considered detrimental, although few experiments on marine invertebrates have been performed to establish the point. However, Barnes and Crisp noticed that many eggs in isolated *C. stellatus* did not develop, possibly because fertilization was unsuccessful. Other abnormalities were found to occur during development. Therefore self-fertilization seems to be less effective in producing normal embryos than cross-fertilization.

The gonads are paired and lie in different parts of the body. The testis consists of numerous lobed seminiferous tubules, which occupy most of the prosoma during the breeding season. Mature spermatozoa collect in two convoluted tubes, the seminal vesicles, which run back on either side of the hindgut. Eventually they join to form a single sperm duct passing through the penis (Fig. 5.14(a)). The penis normally lies between the bases of the cirri, but it can be extended through the opercular aperture and elongated two or three times its resting length. It can probe the surrounding area for another barnacle of the same species in a receptive state (i.e., ready to lay eggs) and, if such an animal is found, the penis is inserted through the opercular aperture, and sperm is passed into the mantle cavity or between the bases of the cirri (Fig. 5.14(e)). Sperm is not normally shed into the sea, possibly because stimuli received during copulation are necessary to evoke ejaculation. The penis of *Semibalanus balanoides* is shed after the breeding season and grows again the following year, prior to the onset of the next breeding season (Crisp and Patel, 1960).

Fig. 5.14 Megabalanus nigrescens. (a) Position of gonads. (b) Oviducal gland. (c) Second instar nauplius. (d) Cyprid of a balanomorph barnacle. (e) *Tesseropora rosea* exchanging sperm. ant. horn., anterior horn; naup. l. 1,2,3, first second and third limbs of nauplius; ov. ap., aperture of oviducal gland; ov. d., oviduct; ov. fluid, fluid secreted by the oviducal gland epithelium (ov. gl. epith.).
Scale bar: (c) = 0.1 mm, (d) = 0.3 mm.

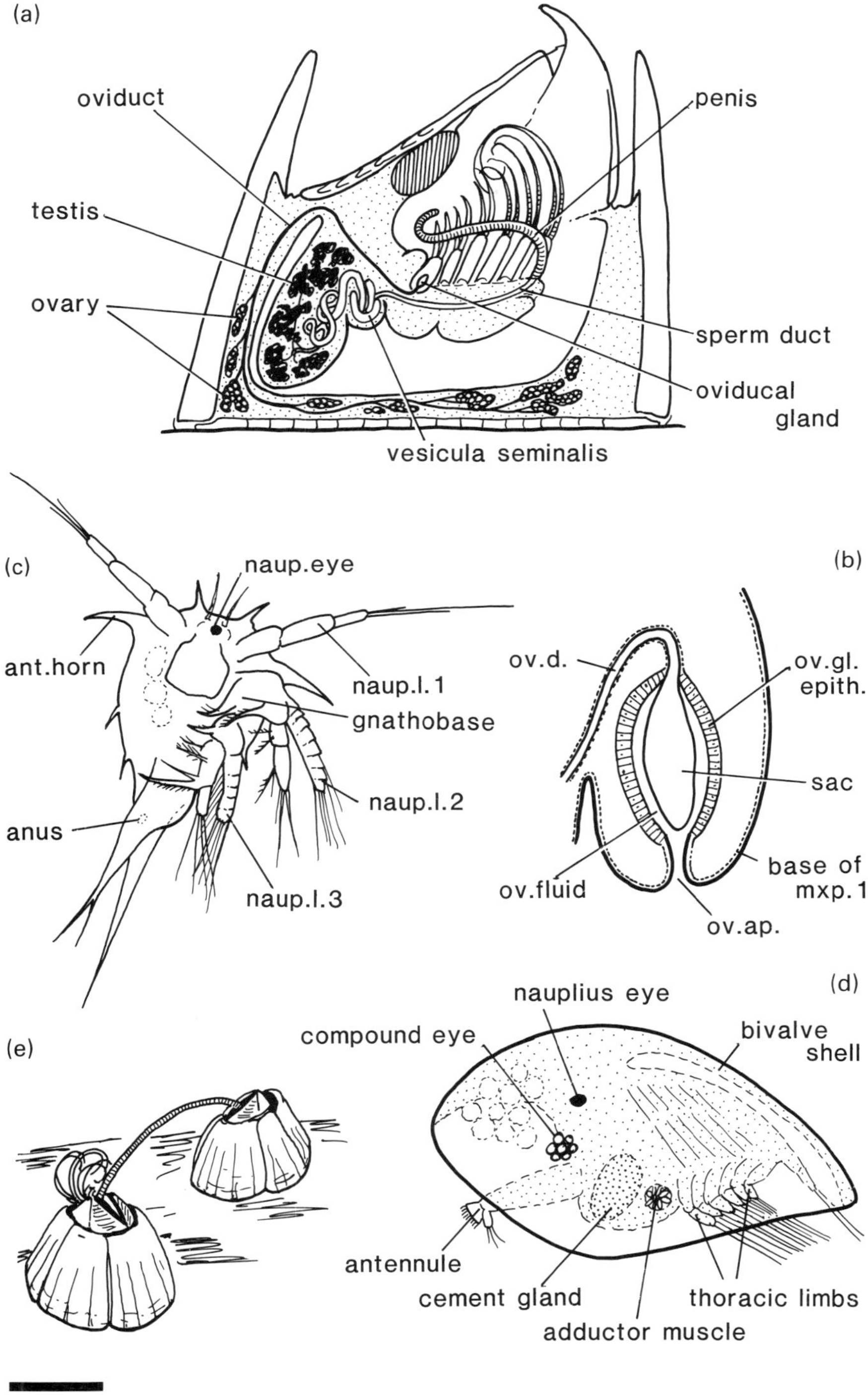
(a)
oviduct
testis
ovary
penis
sperm duct
oviducal gland
vesicula seminalis
(c)
naup.eye
ant.horn
naup.l.1
gnathobase
naup.l.2
anus
naup.l.3
(b)
ov.d.
ov.gl. epith.
sac
base of mxp.1
ov.fluid
ov.ap.
(d)
nauplius eye
compound eye
bivalve shell
antennule
cement gland
thoracic limbs
adductor muscle
(e)

Ovaries consisting of branched ovarian tubules lie within the mantle at the base of the barnacle. The oviducts pass up into the prosoma and finally to the enlarged basal segment of each first maxilliped (Fig. 5.14(a)), where they enter a chamber, called the oviducal gland, that opens to the exterior (mantle cavity) by the genital pore (Fig. 5.14(b)). The oviducal glands play an important role in reproduction. Prior to egg laying, the epithelial lining of the gland secretes an elastic sac, the ovisac, and a fluid secretion that collects between the sac and the wall of the gland (Walker, 1980). Eggs passing down the oviduct accumulate in the ovisac, causing it to expand through the genital pore into the mantle cavity. Oviposition does not take place until after insemination has occurred; indeed, Crisp and Davies (1955) believed that copulation provided the stimulus to ovulation. In *S. balanoides*, multiple insemination occurs and appears to be necessary before the eggs are laid.

When first introduced into the mantle cavity, spermatozoa are inactive and embedded in a jelly-like mass, but the fluid produced by the oviducal glands, and released at the time the ovisacs push into the mantle cavity, causes liquefaction of the sperm mass and activation of the spermatozoa. The fluid also appears to be a chemical attractant, for spermatozoa swim through pores in the greatly distended ovisacs and fertilize eggs. After fertilization, the egg membranes stick together to form a coherent egg mass (Walley *et al.*, 1971; Walker, 1980).

Large numbers of eggs are produced. *Semibalanus balanoides*, for example, may produce up to 8000 eggs (Barnes and Barnes, 1968), while larger specimens of *Balanus balanus* (3 cm) may produce 100 000 or more (Barnes and Barnes, 1954). The eggs are retained in the mantle cavity, where they develop into larvae called nauplii (Fig. 5.14(c)).

Barnacle reproduction falls into two broad patterns, those that produce one brood during the breeding season and those that produce many. *B. balanus* and *S. balanoides* are typical of the first and breed during the autumn and winter. This could be an adaptation to breeding in high latitudes, where there is only a short period of the year during which feeding and dispersal of the larvae can take place.

On British coasts, the gonads of *S. balanoides* grow in size during late summer and autumn, but insemination and egg laying do not take place until November. Embryos are incubated in the mantle cavity during the winter and are set free in spring at the time of the phytoplankton bloom. Breakdown of the ovisac and release of the nauplii is brought about by a proteolytic enzyme (ovicase) secreted by the late-stage nauplii (Barnes and Barnes, 1977a,b). In *S. balanoides*, there is evidence that the adult barnacle releases a chemical substance into the mantle cavity which induces hatching of the nauplii. This 'hatching factor', believed to be a prostaglandin (Clare *et al.*, 1985), is produced when environmental conditions are suitable for larval survival, and appears to be stimulated by the presence of phytoplankton (Crisp and

Spencer, 1958). When first released, the nauplii are positively phototactic and swim towards the surface. The season for this is not clear, but it would ensure that the retreating tide carried them away from the danger of being stranded and killed by desiccation. It would also remove them from the vicinity of adult barnacles that would capture and devour them. The nauplii feed on phytoplankton and eventually change their behaviour to become photonegative and sink to the bottom. Starved nauplii remain photopositive and stay near the surface.

At intervals the nauplii moult. They pass through six nauplius stages, and at the sixth moult undergo a radical metamorphosis into a cypris larva (Fig. 5.14(d)). The nauplii of barnacles generally resemble those of other crustaceans in having three pairs of limbs and a single nauplius eye, but they are immediately distinguished by two antero-lateral horns (Fig. 5.14(c)). All three pairs of limbs (corresponding to antennules, antennae and mandibles) are used for swimming, but the second and third pairs also capture food and pass it to the mouth. Their basal segment is modified as a gnathobase.

The cypris larva is concerned with finding a suitable site for settlement and is specialized for this purpose. The body is protected by a bivalve shell closely resembling that of an ostracod. Besides the nauplius eye, there are two compound eyes, a large antennal or cement gland with a duct opening onto the highly modified antennule, and six pairs of swimming limbs which will eventually be transformed into the adult cirri. The antennules possess sensory organs used to locate a suitable place for settlement. When one is found, the secretion from the antennal gland glues the cyprid to the substratum. A second metamorphosis then takes place, in which the old shell is lost, a new one is secreted, and the body is rotated into the upside-down posture (Crisp, 1976). Dispersal and settlement of the cyprid larvae are discussed in more detail in Chapter 13.

5.7 PHYSIOLOGICAL ADJUSTMENTS OF BARNACLES

The structure of the shell would enable any species of balanomorph barnacle to survive short periods of emersion on the lower shore, but higher levels can be colonized only by those species that have acquired the physiological abilities and tolerances to withstand long periods of exposure to the atmosphere.

Feeding

An animal living at the level of mean high water spring tides (MHWS) will not be submerged for several consecutive days during neap tides; even during spring tides, submersion will seldom be longer than 3 h during the 12 h tidal cycle. The time available for filter feeding (and hence the amount of food caught) diminishes rapidly as one passes up the shore. This is reflected in the smaller size of *Chthamalus stellatus* at its upper limit compared with individuals lower on the shore. Ritz and Crisp (1970) found that specimens of *Semibalanus balanoides* high on the shore had a higher rate of cirral beating when immersed than those living at lower levels, and this presumably compensates for the shorter feeding time. The rhythm of feeding does not persist under constant submersion, suggesting that it is initiated by the tide and is not an endogenous biological rhythm (Naylor, 1976).

As already noted, many barnacles, particularly species of *Chthamalus*, take advantage of runoff from wave splash on exposed shores to supplement the food they collect during submersion. The food these animals catch has to provide for the long periods of emersion as well as the immediate demands of energy and materials for growth. Out of water, barnacles are relatively inactive and so their energy demands will be lower than when submerged. In addition to this, Barnes and Barnes (1957) found that the high-level species *Chthamalus fissus* and *C. dalli* on Pacific shores of North America have a lower metabolic rate than *Balanus glandula* or *Pollicipes polymerus*, which occupy a lower level.

Air breathing

Respiration is another physiological function that must be modified before a marine animal can stay out of water for long periods. Both *Semibalanus balanoides* and *C. stellatus* can respire by taking air into the mantle cavity (Granger and Newell, 1965; Newell, 1979). Soon after being stranded by the receding tide, these barnacles displace some of the water from the mantle cavity by movements of the body, and air takes its place. The opercular valves are held partly open so that a small gap (the pneumostome or micropyle) remains, through which gases and water vapour diffuse. Since oxygen diffuses into the mantle cavity faster than it is used up by the barnacle, it is not necessary for the pneumostome to remain open permanently. This is significant, for water will be lost from the mantle cavity through the pneumostome if there is a large saturation deficit in the atmosphere. Grainger and Newell found that a higher percentage of *S. balanoides* closed their pneumostome at lower humidities than at higher humidities, indicating that the barnacles close the pneumostome to control water loss. Some seawater remains in the mantle cavity, creating a local high humidity, but as water vapour is lost through the pneumostome, the osmotic concentration of the remaining seawater increases and subjects the barnacle to osmotic stress.

Water loss

Although barnacles have a typical cuticle with epicuticle covering the body, limbs and mantle, its permeability to water varies among species and, as might be expected, high-level species liable to severe saturation deficits have the lowest permeability. Foster (1971) found that *Chthamalus stellatus* had a proportionate permeability (proportional water loss / desiccation stress) one-tenth that of the sublittoral *Balanus crenatus*, and *S. balanoides* had half the permeability of the latter species. *B. crenatus* has no pneumostome, and when stranded out of water tends to move the body and keep the operculum open, which results in rapid water loss.

Under experimental conditions using unnaturally low humidities (RH 0% at 19 °C = saturation deficit *c.* 2200 N m^{-2}), *Semibalanus balanoides* had a mean lethal time of 2–4 d (depending on the size of animals) and *Chthamalus stellatus* 7 d. At the higher temperature of 28 °C, *C. stellatus* survived longer (11 d) because the pneumostome was kept closed to reduce water loss (Foster, 1971). This superior ability of *C. stellatus* to survive prolonged exposure helps, in part, to explain its higher position on the shore when both species are present.

In comparison with mussels, which have no cuticular lining to their mantle cavity, these high-level barnacles have evolved considerable control over water loss. Under natural conditions on the shore, they can survive several weeks out of the water. During emersion there is a marked reduction in activity, and aerobic respiration probably satisfies nearly all their metabolic needs. However, if water loss from the body reaches *c.* 20%, the operculum is tightly closed and both *S. balanoides* and *C. stellatus* switch to anaerobic metabolism. The energy-yielding process of glycolysis (Fig. 6.7) proceeds as far as pyruvate, which is then converted to lactate by lactic dehydrogenase (Gäde, 1983; De Zwaan and Putzer, 1985). The lactic acid accumulates in the barnacle's tissues, but unlike that of vertebrates, the oxygen debt incurred is not repaid on resumption of aerobic conditions. Barnes *et al.* (1963) found little evidence of oxidation of the lactic acid or its conversion to glucose when oxygen was again available. They concluded that it was lost to the surrounding seawater by excretion or diffusion. This might appear wasteful of an energy resource, but the waste would be outweighed by the advantage of survival on the rare occasions when animals are forced into anaerobic respiration.

Salinity changes

When *Semibalanus balanoides* is immersed in diluted seawater or fresh water, it closes its operculum tightly and switches to anaerobiosis. This is a valuable adaptation to an animal often washed by rain or living under estuarine conditions where osmotic dilution of the body fluids could occur. It closes its operculum when the salinity drops below *c.* 25‰ (Davenport, 1976), although the exact salinity that triggers the closure response depends upon the rate of salinity change and previous conditions under which the animal has been living (Cawthorne, 1979). Whilst immersed in fresh water they remain tightly closed, and only open when the salinity reaches the same level at which they closed. Nauplius larvae of *S. balanoides* can survive at salinities down to 8‰ due to their impermeable cuticle (Cawthorne and Davenport, 1980). The estuarine barnacle, *Balanus improvisus*, can regulate the osmotic concentration of its body fluids and can tolerate much lower salinities (down to 8‰) before it closes its operculum.

Temperature tolerance

Environmental temperatures affect barnacles in many ways. Each species has a characteristic temperature range in which most activity takes place. Southward (1964) has shown that *Balanus crenatus* and *B. balanus* from Arctic waters are adapted to cirral beating in the range −1 °C to 6 °C, while *Tetraclita squamosa* and *B. amphitrite* from tropical latitudes are adapted to a temperature range from 22 °C to 30 °C. These temperature tolerances also determine to a large extent the geographical range of each species. Within the normal range, temperature changes will affect activities like feeding and pumping, metabolic rates, growth, onset of reproduction, and rate of larval development. Barnacles living in exposed positions on the middle and upper shore will experience the full effect of solar radiation and much wider fluctuations of air temperature than those on the lower shore.

Beyond the normal activity range, there is an upper and a lower limit beyond which further temperature change is lethal. Tolerance to extreme temperatures has been determined for some species under experimental conditions by Southward (1958) (Table 5.3).

Chthamalus stellatus can tolerate high temperatures for much longer periods than *Semibalanus balanoides*; on the other hand, the latter can

Table 5.3 Temperature tolerance of barnacles*

Species	*Time to reach 50% mortality at:* −5 °C	40 °C
Chthamalus stellatus	up to 5 d	29–30 h
Semibalanus balanoides	up to 8 d	3–5 h

* Source: Southward (1958).

withstand lower temperatures. These tolerances are reflected in the geographical distribution of the species. *S. balanoides* is a boreal–Arctic species that becomes frozen during the winter over much of its northern range. It extends as far south as southern England and northern France. Similarly, in the North Pacific *C. dalli* is well adapted to cold conditions and tolerates being frozen in ice for up to 6 months of the year. *C. stellatus*, on the other hand, is a warm temperate species found on Mediterranean and West African coasts as far south as Senegal. Although it penetrates northwards to the west coast of Scotland, it is not found on the colder east coast of the British Isles.

Acclimatization to seasonal changes may occur. Crisp and Ritz (1967) found that *S. balanoides* is more tolerant of low temperatures during the winter than summer. During the summer months, the lower lethal temperature was −6 °C, whereas between December and February it was −17.6 °C. The onset of this cold resistance was found to be induced by several factors, of which photoperiod (short day length), reduced food intake, and falling ambient temperatures were the most important (Crisp and Lewis, 1984). No cryoprotective substances were found in the body fluids, but at −15 °C most of the extracellular water (representing *c.* 80% of total body water) was frozen.

Body temperatures of barnacles exposed to the sun are lower than temperatures of nearby black objects, although higher than ambient air temperatures, which suggests that the white shell reflects some of the sun's radiation (Southward, 1958). *Tetraclita squamosa* could lower its body temperature by up to 4.9 °C below that of surrounding black objects, which suggests that evaporative cooling was taking place (Lewis, 1963); however, this could lead to dangerous loss of water if continued for any length of time.

The rate of cirral beating, frequency of moulting and rate of growth all increase with rise of temperature up to an optimal value characteristic of the species. Above this, rates rapidly decline. *Semibalanus balanoides* reached its maximum beating rate of about one cast per second at 18–20 °C, whereas *Chthamalus stellatus* reached a similar maximum rate at 30 °C, although this was reduced or stopped if the barnacles were starved (Patel and Crisp, 1960). Maximum moulting frequency occurred at about the same temperature as maximum cirral beating.

The dangers of overheating in the sun or desiccating in air of low humidity will be greatest during the period of larval settlement, especially for species living above mid-tide level. After settlement it takes *c.* 24 h for the cyprid to metamorphose and secrete the first protective shell, therefore species living on the upper shore will experience at least one (but probably two) emersions during this period. The potential danger of drying up during emersion would be greatest for cyprid larvae that settled prior to a midday low tide. However, experiments reveal a remarkably high tolerance by cyprids and spat of some

Table 5.4 Median lethal time (h) for cyprids and spat of *Semibalanus balanoides* in air at 15 °C*

P_d *(SW)*† ($N\,m^{-2}$)	*Cyprids* (h)	*Spat* (h)
1667	7	8
834	11	16
416	14	26

* Source: Foster (1971), extrapolated.
† Saturation deficit of seawater.

species to desiccation. Foster (1971) subjected newly settled cyprids and metamorphosed spat to high saturation deficits and measured their median lethal time (MLT). At the relatively high temperature of 25 °C and saturation deficit of 3095 $N\ m^{-2}$, both cyprids and spat of *Semibalanus balanoides* had a MLT a little over 4 h. At temperatures and deficits nearer to natural conditions, survival was much longer (Table 5.4).

In Spring, when the main settlement of cyprid larvae occurs, rock surface temperatures are not likely to exceed 15 °C and saturation deficits will be low (less than 834 $N\ m^{-2}$), therefore, there will be a relatively high probability that newly settled cyprids will survive from one period of submersion to the next. The approach of spring tides would appear to be the best time for settlement, as this would ensure that spat high on the shore are immersed at each tide for several days. Post-larval mortality is highest during the first two or three days after settlement, but much of this can be attributed to destruction of the delicate shell by abrasion and browsing gastropods.

Temperature affects breeding of barnacles in a number of ways. Each species has a critical minimum temperature below which breeding does not occur. Under experimental conditions, the highest percentage of fertilized individuals occurred at or near the optimal temperature for other activities, but fell off rapidly above this temperature. Patel and Crisp (1960) found that *Chthamalus stellatus* and *Balanus perforatus* commenced breeding after 2–3 weeks when kept at temperatures above 15–16 °C. *C. stellatus* living high on the shore benefits from the warmth of the sun during emersion and commences breeding in May, whereas *B. perforatus* living lower on the shore is influenced by water temperatures which do not reach the critical minimum until June or July. *Elminius modestus* has a critical minimum temperature about 8–9 °C and is capable of breeding nearly all the year on British coasts.

The onset of breeding in *Semibalanus balanoides* is more complex; it is influenced by light, temperature and food, and may be subject to an endogenous rhythm (Barnes, 1963). Continuous light inhibits gonad development, both at high and at low temperatures. Breeding was inhibited in specimens subject to normal day/night regime and kept at temperatures above 12 °C. For gonad growth and breeding to take place, animals had to be kept 4–6 weeks below 10 °C and less than 12 h light per day. On British and North American coasts, air temperatures fall below 10 °C long before sea temperatures, and are instrumental in initiating breeding. Individuals low on the shore are influenced more by water temperatures, and commence breeding later, than those high on the shore. The time at which fertilization and egg laying occurs is not influenced by environmental temperatures and may be controlled by an endogenous rhythm.

SUMMARY

Certain species of barnacle occupy the highest position on the shore of any sessile animal. They have evolved a conical shell of low profile composed of four, six or more calcareous plates which can withstand high pressures of wave impact. The barnacle can isolate itself from the surrounding environment by closure of the operculum. The body and tissues of the mantle are covered in cuticle which reduce water loss during emersion. The basis is cemented to the substratum with a proteinaceous material which hardens, but the wall plates themselves are not cemented (this would prevent shell growth at the basal edge). The shell is anchored to the base by attachment fibres. In species of *Balanus* and *Megabalanus* the wall plates of the shell interlock with a calcareous basis.

The shell and cuticle provide protection from water loss, solar radiations, abrasion, predators, micro-organisms and abnormally low salinities.

Feeding is efficient. Planktonic organisms and larger particles of detritus are captured by the cirri, but some species are capable of micro-feeding and collect particles down to 2 μm.

Acorn barnacles are hermaphrodite, but avoid self-fertilization by exchanging sperm with neighbouring barnacles. High level barnacles show a definite breeding season which is initiated by low temperature and photoperiod. An endogenous rhythm may also be involved. Fertilized eggs are retained in the mantle cavity and incubated (brooded). Release of larvae is triggered by temperature rise and the presence of abundant plankton.

During emersion barnacles lower their metabolic rate. The high level species can utilize atmospheric oxygen (a small opening—the micropyle—allows oxygen diffusion but reduces water loss from the mantle cavity). Complete closure of the operculum leads to oxygen shortage and metabolism switches to anaerobic pathways in which lactate is formed, but barnacles do not repay the oxygen debt when aerobic conditions return.

Barnacles are highly tolerant of adverse conditions, some species can withstand weeks of emersion, others can tolerate lowered salinities. Some species living in high latitudes withstand freezing, while others in lower latitudes can tolerate relatively high temperatures out of water.

6

Mussels and oysters (*Mollusca: Bivalvia*)

Mussels and oysters are conspicuous and sometimes extremely abundant members of the midlittoral community around the world. Their success in colonizing the shore can be attributed to their strong protective shell, the ability to withstand strong wave action and a collection of physiological adaptations that help them to withstand exposure to the atmosphere. The ability of some species to tolerate fluctuations in salinity has enabled them to extend their distribution into estuaries.

In favourable localities, closely packed mussels form a black band below the barnacle zone, extending from about mean tide level down to low water spring tides (Fig. 2.6). Rock oysters may be common at about the same level (MLWS).

6.1 REPRESENTATIVE INTERTIDAL SPECIES

A number of different mussels belonging to the family Mytilidae* are found in the littoral zone, but the edible sea mussel, *Mytilus edulis*, has the widest distribution. It occurs in temperate and boreal waters of Europe, the East and West Coasts of North America, Japan and as subspecies in the Southern Hemisphere, *M. edulis planulatus* (Fig. 6.1(a)) in Australia and *M. edulis aeoteanus* in New Zealand. The closely related *M. galloprovincialis* is a Mediterranean species and *M. californianus* occurs on the Pacific Coast of North America.

Other members of the family living intertidally include *Perna canaliculus*, the green mussel of New Zealand, and *Austromytilus rostratus* of southern Australia and Tasmania, both of which form conspicuous bands about mean low water neap tide level on rocky shores. The bearded or hairy mussels,

* See Appendix for classification of mussels.

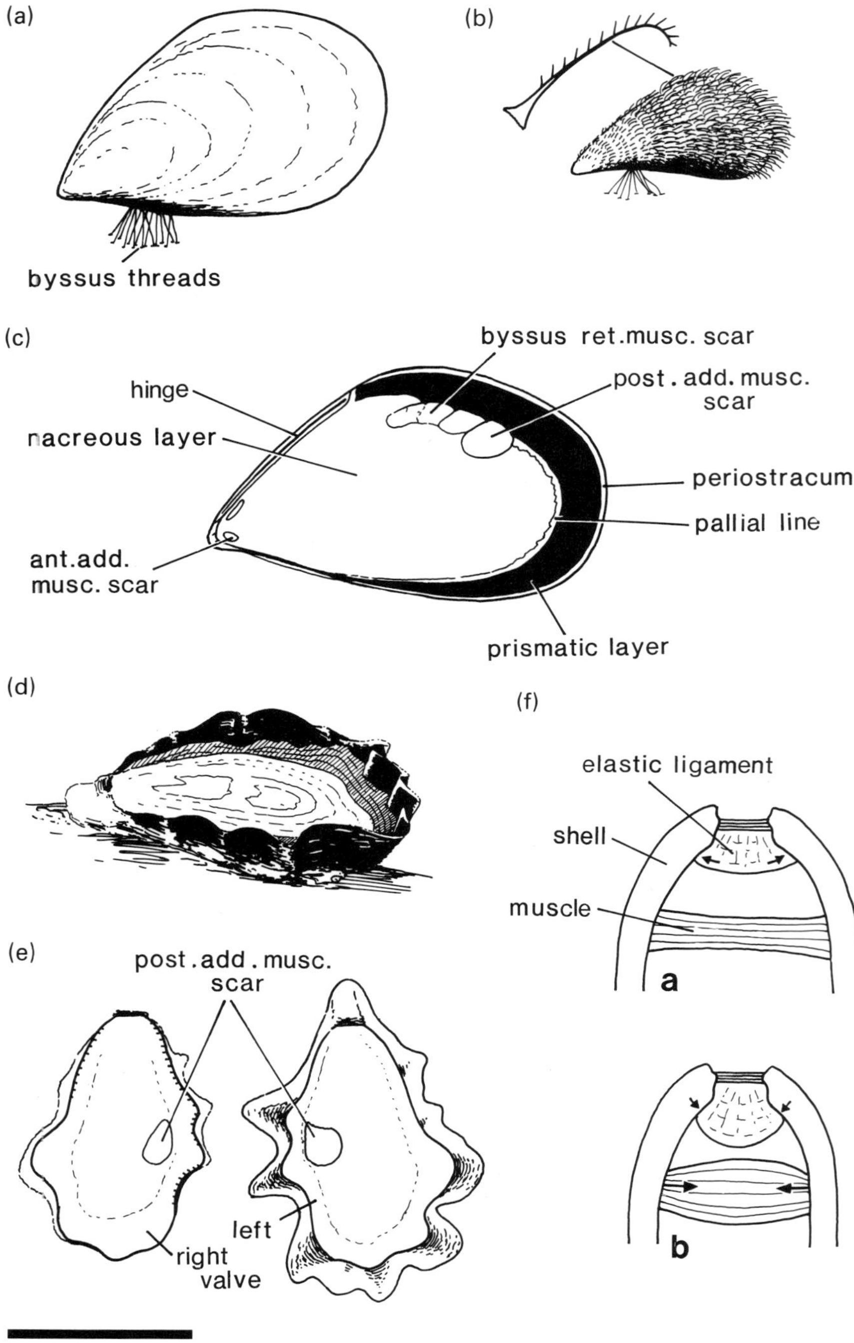
(a)
(b)
byssus threads
(c)
byssus ret.musc. scar
hinge
post. add. musc. scar
nacreous layer
periostracum
pallial line
ant.add. musc. scar
prismatic layer
(d)
(f)
elastic ligament
shell
muscle
a
(e)
post. add. musc. scar
left
right valve
b

Trichomya hirsuta in Australia and *Modiolus barbatus* in England, are other common but less conspicuous members of the family whose shell is covered in horny cuticular spines, which probably prevent other organisms settling on them or prevent predation by sea stars (Fig. 6.1(b)). Both species are found on the underside of boulder rocks on the lower shore.

There are two main groups of oysters in the family Ostreidae: the common European or mud oyster, *Ostrea edulis*, and rock oysters, of which the American oyster, *Crassostrea virginica*, the Sydney oyster, *Saccostrea cucullata* (formerly known as *Crassostrea commercialis*) from Australasia, and the Japanese oyster, *Crassostrea gigas*, are the best known. All four species are cultivated for food, but they may be found growing naturally in suitable habitats. The genus *Ostrea* is distinguished from *Crassostrea* and *Saccostrea* by retaining the eggs and incubating them in the mantle cavity.

6.2 GENERAL ORGANIZATION

In both mussels and oysters, the shell consists of two valves joined together dorsally by an elastic hinge. Adductor muscles close the shell (Fig. 6.1(f) b) and at the same time compress the elastic part of the hinge. When the muscles relax, this elasticity forces the valves open ventrally (Fig. 6.1(f) a). Adoption of a sessile habit has resulted in considerable departure from the more symmetrical shell of species burrowing into soft substrata. The valves of *Mytilus edulis* are equal, but greatly enlarged posteriorly to give a wedge-shaped shell (Fig. 6.1(a)). This results in the anterior adductor muscle becoming much reduced in size compared with the posterior adductor muscle (Fig. 6.1(c)). In some species (*Perna canaliculus*) it is lost altogether. Oysters, on the other hand, have markedly unequal valves. The left valve of *Saccostrea cucullata* is larger than the right, and attached to the substratum so that the animal is lying on its left side. The right valve is flat, and fits into the lower valve as a lid attached by the hinge ligament (Fig. 6.1(d)). An anterior adductor muscle, present in the larva, is absent in the adult, and the shell is closed by the large posterior adductor muscle, lying slightly posterior to the middle of the shell (Fig. 6.1(e)).

Fig. 6.1 (a) The Australian mussel, *Mytilus edulis planulatus*. (b) The hairy mussel, *Trichomya hirsuta*, with single bristle enlarged. (c) Interior of right valve of mussel. (d) The Sydney oyster, *Saccostrea cucullata*. (e) Interior of oyster valves. (f) Hinge ligament of mussel: a, muscle relaxed; b, muscle contracted, compressing elastic ligament. ant. and post. add. musc. scar, anterior and posterior adductor muscle scar; Byssus ret. musc. scar, byssus retractor muscle scar. Scale bar: (a),(b) and (d) = 30 mm.

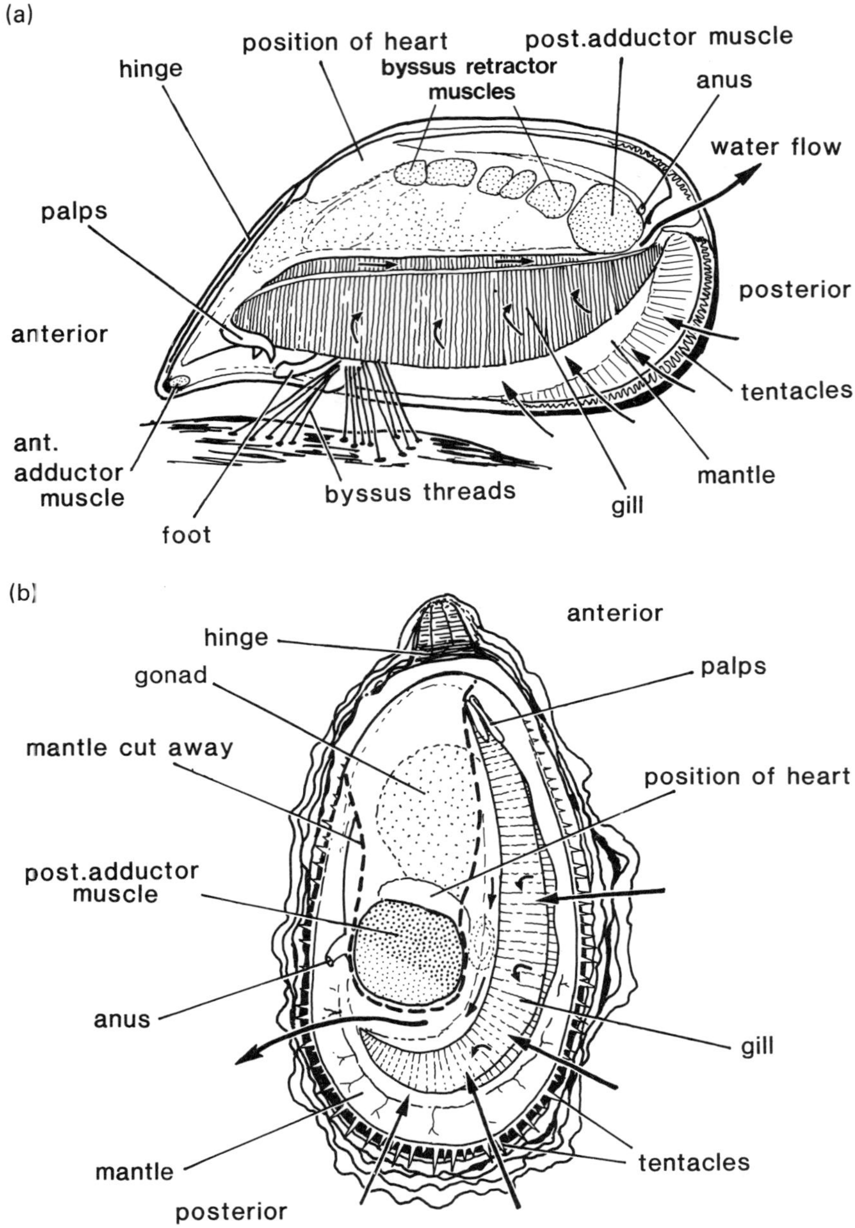

Fig. 6.2 (a) Internal organs of *Mytilus edulis planulatus* (left valve removed and mantle covering gill cut away). Compare with Fig. 6.5 (a). (b) Internal organs of Saccostrea cucullata (right valve and mantle covering gill removed). Arrows show direction of water current.

Some of the internal organs of *Mytilus edulis* and *Saccostrea cucullata* are shown in Fig. 6.2(a) and (b). Lining the whole of the shell interior is a thin sheet of tissue, the mantle, containing blood vessels, nerves and, in the case of *M. edulis*, gonads. Its distal edge bears sensory tentacles, which can be extended to monitor the environment or withdrawn when the shell is closed. The internal lining of the mantle is ciliated, and special tracts are responsible for carrying waste particles entangled in mucus (the pseudo-faeces) out of the mantle cavity via the exhalent opening. The mantle secretes both the shell and hinge ligament. Lying in the mantle cavity on either side of the body are the gills or ctenidia and labial palps which, as their name suggests, lie on either side of the mouth. In *M. edulis*, a tongue-shaped foot and a bundle of brown threads, the byssus fibres, lie between the gills. The byssus threads attach the mussel to the rock surface (Figs 6.1(a), 6.5(c)). The main body or visceral mass contains the gut, digestive gland, heart, kidneys and part of the gonads. Detailed descriptions of the anatomy of *Mytilus edulis* and the American oyster, *Crassostrea virginica*, have been given by White (1937) and Galtsoff (1964), respectively.

6.3 THE SHELL

Although there are considerable differences in shape and structure between the shell of *Mytilus* and *Saccostrea*, both show features that can be regarded as adaptations to sessile life. Each valve has three layers: an outer layer, called the periostracum, which is entirely organic, a relatively thick calcified prismatic layer, and an inner nacreous layer which is calcareous (Fig. 6.3(a) and (b)). The periostracum consists of proteins that have been sclerotized by quinone tanning to give the characteristic horny texture (Beedham, 1958; Gordon and Carriker, 1980), but chitin, reported for the bivalve *Solemya parkinsoni* (Beedham and Owen, 1965), is absent.

In mussels the brown periostracum is relatively thick and persistent. Histological staining shows that it has three layers, each produced by different cells at the edge of the mantle. In *Modiolus barbatus* and *Trichomya hirsuta*, it is greatly developed as a covering of horny spines (Fig. 6.1(b)), but in *Saccostrea cucullata* it is thin and often eroded. The prismatic layer is composed of calcite, with small amounts (2–5%) of a protein matrix called conchiolin that surrounds the prisms. This layer forms the bulk of the shell, and in *Mytilus*, individual prisms are orientated at an angle to the surface (Fig. 6.3(b)). On the internal aspect of each valve there is a distinct line, the pallial line, which marks the origin of muscles to the edge of the mantle (Figs 6.1(c), 6.3(b)), and the boundary of a thin internal layer of mother-of-pearl or nacre. The nacreous layer consists of layers of flattened tablets of aragonite (another crystalline form of calcium carbonate) enveloped in conchiolin. In

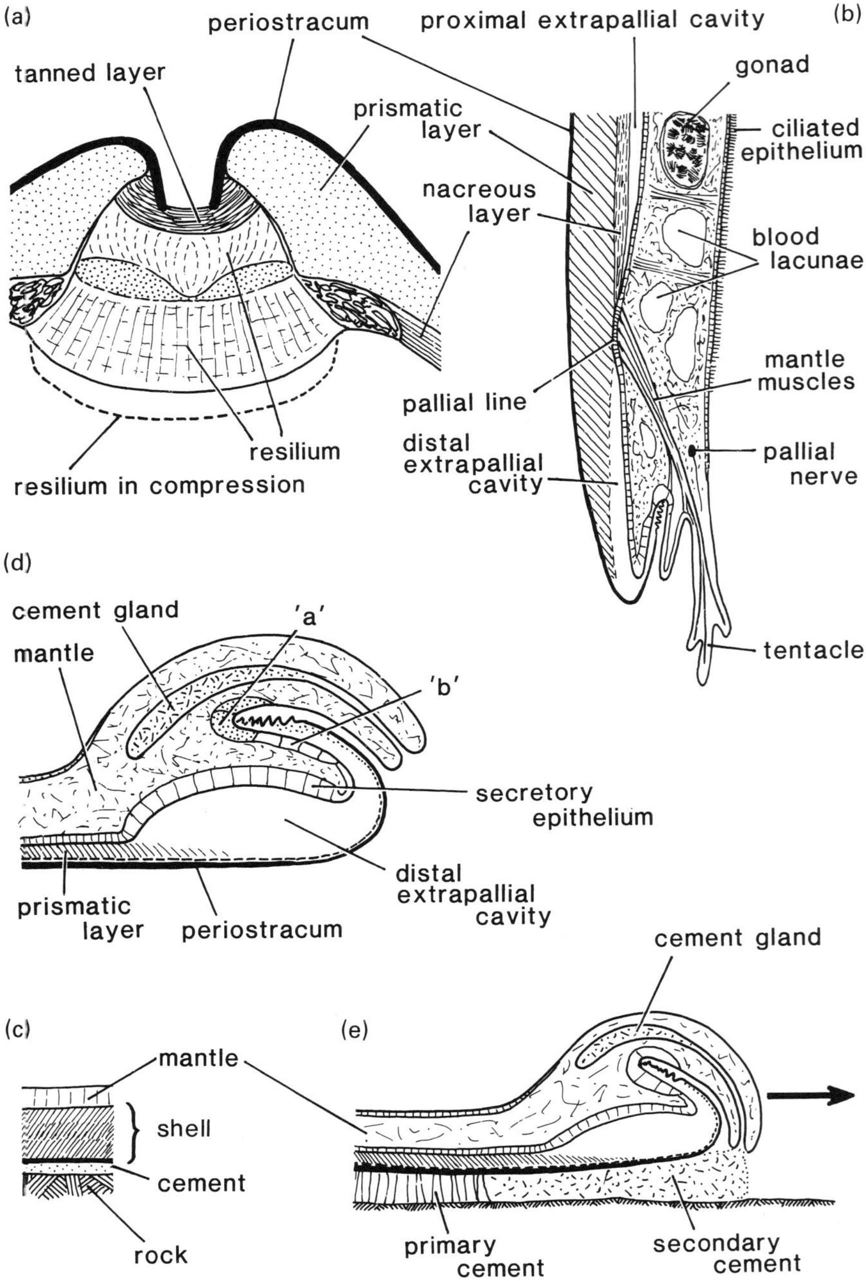
(a)
periostracum
tanned layer
prismatic layer
nacreous layer
resilium
resilium in compression
proximal extrapallial cavity
(b)
gonad
ciliated epithelium
blood lacunae
mantle muscles
pallial line
distal extrapallial cavity
pallial nerve
tentacle
(d)
cement gland
mantle
'a'
'b'
secretory epithelium
distal extrapallial cavity
prismatic layer
periostracum
(c)
mantle
shell
cement
rock
(e)
cement gland
primary cement
secondary cement

the oyster, the corresponding layer (called the calcite-ostracum) is composed of calcite, which is interleaved with oblique layers of amorphous or microcrystalline calcium carbonate containing *c.* 0.6% conchiolin. In both animals, the area to which the posterior adductor muscle is attached, the hypostracum, is composed of prismatic aragonite arranged normal to the surface (Wilbur, 1972).

The shell of *Mytilus edulis* is light and strong. Its shape and smooth surface offer little resistance to water flow, and this, together with the lateral support provided by adjacent mussels, considerably reduces drag forces exerted by turbulent water. The oyster shell is more massive, with characteristically laminated appearance, and fluting that may add to its mechanical strength but will not promote smooth water flow over it.

The entire shell is the product of the external epithelium of the mantle. The relationship between the shell and the mantle in *Mytilus edulis* is shown diagrammatically in Fig. 6.3(b). The internal epithelium of the mantle cavity is ciliated, and probably acts as a respiratory surface to absorb oxygen needed by the mantle tissues, as well as to take up calcium ions from the seawater for shell formation. Much of the thickness of the mantle is taken up by loose connective tissue, blood lacunae, muscle fibres and gonads. Towards the distal edge of the mantle, a distinct band of muscle fibres, the pallial retractor muscles, take origin from the pallial line and insert on the mantle border and tentacles. Their function is to withdraw the tentacles or mantle when the shell is closed. The mantle and tentacles are extended by increased blood pressure in the lacunae brought about by transverse muscle fibres in the mantle. The tentacles are innervated by the pallial nerve and are sensitive to touch, salinity changes, and possibly light intensity. The external mantle epithelium is in contact with the shell along the pallial line and areas of muscle insertion, but elsewhere there is a narrow space between mantle and shell, filled with fluid (the extrapallial fluid) which is similar in ionic composition to blood plus secretory products of the mantle epithelium (Fig. 6.3(b) and (d)). The pallial attachment effectively divides the extrapallial space into two parts, a proximal and a distal cavity, in which the composition of fluid may differ. Periostracum, prismatic layer and nacre are secreted by three different regions of the mantle epithelium.

Fig. 6.3 (a) Transverse section through hinge of a mussel. (b) Diagrammatic section through shell and mantle of mussel. (c) Adhesion of oysters. (d) Diagrammatic section through mantle edge (left valve) of young oyster (modified from Cranfield, 1974): cells marked 'a' produce outer layer of periostracum; cells marked 'b' produce inner layers and tanning substances. (e) Same as (d), showing how the young oyster is cemented to the substratum; arrow shows direction of growth. Scale bar: (a), (b) and (e) = 0.4 mm.

The mechanism for formation of the periostracum in *Ostrea edulis* has been elucidated by Cranfield (1974) and for gastropods by Saleuddin (1979). The outer lamina is produced as small platelets in the Golgi complex of cells at the base of the groove between outer and middle lobes of the mantle edge (Fig. 6.3(d), cells marked 'a'), and these are assembled as a much-pleated sheet in the periostracal groove. Adjacent cells on the outer lobe secrete the inner layers of the periostracum, together with the polyphenol-protein and polyphenol oxidase that sclerotize the proteins (Fig. 6.3(d), cells marked 'b'). Accordingly, the periostracum increases only at its periphery.

The prismatic layer is laid down on the inside of the periostracum, from secretions passed into the distal extrapallial fluid by the mantle epithelium (Fig. 6.3(d)). These secretions contain proteins and mucopolysaccharides that will form the matrix of conchiolin in which calcification takes place. Calcium ions derived from seawater in the mantle cavity pass from the mantle epithelium into the extrapallial fluid, possibly bound to proteins. In the presence of carbonate ions, calcite crystals are formed in the conchiolin matrix. Nacre is formed in a similar manner, but in the proximal extrapallial cavity. Growth in diameter of the valves takes place at the periphery of both the prismatic and nacreous layers, so the pallial line advances centrifugally. Both pallial and posterior adductor muscles also advance centrifugally, by growth of new muscle fibres distally and resorption along the proximal border.

Opinions differ on the exact way in which mineralization of the shell takes place (reviews: Wilbur, 1972; Istin, 1975). It is evident that the mineral layers are derived from the mixture of substances present in the extrapallial fluid (proteins, glycoproteins, mucopolysaccharides, organic acids, and inorganic ions giving a pH of 7.3–7.4), but the problem is to account for the high degree of organization, different patterns of crystallization, and regional differentiation seen in the shell. It is most likely that some organic matrix is laid down first on the inside of the shell, and that this initiates nucleation of calcium carbonate crystals. The first of these appear as separate spheres, which gradually grow in size by further crystallization and finally coalesce to form the prisms or tablets seen in the shell. Whether crystallization takes place in the orthorhombic (aragonite) or hexagonal (calcite) crystallographic system is thought to depend upon the nature of the organic matrix. Its amino acid composition is known to differ between the distal and proximal extrapallial compartments. The conchiolin matrix also appears to determine the orientation of crystal deposition (Grégoire, 1972).

Further crystallization depends upon the presence of calcium (protein-bound or in ionic form) and the enzyme carbonic anhydrase, which promotes the formation of carbonate ions. The latter may be derived from seawater, metabolic carbon dioxide, or bicarbonate in the tissues (Fig. 6.4).

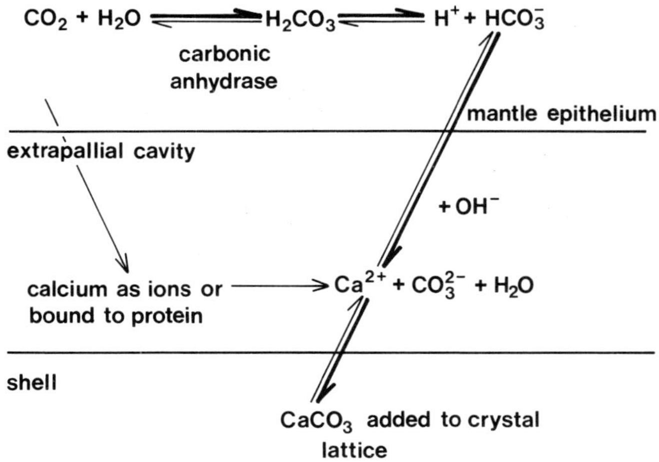

Fig. 6.4 Formation of calcareous shell. Upper compartment, mantle epithelium. Central compartment, extrapallial cavity. Lower compartment, prismatic or nacreous layer of shell.

Crystallization will remove calcium and carbonate ions from the fluid, and so the reaction will proceed in the direction of the heavy arrows. Digby (1968) has shown that the inner side of the thin peripheral periostracum is alkaline, and that the presence of hydroxyl ions in the distal extrapallial cavity would help to promote formation of calcium carbonate.

The reactions are reversible, and under certain conditions calcium may be removed from the shell (light arrows). Lutz and Rhoads (1977) have shown that etching of the aragonite crystals in the nacreous layer occurs during prolonged periods of emersion. This is thought to be due to the accumulation of propionic and acetic acids during anaerobiosis.

Shell growth is incremental, not continuous. This is clearly shown in the frilled appearance of oyster shells. The rate of shell growth (or the balance between crystal deposition and solution) shows seasonal variations which are influenced by environmental temperature (Wilbur, 1964). The various theories of shell calcification and growth have been discussed by Wilbur (1972).

The ligament, which serves the double function of opening the valves and binding them together, may be regarded as part of the shell, for it is produced by a specialized part of the mantle epithelium, the tall secretory cells of the subligamental ridge. The ligament consists of two parts (Fig. 6.3(a)). An outer region, composed of parallel lamellae of fibrous protein hardened by quinone tanning, is well suited to withstanding tensile stress and contains little or no calcium carbonate (5–8% in *Crassostrea*), and an inner or medial resilium consisting of fibrous protein that has undergone little tanning but contains a considerable amount of aragonite (30–67% in *Crassostrea*: Galtsoff, 1964). The resilium is subject to compression when the shell is closed. When the posterior adductor muscle is cut, the valves open further than during normal opening, which suggests that the ligament is maintained under stress*. This poses a problem in explaining how new layers can be added to the ligament in tension or compression. It is possible that tension in the outer region arises from slight shrinkage during tanning, and compression might arise from hydration or the deposition of calcium carbonate in the resilium.

6.4 ATTACHMENT TO THE SUBSTRATUM

Adult mussels attach themselves to the rocks by means of numerous flax-like strands, the byssus threads (Fig. 6.5(a)). These are secreted by special glands in the foot. Yonge (1962) has pointed out that byssal glands are an almost universal feature of bivalve larvae. Their secretion, often as a single thread, is used to anchor the settling larva prior to metamorphosis to the adult form. In most cases (including oysters) they are lost in the adult. The presence of functional byssus glands in adult mussels has been regarded by Yonge as a case of neoteny or retention of larval characteristics by the adult.

The structural complexity of the mussel foot reflects the involved nature of thread formation and the problems of establishing a firm attachment to the substratum. Waite (1983a) has reviewed the subject of biological adhesion and the formation of byssus threads in *Mytilus edulis*.

Although the greater part of the foot of *M. edulis* is composed of muscles, its locomotory function is poorly developed compared with bivalves living in soft deposits. It has become highly specialized for byssus thread formation. The foot is a narrow cylinder capable of being extended three times its resting length. Much of its volume is taken up by glandular tissue. A section through the middle of the adult foot (Fig. 6.5(b)) reveals that two glands (the collagen and accessory glands) occupy about one-third of the volume, and a deep groove runs down the posterior (ventral) side. The solid stem of the byssus, to which all the threads are attached, is deeply embedded in the posterior base of the foot (Fig. 6.5(c)). It is secreted as concentric lamellae by an organ called the byssus generator which consists of muscles and glandular tissue (Tamarin, 1975).

Biochemical and X–ray analysis have shown that the core of each byssus thread and the byssus stem are composed of collagen fibres with an outer sheath or cuticle of protein, similar to β-keratin, which may protect the collagen from microbial attack (Fig. 6.5(d)). Lane and Nott (1975) have

*The very wide gape of dead shells washed up onto the beach is caused by changes in the hinge upon drying out.

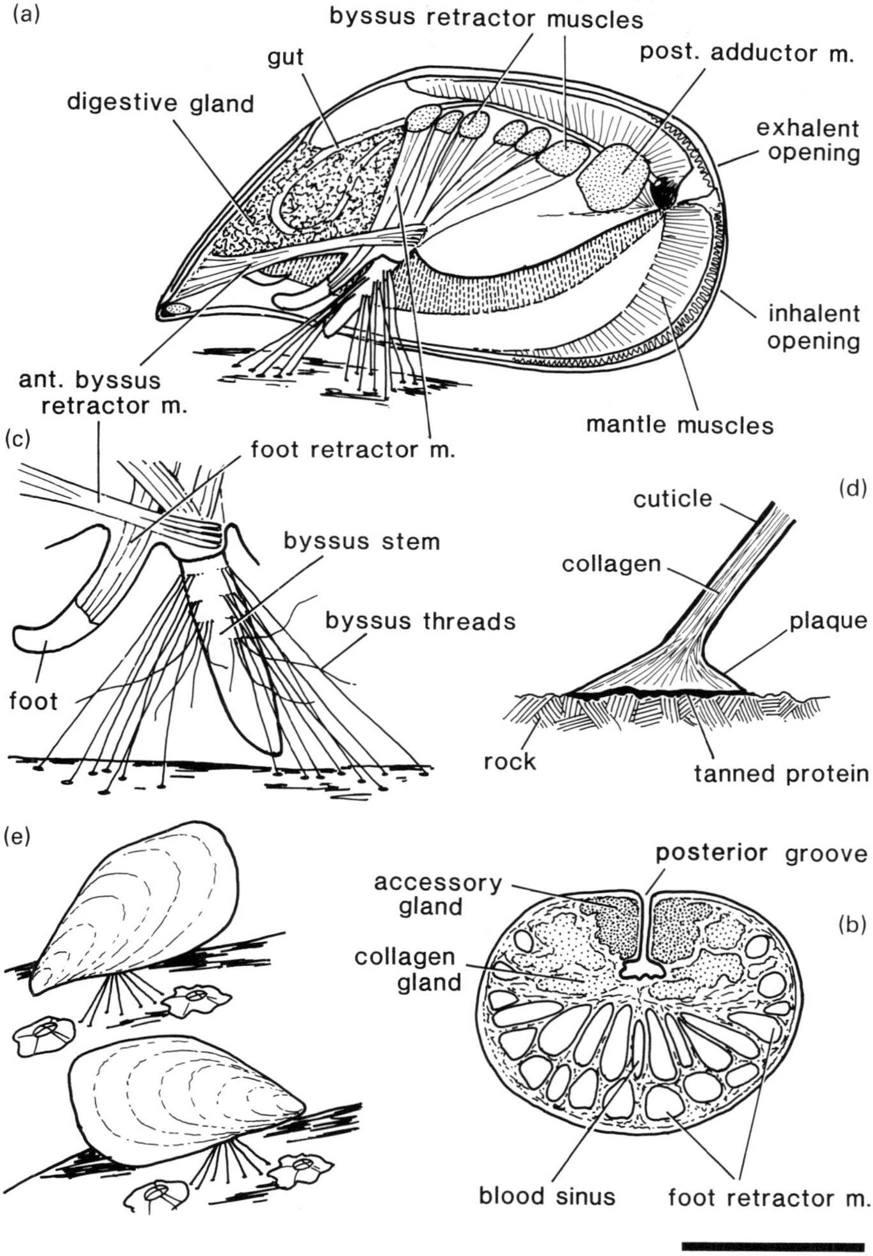

Fig. 6.5 (a) *Mytilus edulis planulatus* dissected to show musculature. Stippled areas mark muscle insertions on inside of right valve. (b) Transverse section through foot. (c) Byssus stem and byssus threads. (d) Attachment plaque of byssus thread. (e) Rotation of mussel. Ant., anterior; post., posterior; m., muscle. Scale bar: (b) = 2.5 mm, (d) = 1.5 mm.

described nine histologically-distinct glands in the foot of the pediveliger larva, but in the adult there are five functionally distinct types. The byssus generator, located at the base of the foot, secretes the byssus stem. The byssus threads are moulded one at a time in the posterior groove, from secretions of the collagen and accessory glands in the foot. The accessory gland secretion contains a protein rich in *ortho*-diphenols which is believed to give rise to the cuticular covering. The byssus threads are anchored to the substratum by a terminal plaque (Fig. 6.5(d)).

When about to secrete a new thread, the foot is extended until the tip touches a firm surface. Mucous glands located at the tip of the foot secrete acid mucopolysaccharides in the depression at the distal end of the groove, and at the same time phenol glands in the region produce a phenolic protein which has surface-active properties capable of displacing water molecules from the surface over which it flows. These two secretions spread over the surface, possibly with the help of paddle-shaped cilia at the tip of the foot, and then harden by a tanning process to form the adhesive bond between substratum and the expanded end of the byssus fibre (Fig. 6.5(d)). While the attachment plaque is being formed, the thread is being moulded in the groove from secretions of the collagen and accessory glands (Waite 1983a,b). It only takes a few minutes for a new thread to be formed, and large numbers (up to 200 or more in *Mytilus edulis planulatus*) are found attached to the stem, although not all of them will be attached to the substratum, for some are broken or detached.

Threads are attached in bundles along the anterior and posterior edges of the massive byssus stem (Fig. 6.5(c)). As the groove runs down the posterior side of the foot, it would be expected that all fibres would be on the anterior face of the stem. Their presence on the posterior side implies that the whole mussel is able to rotate through 180° with respect to the stem, and hence the substratum. This rotation (Fig. 6.5(e)) must be brought about by the anterior byssus retractor muscle, which runs round either side of the stem and inserts on its posterior aspect (Fig. 6.5(c)). Thus contraction of the right anterior byssus retractor muscle will rotate the body clockwise, and vice versa. Contraction of the posterior byssus retractor muscles clamps the mussel down firmly on the rock.

Harger (1970) found that a force of *c*. 60 N was required to tear *Mytilus californianus* from the rocks, whereas *M. edulis* could be dislodged by a little over half this force. However, strength of attachment will depend upon the number and strength of individual fibres. Smeathers and Vincent (1979) found that the fibres break before failure of the plaques, and mussels, particularly young ones, can detach themselves by breaking the threads with strong contraction of the byssus retractor muscles. They can then reattach themselves in a new position. The fibres spread out over a wide area and act as tensile members or guy ropes to take the strain imposed by currents from any direction. In England, Price (1980) found that strength of attachment varied with the season, reaching a maximum in autumn, but whether this is a response to higher wave impact at this time of year or to changes in general activity has not been determined.

Oysters are permanently attached to the rock surface by a process of cementation. There are three phases to this process. When the pediveliger larva is about to settle, it first moves over the surface by means of cilia on its foot. Temporary adhesion may be made by means of mucus secretions of the foot, but eventually this wandering stops and the larva turns onto its left valve. Other glands in the foot now exude a secretion which rapidly hardens and firmly cements the left valve to the rock. This process can only take place once. The glands concerned produce a phenolic protein, sulphated acid mucopolysaccharide and phenol oxidase. They have been homologized with the phenol glands of *Mytilus edulis* (Gruffydd *et al.*, 1975). Cranfield (1975) has shown that after tanning, the secretion forms a highly organized complex of microfibres which firmly cement the periostracum to the substratum.

The second phase of cementation has been described by Cranfield (1974). Along the mantle edge of the metamorphosing larva there are several types of gland cell, some of which have already been noted in connection with production of the periostracum and prismatic layers of the shell (Fig. 6.3(d)). Anothe type of gland cell, the cement glands associated with ciliated sensory organs, are arranged along the border of the inner mantle lobe. They produce mucopolysaccharide, protein and phenol oxidase. By extension of the mantle edge beyond the shell, layers of new cement can be applied to the outer surface of the primary cement to build up the attachment of the shell to the substratum (Fig. 6.3(e)). This new cement hardens by tanning.

As the young oyster grows, cement, periostracum and calcified shell are laid down in contact with the substratum as shown in Fig. 6.3(c) and (e), and firmly attach the oyster to the rock surface (Yonge, 1979). This does not continue indefinitely, and the lip of the valve starts to grow away from the surface.

6.5 NUTRITION

Mussels and oysters obtain their food from the seawater passed through the mantle cavity, so they can only feed whilst immersed. Animals living in the sublittoral could feed at any time and probably do so continuously, but feeding time for those living at mean tide level would be restricted to half the tidal cycle. For such animals, feeding efficiency becomes of prime importance. The abundance of food and its nutritive value, the volume of water pumped, the extraction rate, and the ability to deal with higher levels of food intake over short periods are all factors that will influence survival and for which various adaptations would be expected. A detailed review of feeding and digestion in mussels has been given by Bayne *et al.* (1976a).

The current of water drawn through the mantle cavity by activity of the lateral cilia on the gills has a dual function: it brings both oxygen and food to the animal. In mussels, the gills are made up of numerous W-shaped filaments, which are not fused together and are attached to the body (Fig. 6.6(a)). They are held together by interlocking discs of cilia, the ciliary junctions, which occur at intervals down the filament (Fig. 6.6(b)). On either side of each filament, a band of lateral cilia pumps water between adjacent filaments, into a central cavity, and then upwards to the suprabranchial cavity, which leads posteriorly to the exhalent siphon. This creates a current of water which is drawn in ventrally past the mantle tentacles and exits posteriorly through the exhalent aperture (Figs 6.2(a), 6.6(a)).

Other cilia are concerned with collecting food and moving it to the mouth. The latero-frontal cirri (Fig. 6.6(b)) are composed of about 58 cilia fused together except for their distal ends, which project stiffly at right angles to the main axis and across half the space between adjacent filaments (Fig. 6.6(c)). Adjacent cirri beat alternately forming a mesh *c.* 1.2 μm by 4.8 μm, through which the water current has to pass (Owen, 1974). Silvester and Sleigh (1984) pointed out that, if the mesh of cirri is to act as a sieve, it will impose a resistance to flow: lateral cilia producing a flow rate of 1.02 $\mathrm{l\,h^{-1}}$ must produce a pressure difference of 3 $\mathrm{N\,m^{-2}}$ between inhalent and exhalent sides of the gill, plus *c.* 4 $\mathrm{N\,m^{-2}}$ to overcome drag forces induced by the lateral cirri. They calculated that the lateral cilia are capable of producing this pressure difference. They used this evidence to argue against the idea that the latero-frontal cilia are sticky or the suggestion that shear forces caused by velocity gradients were responsible for extracting particles from the water flow (Jørgensen, 1981a, 1983). They considered that the latero-frontal cirri could act as a sieve capable of retaining particles down to *c.* 1 μm in diameter. Fine particles retained by this sieve are transferred by the beat of the cirrus to the frontal cilia, which carry them to the ventral groove and thence towards the mouth. Particles too large to enter the ventral groove will fall onto the mantle, where ciliated tracts remove them, bound with mucus, as pseudo-faeces. Particles in the ventral groove pass to the labial palps, where further sorting takes place on ciliated ridges. Larger inert particles are rejected and passed to the mantle to form pseudo-faeces, while small particles are passed to the mouth.

It is usually held that bivalves are mucociliary feeders, collecting food particles on the gills and conveying them to the mouth entangled in mucus strings (Morton, 1963; Purchon, 1968; Bayne *et al.*, 1976a). But this idea has always created problems in explaining how the labial palps sort the particles according to size or density if they are entangled in a viscid thread. The common observation that the labial palps do sort particles by means of ciliated tracts, and can regulate the size and quantity of food particles passed

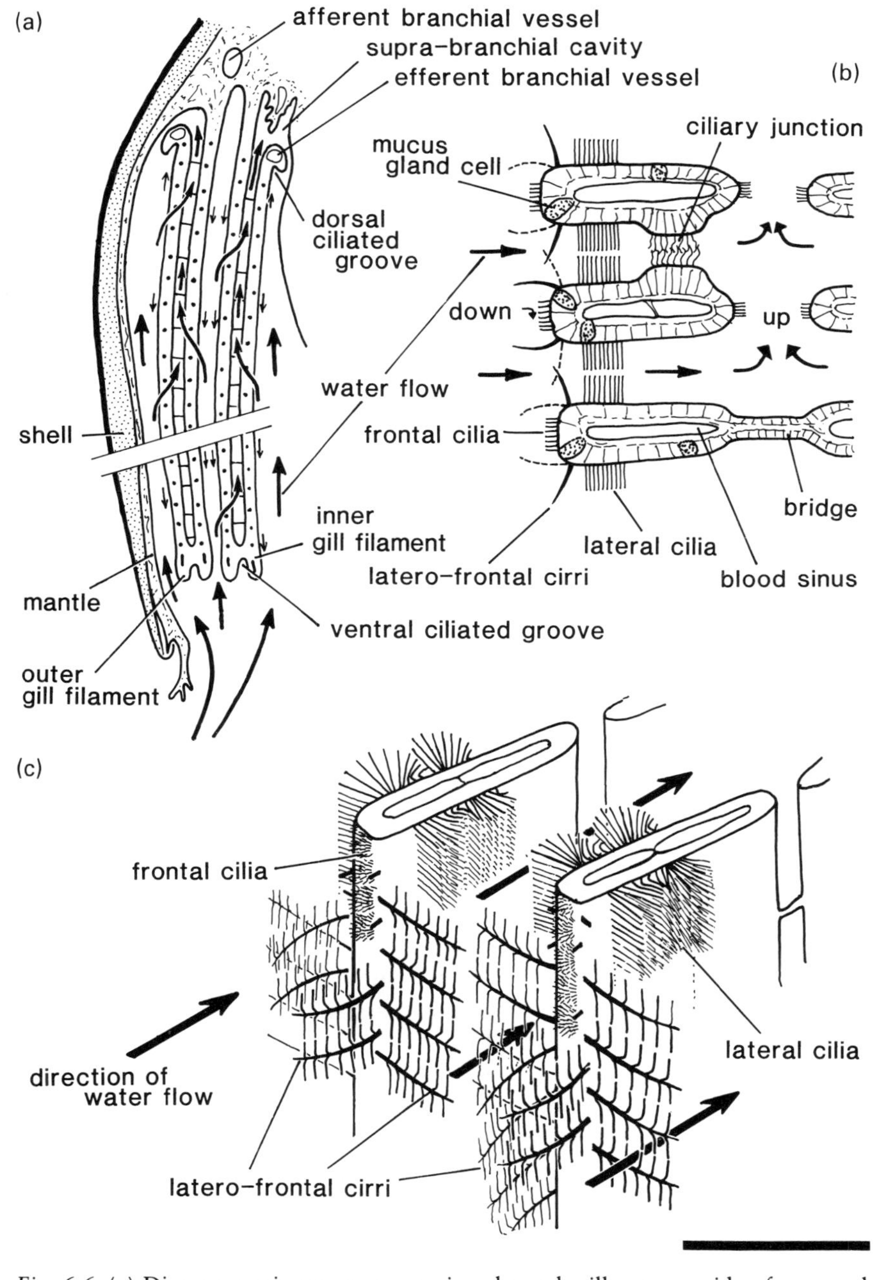

Fig. 6.6 (a) Diagrammatic transverse section through gills on one side of a mussel, showing water flow (heavy arrows) and transport of food to the ciliated grooves (fine arrows). (b) Transverse section through filaments. (c) Stereogram illustrating alternate beating of latero-frontal cirri and water flow through the sieve created by the cirri. Scale bar: (b) = 0.5 mm.

to the mouth, is incompatible with the concept that food is bound in mucus. The familiar classroom experiment of tracing the ciliary tracts on mussel gills with a carmine suspension after the mussel has been cut open gives a false idea of how filter feeding takes place. The trauma of being cut open causes a mussel to produce abnormally large quantities of mucus. This excess mucus may partially block the gills and cause the carmine to be passed to rejection tracts.

A much more plausible explanation is that food particles collected by the latero-frontal cirri are not caught in mucus but remain free, and are passed in a current created by the frontal cilia to the food groove, and so to the labial palps where, unhindered by mucus, the cilia sort or regulate the particles that are passed to the mouth. Particles that are too big to pass into the food groove or are rejected by the palps are entangled in mucus and passed to the rejection tracts on the mantle.

Jørgensen (1981b) has made observations on *Mytilus edulis* which support this suggestion. He noted that profuse secretion of mucus occurred when the adductor muscle was cut. This may be a protective reaction. During normal undisturbed feeding, particulate matter in the stomach was freely suspended, with little evidence of mucus being present. He postulated that food particles filtered off by the latero-frontal cirri are prevented from being lost, as they pass along the frontal cilia, by hydromechanical shear forces created by the water current passing through the gill. Larger particles are trapped by the mucociliary system in small amounts of mucus and are passed to rejection tracts as pseudo-faeces. In this view, the system usually held responsible for feeding is really concerned with cleaning the gills and mantle cavity. A function of the labial palps may be to concentrate food particles before passing them to the mouth, because a comparison of particle concentration in the seawater and stomach suggests that particles may be concentrated by a factor of up to 200 000 (Jørgensen, 1981b). This indicates that the gills of bivalves are extremely efficient filtering devices.

It has been found experimentally that many inert substances (carbon, carmine, titanium oxide particles) are ingested, and this has led to the widespread belief that bivalves do not select or discriminate among the nutritive value of particles. However, *Mytilus californianus* was found to feed almost exclusively on diatoms in a mixture of diatoms and dinoflagellates where the latter were far more abundant (Newell, 1979). Not all potential food organisms may be digested. Galtsoff (1964) found that yeast cells passed through the gut of *Crassostrea virginica* unharmed and were of no nutritive value.

Mussels can control the velocity of the current passing through the mantle, the rate of extraction, and the rate of ingestion of food particles. Current velocity is controlled by the size of inhalent and exhalent apertures (degree of shell opening), muscular movements of the velum and gill filaments, the rate of beating of the lateral cilia, and possibly the activity of the latero-frontal cirri. Beating of the lateral cilia is under nervous control (Paparo, 1972) and can be stopped (i.e., when the shell is closed on emersion), but control of metachronal rhythm appears to be mechanical (Aiello, 1974). Stimulation of the branchial nerve increased the rate of lateral ciliary beating from 7 to 19 beats per second (Paparo, 1985). The rate of ciliary beating is accelerated when 5-hydroxytryptamine (serotonin) is administered in physiological concentrations, and this substance may be the neurotransmitter released at nerve endings in ciliated epithelium of gills (Clemmesen and Jørgensen, 1987).

Filtration efficiency or extraction rate can be adjusted by movements of the filaments which increase the space between them, or by controlling the latero-frontal cirri and frontal cilia. Particle collection can be stopped without altering the current of water passing through the gills. This may occur if the concentration of food particles in the water falls below a critical value, or if it is so high that clogging of the gills is likely. It is clear that the maintenance of the respiratory current is independent of food collection. The rate at which food is ingested can be adjusted by increasing the rejection rate. When mussels are fed on high concentrations of algal cells, a greater proportion are rejected by the palps and appear in the pseudo-faeces. Extraction efficiency can be high. Motile flagellates 1–2 μm in size can be trapped, and in the size range 2–8 μm, extraction efficiency ranges from 80% to 100% (Vahl, 1972).

Only fine particulate food is taken into the stomach, thus no trituration is required. The stomach possesses a crystalline style which is constantly rotated by cilia. It is secreted in the first part of the intestine, which is more or less divided into a style sac and intestine by a fold, the typhlosole. In *Mytilus edulis* the style contains carbohydrases and an endogenous cellulase. Changes in pH of the stomach contents cause liquefaction of the style and the release of these enzymes, so that some digestion of food takes place in the stomach. A complex ciliary sorting apparatus passes the finest particles into the digestive diverticula, where digestive cells in the epithelium take them up by phagocytosis for intracellular digestion.

Endogenous biological rhythms entrained to environmental oscillations are widespread among animals and plants. Both circadian and circatidal rhythms (following daily and tidal cycles) have been demonstrated for many shore-living animals, particularly crustaceans, where anticipation of the onset of changes in environmental conditions (i.e., submersion or emersion) is of adaptive significance (Naylor and Williams, 1984; Naylor 1982, 1985). Little attention, however, has been paid to sessile shore animals, probably because of the lack of overt activities and the difficulty of following internal physiological processes without disturbing the animal.

Beentjes and Williams (1986) have demonstrated an endogenous circatidal rhythm of shell gaping, siphon protrusion and adductor muscle activity associated with feeding in the New Zealand cockle, *Chione stutchburyi*. Evidence of a tidal periodicity in other littoral bivalves comes from cyclic changes in the digestive system, although it has not been demonstrated that these follow an endogenous rhythm. The clearest demonstration of such a periodicity is in the minute bivalve, *Lasaea rubra*, found living up to mean high water spring tides (MHWS) on British and Australasian shores (Morton, 1956). Submersion may last for only 3 h or 4 h. During this period, a crystalline style and food are present in the stomach, and particles are absorbed by digestive cells. During the first hour or two of emersion, intracellular digestion takes place, followed by absorption of nutrients and the final breakdown of the digestive cells. The crystalline style disappears, and within 8 h of emersion new digestive cells are forming. Faecal matter consisting of undigested material from the stomach is voided, thus completing the cycle. A new cycle commences with the secretion of a new style about the time of immersion on the next tide. It is clear that this well-defined digestive cycle follows the tidal cycle, but how the periodicity is phased, and whether there is an endogenous rhythm, has not been determined. In view of the fact that endogenous rhythms are so widespread, it is not unlikely that *Lasaea rubra* has a biological rhythm entrained to a semilunar periodicity.

Less clearly-defined cycles occur in the digestive system of other intertidal bivalves. In *Cerastoderma* (= *Cardium*) *edule*, Morton (1970) has shown that secretion of the crystalline style (which is not lost) only takes place during periods of submersion, and digestive cells show cyclic changes. Langton and Gabbott (1974) have shown a tidal rhythm of extracellular digestion in *Ostrea edulis*, and Langton (1975) found that in *Mytilus edulis*, the proportion of digestive gland cells carrying out intracellular digestion increased during submersion. It seems highly probable that mussels possess biological rhythms controlling physiological processes that are adaptive to intertidal life.

Feeding on particulate matter is not the only way that mussels can obtain nourishment, for it has been established that they can absorb the traces of dissolved organic matter present in the sea. Péquignat (1973) was the first to show that [^{14}C] glucose and [^{3}H] glycine were absorbed through the gills and mantle of *Mytilus edulis*. *M. californianus* can extract certain amino acids more readily than others. Serine and glycine, for example, can be absorbed from ambient concentrations as low as 0.01 g m^{-3}. As these amino acids were not lost from the mussel to the water, absorption must represent a net gain of nutrients. The amount of amino acid taken up from the sea accounted for 6–60% of the mussel's metabolic requirements and could constitute a significant nutritional supplement (Wright, 1982).

Manahan and Crisp (1982) showed that the veliger larvae of oysters and mussels can absorb amino acids through the velum. This nutritional supplement could be of great importance to early larval stages before they start feeding on phytoplankton, as the yolk reserves are often minimal and insufficient for proper development.

6.6 REPRODUCTION

The gonads of mussels are located in the mantle and dorsal part of the body (Fig. 6.3(b)). The sexes remain separate (diœcious), and sex reversal does not occur in *Mytilus edulis* or *M. californianus*. In Britain and North America, *M. edulis* tends to spawn in spring and late summer, but *M. californianus* does not show such a marked reproductive season; spawning occurs throughout the year, although it tends to peak in autumn or winter. Near Sydney, *M. edulis planulatus* spawns during the colder months of winter. Fecundity is high. About 10^7 eggs are shed during the season into the sea, where they are fertilized and undergo development into planktonic larvae.

Oysters have a less diffuse gonad located in the visceral region (Fig. 6.2(b)). *Ostrea edulis* regularly changes sex in a process referred to as rhythmic consecutive hermaphroditism. After functioning as a female, an individual becomes a functional male in the next season and vice versa. *Crassostrea virginica*, however, does not change sex regularly. In their first season, 70% of young were found to be functional males. Some of these must subsequently have changed sex, for there was an equality of males and females in the second year. Old individuals showed a bias towards the female sex, thus sex change is an erratic phenomenon in this species (Galtsoff, 1964). There is a marked difference in fecundity of *C. virginica* and *O. edulis*. The latter produces fewer eggs (*c.* 10^6), which are fertilized and retained in the mantle cavity for incubation to the shelled pediveliger stage. Maximum spawning was found to take place on spring tides following full moon, and to a lesser extent after new moon. *C. virginica* sheds its eggs into the sea, where they are fertilized and develop into planktonic larvae. Eggs are produced in prodigious numbers. Galtsoff found that the gonad of an oyster that had just spawned more than 10^8 eggs still contained vast numbers of eggs.

Spawning of eggs is accompanied by rapid rhythmic contractions of the adductor muscle, forcing the eggs out of the mantle cavity in a jet of water, but there are no corresponding contractions in oysters functioning as males and shedding sperm. Not all the eggs are released at one time. *Crassostrea virginica* spawns two or three times during the spring or summer reproductive period. Environmental factors such as temperature and salinity influence spawning, but the presence of sperm in the water is a strong inducement to egg release.

Cleavage of the fertilized egg gives rise to a trochophore and then a veliger larva. The latter develops into a veliconcha with a larval shell and then, after the development of a foot, becomes a pediveliger. The pediveliger undergoes a metamorphosis in which the velum is lost, gill filaments and labial palps are formed, and byssus glands become functional (Bayne, 1976). This stage is called a plantigrade larva; its function is to find a suitable site for attachment and growth to an adult.

6.7 METABOLIC ADAPTATIONS TO SHORE LIFE

Three areas of metabolic adaptation are essential to survival out of water: reduction in energy demand, changing metabolic pathways to overcome oxygen shortage, and modification of excretion and ionic regulation. Because of their suitability for experimental investigation, considerably more is known about mussel metabolism that that of other sessile shore animals.

Reduction in energy demand

Upon complete closure of the shell, ciliary action of the gills comes to a halt and heart rate slows or stops, except for occasional short bursts of normal beating. This shutdown of activity significantly reduces energy expenditure, but keeping the shell closed requires the adductor muscles to do work against the elasticity of the hinge ligament. Prolonged muscular contraction is normally an energy-consuming process, but in *Mytilus edulis* and other bivalves a so-called 'catch mechanism' enables the muscle to maintain tension with very little energy expenditure. The nature of the mechanism enabling byssus retractors and adductor muscles to exert a constant tension for very long periods (many hours) has been the subject of considerable debate (reviewed: Hoyle, 1964).

Plain and striated muscle fibres are intermingled in the byssus retractor and posterior adductor muscles of *Mytilus edulis*. The plain fibres contain paramyosin as well as actinomyosin and are capable of maintaining tension for long periods following a single stimulus. Such tonic contraction would normally be initiated by motor impulses from the central nervous system and is probably brought about by an actin–myosin sliding-filament mechanism similar to that proposed for vertebrate muscle. Tension does not appear to be maintained by repetitive stimulation, but could be due to the formation of linkages between the actin and the myosin-paramyosin in the plain muscle. Alternatively, tension could be maintained by a change in state of the paramyosin (i.e. crystallization or increased viscosity). Relaxation would not take place until the linkages were broken or the normal state of paramyosin was restored. The rate of decline in tension depends upon the presence of a relaxing substance, 5-hydroxytryptamine, which is released into the muscle when relaxing nerves are stimulated (Hoyle, 1964).

Contractions of short duration (phasic), brought about by the striated fibres, could be in response to brief motor discharges or could be caused by simultaneous stimulation of motor and relaxing nerves. The 5-hydroxy tryptamine would prevent prolonged contraction of the plain muscle fibres without inhibiting striated muscle contraction. Tension in the latter could be maintained for short periods by repeated impulses (tetanus) from the central nervous system. The neural control of muscular activity in molluscs, involving motor and relaxing nerves, is in contrast to that of crustaceans, which have two types of motor nerves (fast and slow), plus an inhibitory nerve which can block action of motor nerves to bring about relaxation.

When the shell is completely closed during emersion, the metabolic rate may drop to *c.* 5% of the normal value of active submerged mussels, so the energy demands during emersion are extremely low.

Modified metabolic pathways

During vigorous activity, the energy demands of muscle tissue may exceed the rate at which aerobic metabolism can proceed. This has been overcome in crustaceans and vertebrates by diverting metabolic pathways to produce lactic acid which accumulates in the tissues. A similar diversion in molluscs results in the formation and accumulation of opines*. Opines are formed during recovery from a long period of anaerobiosis when energy demands are high. This has been called functional anaerobiosis. When environmental conditions cause closure of the shell, oxygen tensions in the body drop to a level at which aerobic metabolism cannot be sustained, and metabolism switches to anaerobic pathways. This has been called environmental anaerobiosis. For barnacles, functional and environmental anaerobiosis follows the same pathway, but for mussels there is a difference.

When first exposed, *Mytilus edulis* and *M. californianus* are known to utilize atmospheric oxygen by keeping the shell slightly open and taking air into the mantle cavity. This will allow oxygen to diffuse into the tissues of the mantle and gills, but oxygen tension in the deeper tissues rapidly falls below the level at which aerobic metabolism can be sustained. The blood does not have any oxygen-carrying pigment which would improve oxygen delivery to deeper tissues. The rate of oxygen uptake from air is *c.* 0.74 times the rate of extraction from seawater. When the shell of *M. californianus* is completely closed, the oxygen tension drops from $1.6 \times 10^4\,\mathrm{N\,m^{-2}}$ to $1.06 \times 10^4\,\mathrm{N\,m^{-2}}$ within *c.* 5 min and to $0.53 \times 10^4\,\mathrm{N\,m^{-2}}$ in 15 min. Oxygen tensions then remain constant at this level throughout anaerobiosis (Bayne *et al.*, 1976b).

* Opines are condensation products of pyruvate and an amino acid formed by enzyme action of an opine dehydrogenase, for example, strombine (2,2′-iminopropioacetic acid) is formed from pyruvate and glycine in the presence of strombine dehydrogenase.

Mussels and oysters can survive out of water for many days by utilizing their stores of glycogen, adenosine phosphate and aspartate in anaerobic metabolic pathways. They can survive emersion much longer than their position on the shore would suggest, but it is likely that regular immersion (at last once every 24 h) is normally necessary to excrete the waste products that accumulate whilst the shell is closed.

Mytilus edulis has been the subject of many biochemical studies on environmental anaerobiosis. The following summary is based on the reviews by De Zwaan (1977, 1983) and De Zwaan and Putzer (1985).

Two phases of anaerobiosis can be distinguished: an initial or transitional phase, lasting 5–10 h after closure of the shell, during which aspartate is used in conjunction with normal glycolysis, and a steady-state anaerobiosis that sets in when the aspartate pool has been exhausted.

The end product of glycolysis is pyruvate, with the liberation of energy in the form of energy-rich adenosine triphosphate. During aerobic metabolism, pyruvate is degraded in the citric acid cycle to carbon dioxide, and further energy is released in electron transfer through the cytochromes with the formation of water (Fig. 6.7, pathway at far left). At low oxygen tension, electron transfer cannot take place through oxidative phosphorylation, and alternative pathways are followed. In the initial phase of anaerobiosis, pyruvate undergoes transamination with aspartate to form alanine and oxaloacetate; the latter is then converted in the mitochondria, via malate and fumarate, into succinate, with further energy release (Fig. 6.7, pathway in centre). Succinate, alanine and small amounts of the opine, strombine, accumulate.

When the aspartate pool has been used up, the glycolytic pathway passes from phosphoenolpyruvate by carboxylation to oxaloacetate, and then through malate and fumarate to succinate, but succinate is now converted

---

Fig. 6.7 Scheme outlining major steps in metabolic pathways of mussels and barnacles. Upper half — reactions in cytoplasm, lower half — reactions in mitochondria.

Left: normal pathway during submersion (aerobic conditions) for mussels and barnacles.

Middle: anaerobic pathways when shell is closed during emersion or lowered salinity. Mussels produce succinate during short periods of anaerobiosis, but on prolonged emersion they switch to a pathway producing propionate as the end product.

Right: pathway followed in mussels during recovery on re-submersion. (See text for explanation).

– – – – E, energy release in form of ATP; ♯, transamination; ♯ ♯, coenzyme A transfer.

METABOLIC PATHWAYS

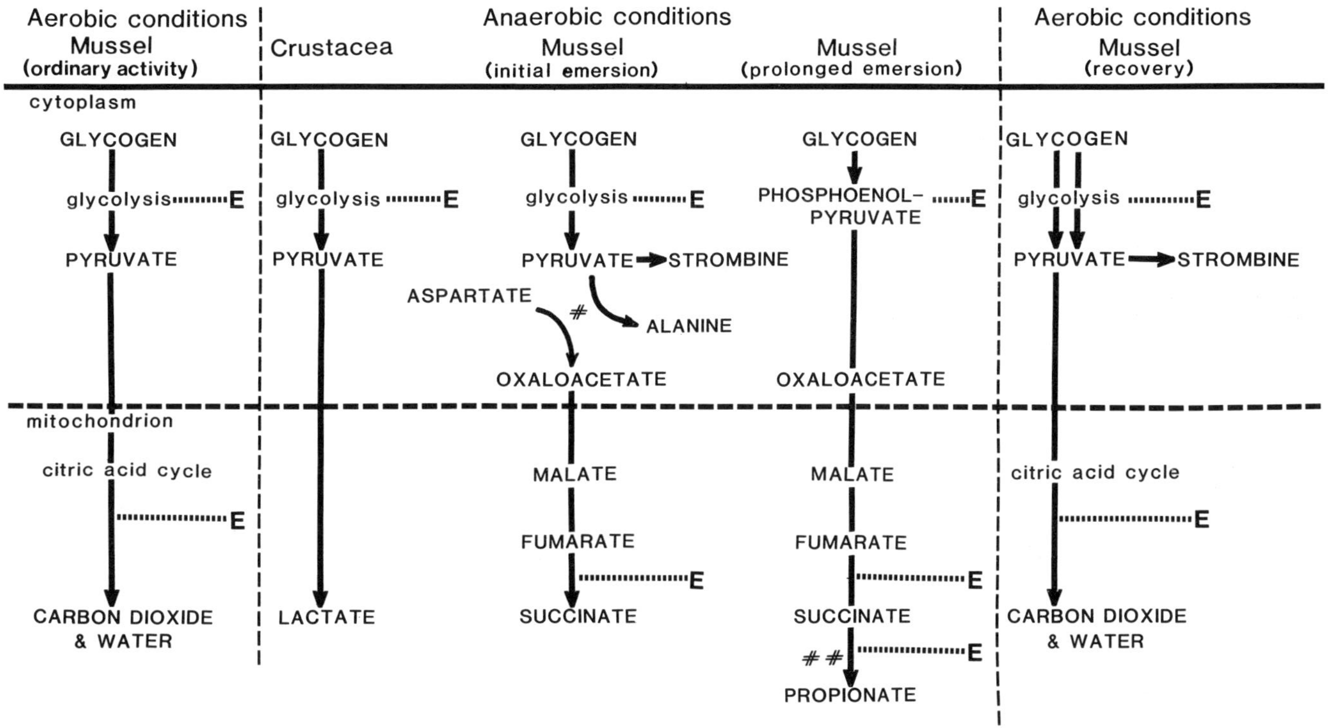

into propionate by coenzyme A, with further energy release (Fig. 6.7, pathway first from right). Thus during prolonged anaerobiosis, propionic acid is the main end product. Lutz and Rhoads (1977) have suggested that the calcium carbonate of the shell plays a part in buffering the acidity of rising propionate levels in the blood during prolonged anaerobiosis.

In comparison with the lactic acid pathway followed by crustaceans (Fig. 6.7), the production of succinate or propionate increases the yield of energy-rich adenosine triphosphate per glycosyl unit metabolized, thus allowing a much lower rate of metabolism to supply energy needs.

When resubmerged by the incoming tide, the mussel opens its shell and resumes pumping water through the mantle cavity. There is an immediate need to clear the tissues of accumulated waste products, raise the oxygen tension so that aerobic respiration can resume, and replenish the depleted arginine phosphate and aspartate pools. The course of events has been studied by De Zwaan *et al.* (1983). They found that blood ammonia dropped to normal levels within 3 h, partly by direct diffusion into the seawater and partly by conversion to glutamate. The high energy demands during this early period cannot be adequately supported by aerobic pathways, and functional anaerobiosis takes place. A surge in the rate of glycolysis is achieved by diverting some of the pyruvate to form opines (strombine and small amounts of octopine) which temporarily accumulate in the muscle (Fig. 6.7, pathway at right). Later they are converted back into pyruvate, but this may take 12–24 h.

The formation of opines in the presence of a dehydrogenase has a number of advantages over lactate production, both during anaerobiosis and during the recovery period. They will not upset the pH as they are not as acidic as lactic acid, nor will they cause changes in osmolarity. A number of other metabolic functions have been proposed for octopine in cephalopods (Storey and Storey, 1983).

Excretion and ionic regulation

Closure of the shell for long periods, which occurs during submersion at low salinities as well as emersion, requires physiological adjustment of the closely related functions of excretion and ionic regulation (reviews: Bayne *et al.*, 1976b; Davenport, 1985).

Excretory organs are the paired kidneys and pericardial glands, and the main nitrogenous end products are ammonia and amino acids. *Mytilus edulis* excretes 41% of its nitrogenous waste as ammonia and 55% as amino acids during winter, but in summer, ammonia rises to 68% and amino acids drop to 28%. This reflects the fact that protein is more important as a substrate for metabolism in winter, and glycogen in summer (Bayne *et al.*, 1976b). *Crassostrea virginica* excretes 68% of nitrogenous waste as ammonia, 21%

as amino acids, 8% as urea, and *c.* 3% as uric acid (Galtsoff, 1964). Amino acid loss from the body is a drain on energy resources and may account for up to 11% of energy output. Amino acids must be actively excreted, not lost by diffusion, for some are more readily eliminated than others. The total amount lost by a dense mussel population is considerable, and will contribute a significant proportion of the energy budget of the microbial population of inshore waters.

Out of the water, the rate of ammonia excretion in *Mytilus californianus* is reduced to only 5% of the rate when immersed. This would prevent the rapid buildup of ammonia in the mantle cavity. The mechanism for this reduction is not clear. It would reflect the fact that glycogen is the sole substrate for metabolism during anaerobiosis; however, higher amino acid excretion or the fixation of ammonia as alanine could offset ammonia production. These alternatives are an important adaptation to shore life because they reduce accumulation of toxic substances.

Mussels and oysters may be subjected to lowered salinity when heavy rain falls at low tide, or in estuaries where significant quantities of fresh water enter the sea. The osmotic concentration of the blood is identical or very slightly hyperosmotic to the surrounding water. The ion species are present in the same proportions except for a slight increase in potassium and a very small amount of protein, but the ionic composition and osmotic concentration of the tissue fluid is very different from that of the blood. In the muscle tissue of *Mytilus edulis*, sodium, magnesium and chloride ions are reduced in amount, while calcium is slightly increased and potassium greatly increased compared with blood. In addition, phosphate and considerable amounts of organic solutes are present. The ionic constituents contribute *c.* 60% and the organic solutes 40% of the total osmotic concentration. *M. edulis* is an osmo-conformer; the osmotic pressure of its blood is rapidly adjusted to changes in the salinity of the surrounding medium. Lowered salinities would normally cause cells of the tissues to take up water and so increase in volume, but mussels can carry out intracellular ionic regulation. This is achieved by active regulation of the intracellular concentration of osmotically active organic compounds. At salinities down to 50% of normal seawater, the shell may not close, consequently dilution of the blood occurs. Intracellular ionic regulation causes amino acids to be lost from the tissues to the blood, thus reducing the osmotic concentration of the tissues and preventing the cells increasing in volume. At salinities below 25% of normal seawater, mussels close their shell tightly to avoid excessive dilution of their blood (Davenport, 1979, 1981). They are now forced to enter anaerobic metabolism. Under experimental conditions, Bayne (1975) showed that *M. edulis* more than doubled its ammonia output when placed in water with a salinity of 14.5‰, but gradually reverted to normal levels within 14 d. At the same time there was a sevenfold increase in the rate of amino acid loss, which reverted to normal after *c.* 20 d. This shows that reduced salinity increases the breakdown of amino acids and the excretion of ammonia. However, the total amount of ammonia nitrogen plus amino nitrogen lost in the first day accounted for less than 10% of the change in free amino acid concentration in the mantle. This suggests that free amino acids may be bound to less soluble or less osmotically active molecules, possibly proteins, when the blood is diluted. Taurine, betaine, glutamate and glycine are the principal nitrogen compounds involved in intracellular ionic regulation (Bayne *et al.*, 1976b).

These adjustments to external change are examples of acclimatization. Almada-Villela (1984) found that *Mytilus edulis* taken from normal salinities could be acclimatized to salinities down to 13‰, although dilution beyond 9.5‰ proved fatal. However, *M. edulis* occurs in the Baltic Sea at salinities of 15‰, and is even found in Finnish waters where salinities may fall as low as 6‰. It is likely that the Baltic mussels are a distinct physiological race adapted to low salt concentrations.

Reduced salinity may affect reproduction. Salinities below 25‰ adversely affect fertilization in *M. californianus*, but in *M. edulis*, fertilization is successful over the range 5–22 °C and 15–40‰ salinity, although normal cleavage of the fertilized eggs only occurs in the narrow salinity range of 30–40‰.

Lowered salinity also affects shell growth, by altering the amino acid composition of the conchiolin matrix. Increased salinity reduced the aragonite content of the shell (Grégoire, 1972).

SUMMARY

Mussels and oysters are well adapted to life at lower levels of the middle shore. The calcareous shell gives protection from water loss, abrasion, lowered salinity and predators, but the body, gills and mantle do not have a cuticular covering. Water is retained in the mantle cavity during emersion. When the valves are closed the animal is completely isolated from its external environment.

Oysters attach permanently by means of a proteinaceous cement substance. Mussels attach by means of byssus threads secreted by glands in the foot. The distal ends of the threads are cemented to the substratum by a tanned protein. The threads can be broken and new ones formed in such a way that the animal can move to a new position.

Feeding is efficient, particles down to 2 μm can be collected, and there is evidence that feeding is selective. In mussels specialized latero–frontal cirri act as a sieve. Mucus appears to be involved in removing unwanted particles; not food capture.

Mussels have evolved efficient adaptations to emersion. Out of the water, metabolic rates are very low. Some species are able to utilize atmospheric oxygen, but the shell cannot remain open for long periods without the danger of water loss. When the shell is closed, the oxygen supply is depleted and metabolism switches to anaerobic pathways. At first the end product is succinate, but on prolonged emersion, propionate is formed. During recovery, on returning to aerobic conditions, opines may be formed, but the oxygen debt is eventually repaid as metabolic pathways return to normal. The pathways for nitrogenous excretion are modified to minimize the toxic effects of ammonia accumulation during anaerobiosis and during recovery.

Mussels show wide tolerance to shore conditions. In high latitudes they can tolerate freezing. Freezing of extracellular water increases the osmotic concentration of solutes and prevents further freezing. Some of the intracellular water may be bound to macromolecules and is not free to crystallize. Mussels are osmo-conformers and tolerate lowered salinities down to a certain level when further dilution is prevented by closure of the shell.

Mussels and oysters do not show reproductive specializations for shore life.

7

Tube-dwelling worms (*Annelida: Polychaeta*)

A few tube-dwelling worms are found in the intertidal zone; they represent four families: Sabellariidae, Sabellidae, Serpulidae and Spirorbidae*. Although none show structural features than can be regarded as specific adaptations to periods of emersion, some physiological adaptations appear to have evolved. The calcareous tube of an intertidal serpulid worm is well suited to strong wave action, for it is firmly cemented to the substratum throughout its length and has a low profile. Its hardness and rigidity provides excellent protection for the delicate worm inside. The number of species increases in the sublittoral zone.

These worms differ from all other animals described in this book by not being directly attached to the substratum. Imprisoned by the very tube they construct, they are free to move within it, but cannot survive for long without its protection. Few species are able to produce a new tube if removed from their existing tube, and without it would soon fall prey to predators.

A number of profound modifications related to the tubicolous mode of life have evolved in these families, and yet they still possess the main features of annelid organization. They have an elongated, segmented body, a straight gut with terminal anus, a large perivisceral coelom that may be filled with developing gametes in the breeding season, and groups of bristles or chaetae, which project from the sides of each segment.

7.1 REPRESENTATIVE SPECIES FROM ROCKY SHORES

On British shores, two species of spirorbid tube-dwelling worms are likely to be the first noticed. They secrete small white calcareous tubes coiled in a flat

* See Appendix for classification of tube-dwelling worms.

helix 1–3 mm in diameter. *Janua* (= *Dexiospira*) *pagenstecheri* (also found on the southern shores of Australia) may extend up the shore to high water neap tide level on stones and various seaweeds. Its tube coils anticlockwise (dextral when viewed from the opening, Fig. 7.1(a)). The tube of *Spirorbis spirorbis* (= *borealis*) coils clockwise (sinistral Fig. 7.1(c)) and it is found lower on the shore, where it is particularly conspicuous on *Fucus serratus* and *F. vesiculosus* in sheltered places. In both, the body is short, with only five feeding tentacles or radioles, an operculum, and three thoracic segments (Fig. 7.1(b)). Ova form in the first two abdominal segments, and sperm in the remainder. The two species differ in the way the eggs are incubated; *Spirorbis spirorbis* retains them in the tube, while *Janua pagenstecheri* incubates a few at a time in the operculum (Fig. 7.1(b)). A new cap to the operculum regenerates when the batch of embryos has been released.

Forming a conspicuous white band some 30 cm wide below the barnacle zone on Australian shores, vast numbers of the serpulid worm *Galeolaria caespitosa* cover the rocks (Fig. 7.1(d) and (e)). Their calcareous tubes 20–30 mm long intertwine and tend to fuse, but isolated tubes are straight or slightly S-shaped, with a prominent keel that projects beyond the mouth of the tube (Fig. 7.1(e)). The worms have 23 pairs of dark blue or brown radioles arranged in a double spiral, and an operculum bearing elaborate tooth-like spines (Fig. 7.2(b)). The operculum is not used as an incubation chamber. Vast numbers of orange-coloured eggs are produced in the abdomen and shed directly into the sea. The abdomen of a sexually mature male is white. *Pomatoceros triqueter* is a very similar species found on British shores on stones near low water.

Another serpulid worm, *Salmacina dysteri*, forms masses of fine, tangled, white calcareous tubes under boulders below the *Galeolaria caespitosa* band and into the sublittoral zone. It is of interest here because it reproduces asexually. The body undergoes fission to produce two worms, one of which secretes a new tube. In this way a clone is rapidly formed which spreads out over the surface. The worm itself is only 1 mm long. It does not have an operculum and there are only eight colourless radioles; the body is orange.

The feather duster worm *Sabellastarte indica*, belonging to the Sabellidae, is a common species on Australian shores, where the expanded feeding crown makes it a conspicuous animal in tide pools or between boulders in the sublittoral zone (Fig. 7.2(c)). It closely resembles *Sabella pavonina* found in soft deposits on British shores. *Sabellastarte* lives in a tube of parchment-like material formed from the secretion of glands on the ventral side of the body. Exposed parts of the tube are heavily impregnated with silt and only loosely attached to the rocks, but it is anchored deeply in crevices. Large specimens (up to 12 cm contracted length) have a feeding crown of 200 or more brown radioles arranged in a double spiral. Each radiole is fringed with ciliated white pinnules which create the feeding current (Fig. 7.3(a)). There are two short palps, but no operculum or lateral folds to the thorax.

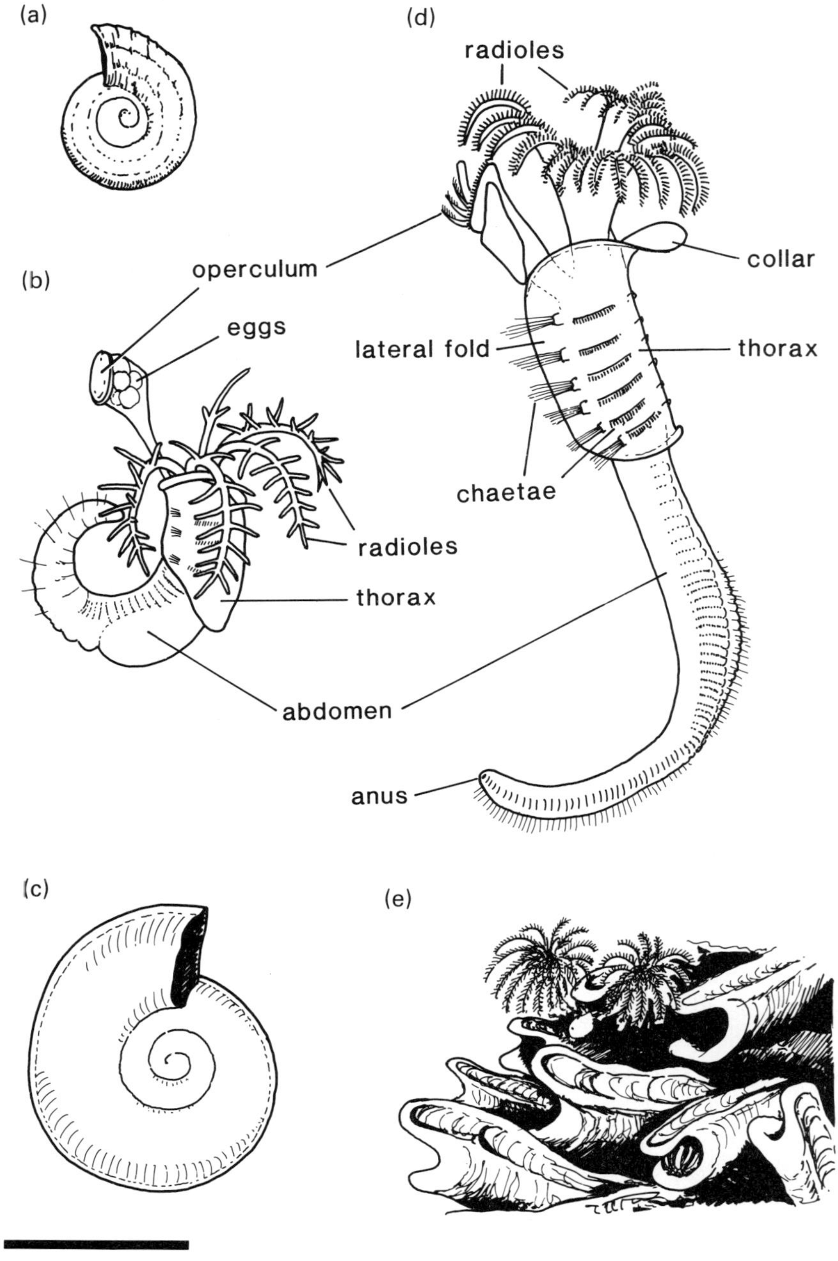

Fig. 7.1 (a) Calcareous tube of *Janua pagenstecheri*, × 10. (b) Worm removed from its tube, showing eggs in operculum. (c) Tube of *Spirorbis spirorbis*, × 6. (d) *Galeolaria caespitosa*, removed from its tube, × 5. (e) Tubes of *Galeolaria*.

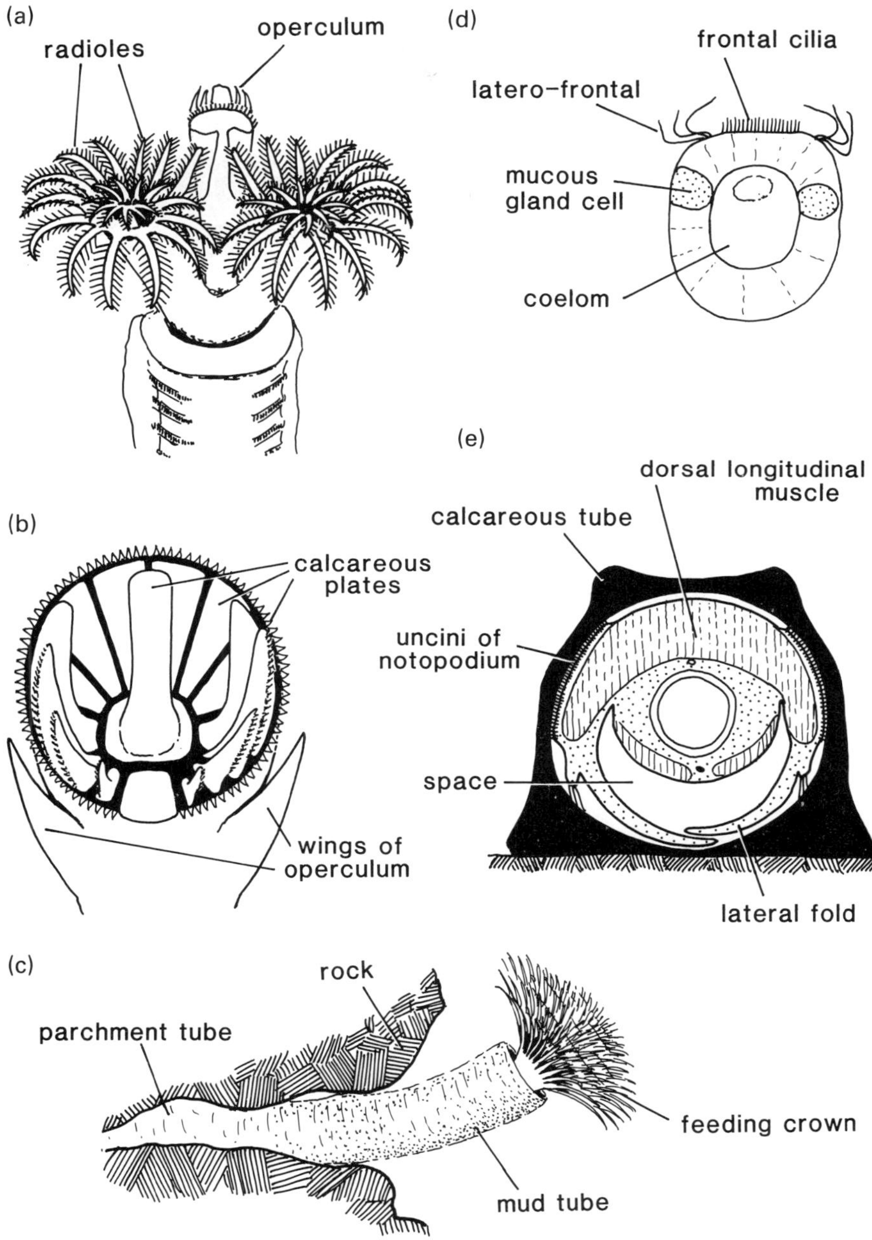

Fig. 7.2 (a) Dorsal view of *Galeolaria caespitosa*, showing feeding crown. (b) Calcareous plates of operculum. (c) The fan worm, *Sabellastarte indica*, about half natural size. (d) Diagrammatic section through pinnule of radiole. (e) Diagram showing probable role of lateral folds (see text).

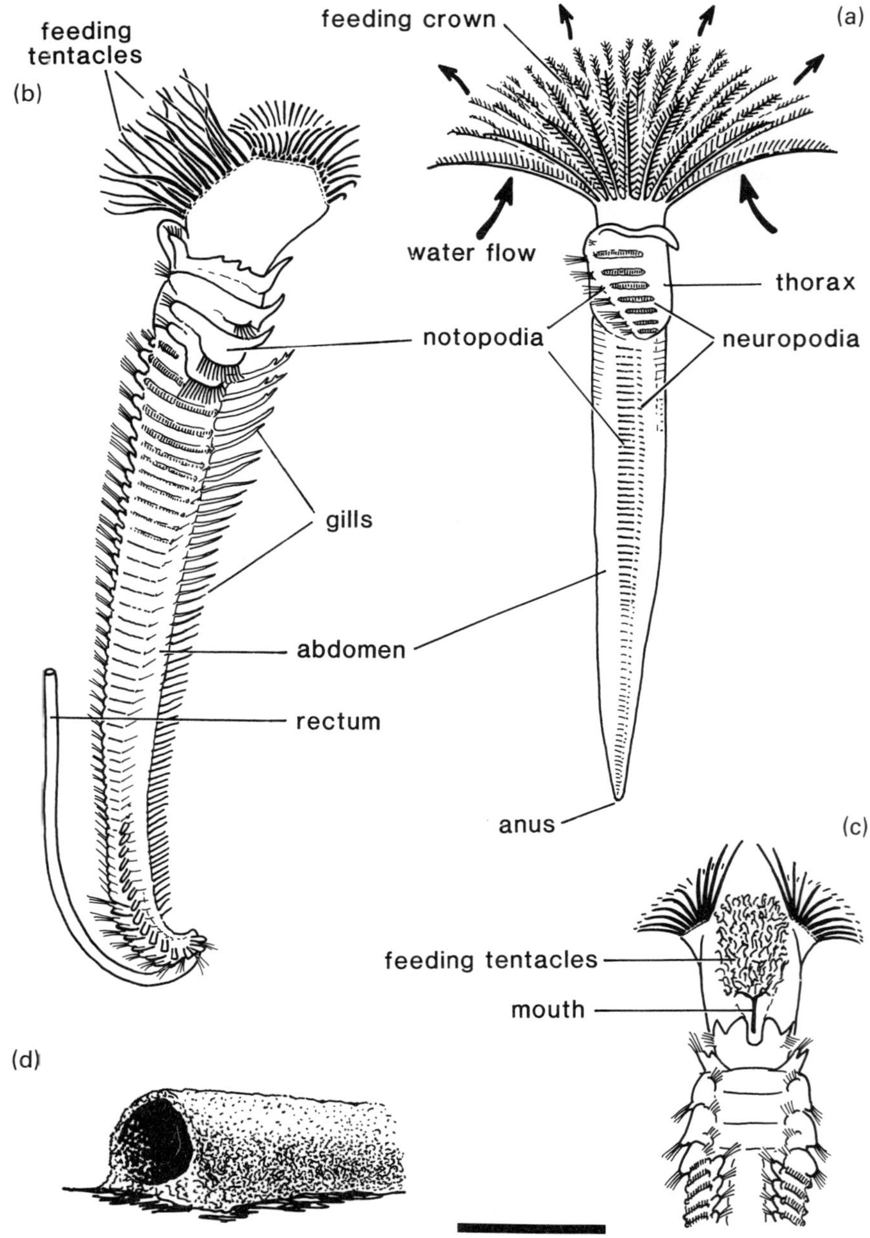

Fig. 7.3 (a) *Sabellastarte indica* removed from its tube (side view). Arrows, direction of water flow. (b) to (d) *Idanthyrsus pennatus*; (b) removed from its tube; (c) Ventral view of anterior showing mouth; (d) Tube constructed from sand grains cemented together. Scale bar: (a) = 20 mm, (b) = 8 mm.

Another worm, *Idanthyrsus pennatus*, belonging to the Sabellariidae, is found in the same situations as *Sabellastarte indica*, but it lacks a spiral crown of radioles, and builds an extremely hard tube of sand grains cemented together with organic material and firmly fixed to the underside of rocks (Fig. 7.3(d)). *Sabellaria alveolata* is a similar species found on some British shores. Two clusters of feeding tentacles without pinnules lie on the ventral side of the head region (Fig. 7.3(b) and (c). Above them there is a palisade of very strong golden bristles that guard the entrance to the tube when the worm withdraws (Fig. 7.4(d) c). Segments three to five bear stout dorsal parapodia with strong paddle-shaped chaetae (Fig. 7.4(d) b), while the remainder of the segments have prominent ridges bearing numerous hooked chaetae called uncini (Fig. 7.4(d) a) which are characteristic of all these tube-dwelling species. Each segment of the body bears a pair of dorsal gills, except for the long rectal region which shows no sign of segmentation or chaetae. The rectum is bent upwards ventrally and can be extended to discharge faeces out of the tube (Fig. 7.3(b)).

7.2 STRUCTURES RELATED TO A TUBICOLOUS LIFE STYLE

The repertoire of movements made by tube-dwelling worms is limited. In species that close the mouth of the tube with a stopper or operculum, movement back along the tube is restricted by that structure, but species lacking an operculum are able to move back along the tube, if necessary, by using their chaetae to gain purchase on the wall. Many species, such as *Sabella pavonina*, perform peristaltic movements or undulate the body to circulate water in the tube (Mettam, 1969), and all species must have the ability to rotate in order to construct the tube. A characteristic reaction of all sabellid and serpulid worms is the instantaneous withdrawal of the feeding tentacles into the tube when the environment is disturbed, as for example when a shadow falls on an expanded worm. Many have photoreceptors on the radioles that respond to changes in light intensity, and giant nerves in the ventral nerve cord that co-ordinate the instantaneous contraction of the large segmental longitudinal muscles and the extension of chaetae to anchor the worm in the tube.

A hard cuticle to protect the body of the worm is absent. The epidermis is a simple epithelium which contains an abundance of mucous cells and is covered by a very thin flexible pellicle of mucopolysaccharide. The chaetae are secreted by special epidermal cells, and consist of chitin or a chitin–protein complex. They are of two types: short hook-shaped chaetae called uncini (Fig. 7.4(b) b, (c) b and c, (d) a), arranged in rows, and elongated bristles which may be plain or comb-like (Fig. 7.4(a), (b)a, (c)a, and (d)d). The uncini function as crampons to grip the tube when the worm rotates, moves

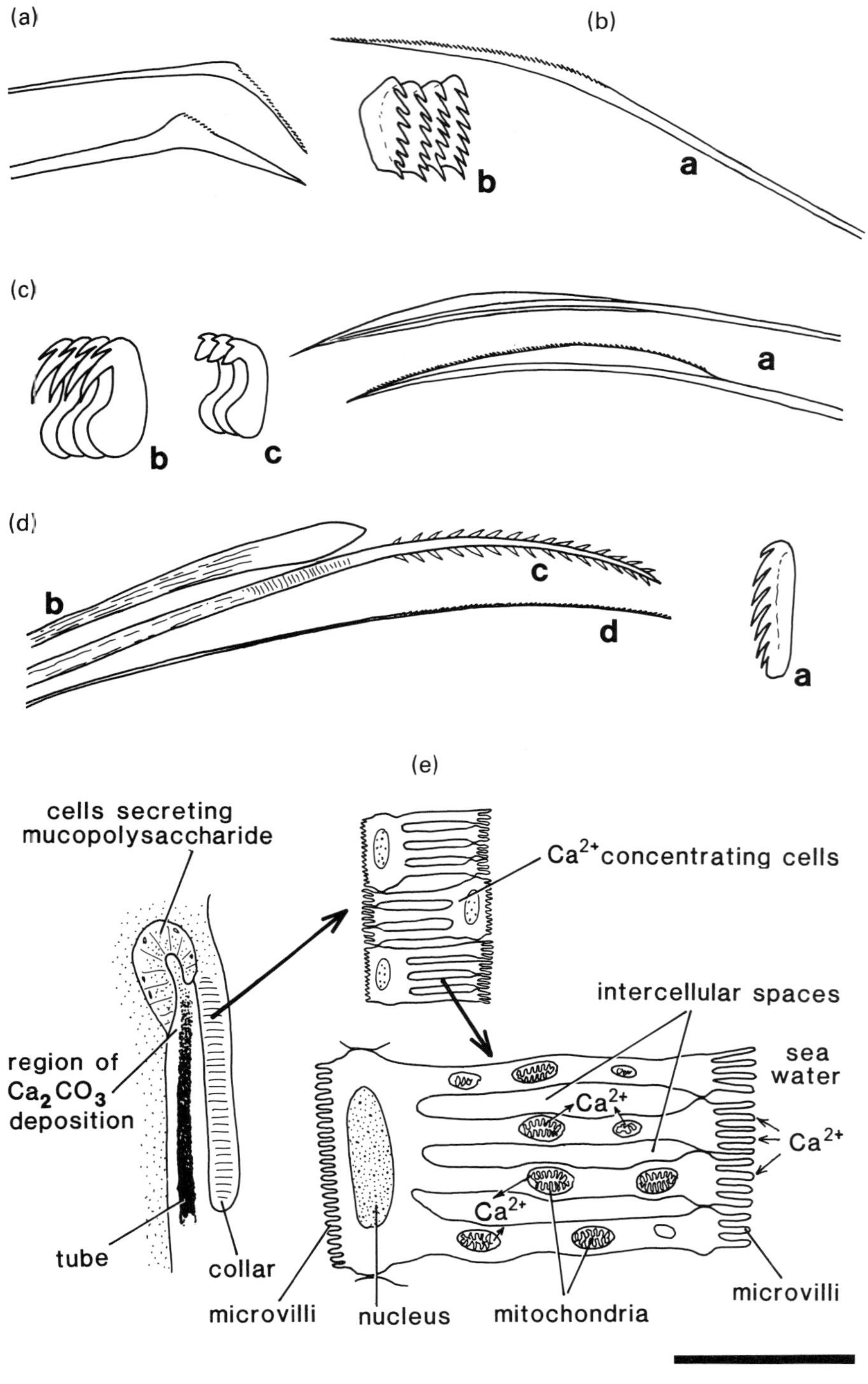
(a)
(b)
b
a
(c)
b
c
a
(d)
b
c
d
a
(e)
cells secreting mucopolysaccharide
region of Ca_2CO_3 deposition
tube
collar
Ca^{2+}concentrating cells
intercellular spaces
sea water
Ca^{2+}
Ca^{2+}
Ca^{2+}
microvilli
nucleus
mitochondria
microvilli

along the tube, or maintains its position during pumping movements of the body. The function of the long comb-like bristles is less clear, but their shape and ornamentation suggest that they are used to keep the interior of the tube clean or free from peritrichous ciliates and other micro-organisms, as well as for movement in the tube. The first three to seven bundles of stout ventral chaetae can be moved in and out and may be responsible for rotation of the worm in the tube.

The anterior segments of sabellid and serpulid worms are modified in two important ways: the head has a crown of pinnate radioles arranged in two spirals or semicircles above the mouth, and some of the anterior segments form a thoracic region (Fig. 7.1(d); 7.3(a)).

In the Serpulidae, one (sometimes two) of the radioles is modified as an operculum, which blocks the mouth of the tube when the worm retreats. The heavily armoured operculum of *Galeolaria caespitosa* appears to be a defence against small carnivorous gastropods. Another adaptation of the operculum in certain species is for incubation of the eggs (Fig. 7.1(b)). The crown of extended radioles forms a funnel through which a current of water passes slowly from the outside, up through the pinnules to the middle of the funnel, (Fig. 7.3(a)). Lateral cilia like those found on *Mytilus* gill filaments (Fig. 6.6(b)) or bryozoan tentacles (Fig. 9.5(e)) are absent, and the current is generated by long latero-frontal cilia on the pinnules (Fig. 7.2(d)). Frontal cilia on the oral side of the pinnules beat towards the radiole, down which a ciliated gutter runs to the mouth.

Tube-dwelling worms are suspension feeders. Particles of food are collected from the water current by the pinnules and passed to the mouth. The classic studies of Nicol (1930) on *Sabella pavonina* showed that a ciliary mechanism sorted particles at the base of the radioles, and only the smaller particles were passed to the mouth. Large particles were rejected, while medium-sized particles were retained and mixed with mucus to be incorporated into the tube. *Sabellastarte* appears to have the same sorting mechanism. Serpulid worms also feed on small particles, but there is evidence that selection may be done at the pinnules rather than at the base of the radioles.

The method by which particles are captured is not well understood. Abundant mucous gland cells are present in the epithelium of the pinnules

---

Fig. 7.4 (a) Chaetae of *Spirorbis*. (b) Chaetae of *Galeolaria*: a, distal end of long chaeta, b, uncini. (c) Chaetae of *Sabellastarte*: a, distal end of long chaetae; b and c, uncini. (d) Chaetae of *Idanthyrus*: a, uncinus; b, distal end of paddle from thoracic notopodium; c and d, distal end of long chaetae from thoracic neuropodium. (e) Tube secretion (see text for explanation). Scale bar: (a) = 0.02 mm, (b), (c) and (d,a) = 0.05 mm, (d,b,c,d) = 0.2 mm.

and radioles, and it is usually held that particles are trapped by mucus (Dales, 1967), however, the same problem arises concerning ciliary sorting that has already been discussed for *Mytilus* (see page 98). It is difficult to explain how particles entangled in mucus can be sorted according to size by ciliary currents. The latero-frontal cilia of the pinnules create the water current, but they do not act as a sieve like the latero-frontal cirri of mussels, nor as sensory organs as in bryozoans. It is possible that eddies, caused by beating of the C-shaped latero-frontal cilia, direct the smallest particles to the frontal tract, where local currents carry them to the ciliated gutter on the radiole. It is clear that the mechanism is very different from that of *Mytilus* or the bryozoans. Dales (1957) found that flagellate algae passed through the feeding crown of sabellid and serpulid worms, and only inert particles, down to 1–2 μm, were retained. This finding implies that these worms are detritus feeders. It also supports the idea that food is not retained by mucus.

According to the species, between three and seven of the anterior body segments are modified as a thorax, on which there is a row of ventral uncini and a dorsal bundle of long chaetae. On the abdomen, the arrangement of the chaetae is reversed, with the uncini in the dorsal position (Figs 7.1(d), 7.3(a)). The functional significance of this reversal has not been satisfactorily explained; it might be concerned with maintaining a firm grip on the tube to stop the worm being sucked out by the sudden reduction in pressure after a wave has broken heavily.

A sabellid worm can wedge itself in the tube by increasing the diameter of a few segments of the body, but this would seal the tube and cut off the posterior end from oxygen and removal of faecal waste. Wing-like expansions of the thorax in serpulid worms overcome this difficulty. They allow the worm to wedge itself in the tube while providing a space through which water can pass (Fig. 7.2(e)). This arrangement has the added advantage that water can escape when the worm rapidly retracts into the tube, or enter easily when the feeding crown is protruded.

Water cannot flow through the tube because it is closed at one end. This poses two problems, disposing of faecal waste and obtaining oxygen. Faecal pellets are carried forward in a ciliated groove on the ventral surface of the abdominal region as far as the thorax, and are then passed to the dorsal side, where they are carried forward and out of the tube. Tube-dwelling worms have a well-developed blood system, with blood pigments (chlorocruorin or haemoglobin) capable of carrying oxygen from the radioles to all parts of the body. Oxygen can also be taken through the body surface or dorsal gills in the Sabellariidae (Fig. 7.3(b)). Undulatory movements or peristalsis circulate water in the tube for this purpose. Serpulids like *Galeolaria caespitosa* may not close their tubes completely during emersion, so air is in contact with at least part of the retracted radiole surface, and thus aerobic respiration probably continues during this period.

Other physiological and behavioural adaptations for shore life have been noted. De Silva (1967) found that *Spirorbis spirobis* showed a clear semilunar rhythm of egg release during the breeding season from May to October. Larvae were released a few days before the first and last quarters of the moon (i.e., at the approach of neap tides). It is possible that the quieter waters at this period would favour larval settlement on fucoid seaweeds, for settlement at neap tides gives the young time to secrete a protective tube before the danger of exposure during the following spring tides. De Silva also found evidence of physiological adaptation to temperature stress. *Janua pagenstecheri* living relatively high on the shore showed a wider range of temperature tolerance than *Spirorbis corallinae* living on the lower shore.

7.3 TUBE SECRETION IN SERPULIDAE AND SPIRORBIDAE

The tube is not part of the worm. It is formed from a mucopolysaccharide matrix hardened by deposits of calcium carbonate. Both components are secreted by glands in the collar, which also moulds the shape of the tube. The proportion of organic matter in the tube is high. Dixon (1980) found that it contributed 4.5% of the tube weight in *Mercierella*, compared with *c.* 0.5% in bivalve shells, and represented a considerable expenditure of energy by the worm. Dixon calculated that 68% of the energy used in growth and tube formation went to production of the matrix, 20% to somatic growth and 12% to gamete production.

The early stages of tube formation have been studied by Nott and Parkes (1975), who worked with settling larvae of *Spirorbis spirorbis*. Initial attachment is effected by a secretion from the attachment gland which accumulates in the larval gut prior to settlement. At settlement it is released through the anus. The tube itself is produced by activity of the collar epithelium, which has two types of cell: cells of the tube-forming gland that secrete mucopolysaccharides, and collar cells with fine processes ending in microvilli and packed with mitochondria (Fig. 7.4(e)). A large intercellular space exists among the fine processes. Nott and Parkes considered that these cells are the site of cation pumps, which move calcium ions from the seawater into the intercellular spaces and so increase their concentration. Crystallization of calcium carbonate takes place outside the cells, presumably when secretions of the tube-forming gland come into contact with the high concentration of calcium ions. Calcium carbonate and organic matrix accumulate under the collar (Fig. 7.4(e)). The collar acts as a mould in which the tube material solidifies.

The larvae of *Spirorbis* figure prominently in studies on larval settlement (pages 306–309).

SUMMARY

Tube-dwelling worms living at lower levels on the shore will not be subjected to very long periods of emersion or harsh conditions, consequently there are few direct adaptations to shore life. The hard tube of serpulid worms provides effective protection and can be closed by the operculum. Water is retained in the tube during emersion, and the feeding tentacles may not be completely withdrawn, suggesting that some species may be able to utilize atmospheric oxygen. Littoral species differ little in structure or mode of reproduction from sublittoral species.

8

Sea squirts (*Tunicata: Ascidiacea*)

In comparison with barnacles and mussels, the majority of sea squirts are not well suited to living on the shore and are only found at lower levels. Water loss from the body during emersion at low tides may not be a problem for them, as they retain a relatively large volume of water in the body cavity; however, the majority of species appear to be sensitive to light, and live in places where they avoid direct rays of the sun, like crevices, caves and under rocks or seaweed. There are some notable exceptions, such as the large surf sea squirt, *Pyura stolonifera*, living on rock platforms at low tide level in South Africa and Australia, which is exposed to intense sunshine for 2–3 h during spring tides, or certain species of *Didemnum* and *Lissoclinum*, which contain symbiotic algae and live in exposed positions.

Few specific adaptations to intertidal conditions are shown, although many of the species in the sublittoral zone appear well suited to turbulent wave action. Sea squirts display many features of biological interest, some of which will be discussed here even though they are not directly connected with shore life.

The larval stage of sea squirts is a free swimming tadpole. Its locomotory organ is a post-anal tail which contains muscle blocks arranged round a rod of vacuolated cells–the notochord. During development, slits appear in the wall of the gut which grows into the gill slits or feeding apparatus of the adult. These features are characteristic of the phylum Chordata. Sea squirts are placed in the class Ascidiacea of the subphylum Tunicata.* Two different types of ascidian occur on the shore: compound ascidians that form a colony in a common gelatinous mass, and large solitary ascidians enveloped in a tough outer tunic and possessing two distinct siphons.

* See Appendix for classification of sea squirts.

8.1 REPRESENTATIVE SEA SQUIRTS FROM THE SHORE

Compound ascidians

Compound ascidians are amongst the most abundant sessile animals to be found on the lower shore and in some places completely dominate the rock surface. Many are cryptic and form sandy or chalky incrustations that give no indication of containing living animals. The individual members of the colony (or zooids) are very small (seldom more than 5 mm), nevertheless most species can be squeezed out of the gelatinous tunic and examined under a microscope. All are hermaphrodite and retain fertilized eggs within their body (or the tunic) for development, consequently tadpole larvae are easily obtained. A characteristic feature of the compound ascidians is their ability to reproduce by budding to form a colony or clone in which zooids show various degrees of interdependence.

Eudistoma elongata (Fig. 8.1(a) and (b)) appears as a transparent gelatinous mass, enclosed in a thin outer cuticle, in which sand grains and hundreds of minute zooids are embedded. Zooids are independent of each other. Those towards the periphery are in various stages of development from buds, while those in the centre will show various stages of sexual reproduction.

The structure of a typical zooid is shown in Fig. 8.1(a). The thin outer body wall or mantle has longitudinal and circular muscle strands, but is otherwise transparent, so other internal organs can be seen (Fig. 8.1(b)). An oral or inhalent siphon leads to the greatly enlarged anterior part of the alimentary canal, called the branchial sac or pharynx, which is perforated by numerous openings, the gill slits or stigmata. These lead into the surrounding atrial cavity, which opens to the exterior by an atrial or exhalent siphon. The endostyle, a conspicuous structure concerned with feeding, lies on the ventral wall of the pharynx. The remainder of the digestive system consists of a relatively simple U-shaped tube which terminates in an anus near the atrial siphon. A current of water, generated by cilia in the wall of the pharynx, is drawn in at the oral siphon, passed through the stigmata into the atrium, and expelled through the atrial siphon. Food particles trapped in the pharynx by mucus are carried through the digestive system by cilia. An ovary and scattered lobes of the testis lie in the posterior or abdominal region of the body.

A characteristic feature of this genus is the presence of two similar siphons. When the siphons of a zooid are touched, the whole zooid is pulled down into the gelatinous tunic by the retractor muscle. Restoration to the normal condition is probably dependent on the elasticity of the tunic and contraction of the circular muscles.

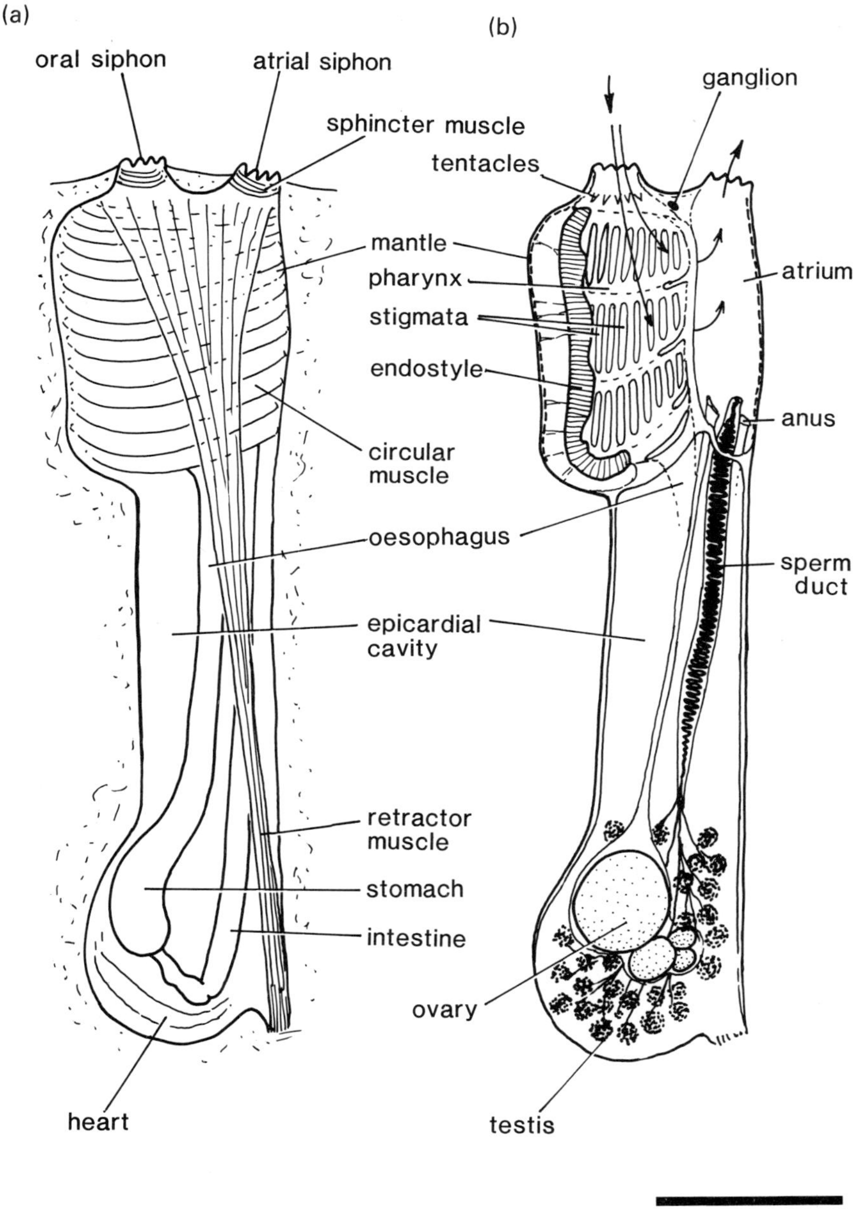

Fig. 8.1 Eudistoma elongata. (a) Zooid removed from its jelly-like tunic to show musculature. (b) Internal structure. Arrows depict direction of water flow. Scale bar: (a),(b) = 1 mm.

This species produces a few very large, yolk-laden eggs, which are retained in the atrial cavity during their development. Usually three embryos are found in the branchial region, each at a different stage of development, indicating that the ova mature one at a time. A tadpole with black eye spot can be seen in the first egg ready to be released through the atrial siphon. The large eggs distend the atrium and compress the pharynx, but feeding appears to continue.

The body of the tadpole is oval and has three anterior adhesive papillae in the medial plane, plus four bi-lobed ampullae between them (Fig. 8.2(e)). Other structures may be obscured by the large amount of yolk that is present, but in well-developed tadpoles the siphons, endostyle, three rows of protostigmata, and the digestive region of the gut can be seen. The tail has a central row of chordal cells with longitudinal muscles on either side and a thin dorsal and ventral fin.

The genus *Aplidium* is well represented around the world and more than 20 species are known from Australia alone (Kott, 1963). Many of these species live on the lower shore. Elongate zooids *c.* 3 mm long are embedded in a gelatinous tunic together with sand grains, similar to that of *Eudistoma*, but with only one siphon, the oral siphon, projecting through the cuticle of the tunic. The exhalent aperture opens into a cloacal cavity common to several zooids (Fig. 8.2(a) and (b)).

The body is relatively simple and characteristically divided into three regions (Fig. 8.2(a)): a branchial region comprising the oral siphon, a pharynx with several rows of stigmata (8–20 rows according to the species) and the surrounding atrium; an abdominal region containing the digestive tract; and a post-abdominal region in which the gonads and heart are located. The spherical stomach has about a dozen deep longitudinal ridges, which are characteristic of the genus. The testis occupies nearly all of the post-abdomen except for the terminal heart and the posterior extension of the epicardium. During spring and summer, the branchial region may be greatly distended by a single developing embryo that grows into a tadpole.

The tadpole (Fig. 8.2(d)) has three torch-shaped adhesive papillae (some species have up to 21), and numerous epidermal vesicles. Much yolk is present, but usually the endostyle, future oral and atrial apertures, pharynx and a larval eye spot can be seen.

Didemnum moseleyi (Fig. 8.3(a) and (b)) is a common littoral species that forms thin white, pink or pale lemon incrustations with faint star-like areas marking the position of the oral siphon of the zooids. The white colour results from the vast numbers of minute calcareous spicules (0.02 mm in diameter, Fig. 8.3(c)) which are scattered through the gelatinous tunic, particularly near the surface. Individual zooids are very small (less than 1 mm) and are divided into two regions, a branchial region and an abdominal region. The former has only four rows of stigmata, and the large atrial opening discharges into a

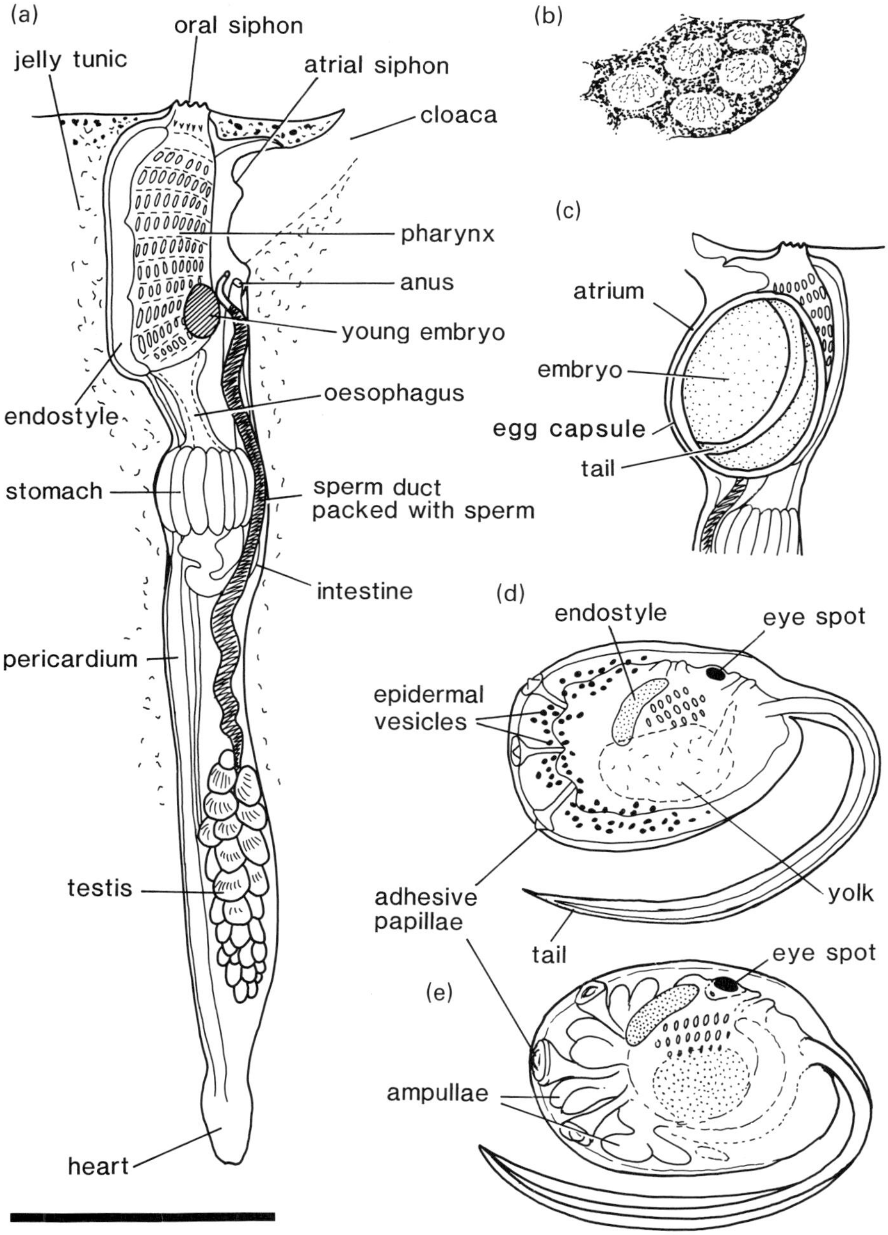

Fig. 8.2 (a) to (d) *Aplidium sp.*: (a) Zooid, showing internal structure. (b) Natural appearance of small colony, × 2. (c) Embryo developing in atrium. (d) Tadpole larva. (e) Tadpole larva of *Eudistoma elongata*. Scale bar: (a) = 1 mm, (d) and (e) = 0.7 mm.

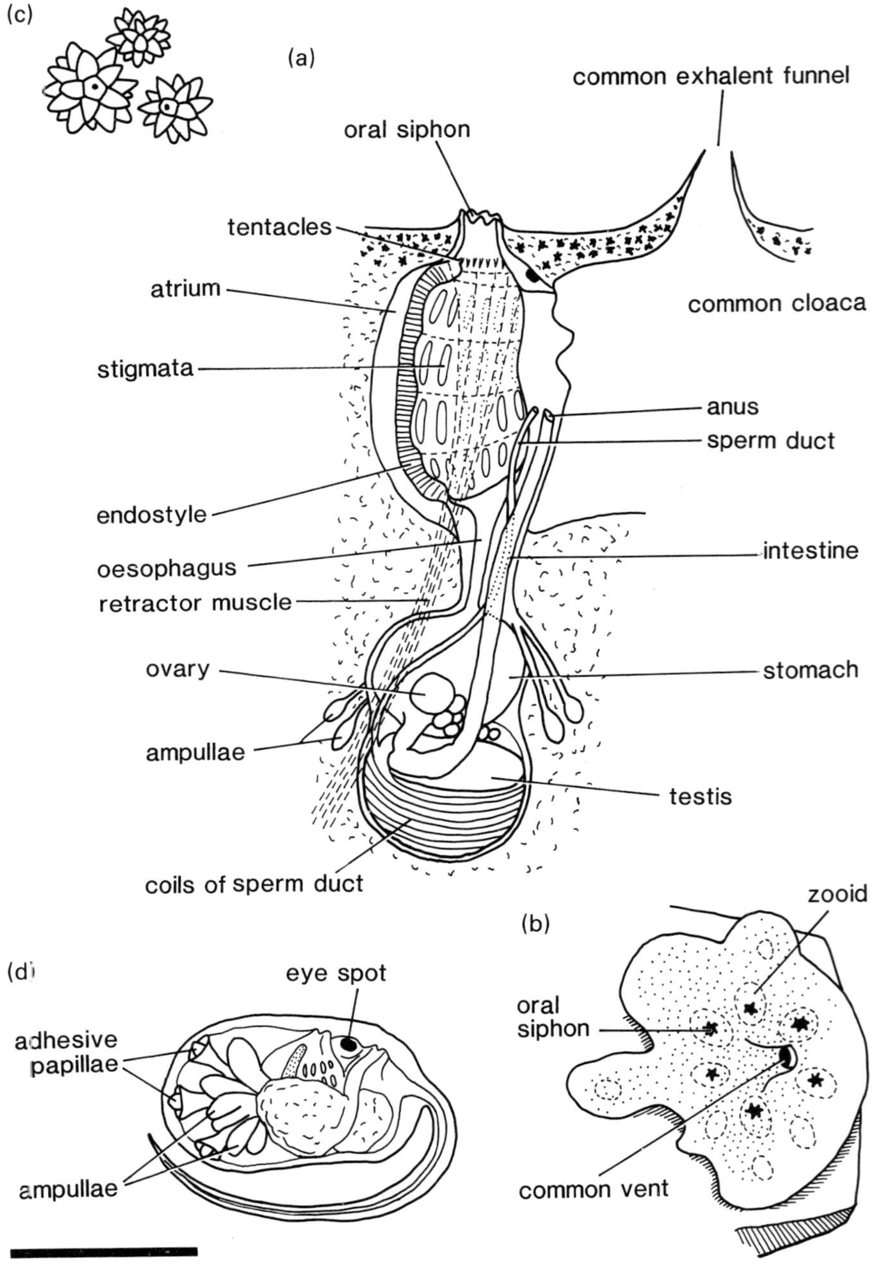

Fig. 8.3 Didemnum moseleyi. (a) Zooid, showing structure. (b) Small colony, × 8. (c) Spicules removed from gelatinous tunic. (d) Tadpole larva. Scale bar: (a) = 1.2 mm, (c) = 0.05 mm (d) = 0.5 mm.

cloaca shared by other zooids (Fig. 8.3(a)). The abdominal region is firmly embedded in the gelatinous tunic, so it is difficult to extract the whole zooid without damage. A long retractor muscle inserting on the oral siphon can pull the whole zooid down into the common tunic.

The large hemispherical testis, with its sperm duct wound spirally round the outside, is the most conspicuous structure in the abdominal region. The stomach does not have longitudinal ridges. Four elongate ampullae project posterio-laterally (Fig. 8.3(a)). The tadpole has three large adhesive papillae and four pairs of ampullae, but no epidermal vesicles (Fig. 8.3(d)).

Botryllus schlosseri (Fig. 8.4(a)) is another common compound ascidian, forming flat gelatinous colonies under boulders, on wharf piles and even on algae below low tide level. It occurs on most shores round the world. Its colour is striking. Colonies are brown, purple, violet, pink, yellow or green, and the zooids, which are characteristically arranged in rosettes of six to twelve like the petals of a flower, may be outlined in white. Each ring of zooids, called a system, has its own central exhalent or cloacal opening. The zooids of a colony are linked together by a common network of blood vessels which ramify through the tunic and connect with hundreds of contractile, flask-shaped ampullae, looking like bunches of grapes, round the perimeter of the colony (Fig. 8.4(d)). The zooids of a system are primarily concerned with feeding and are called trophic zooids.

Although colonies may appear stable, individual zooids are short-lived. Every 5–7 d, all the trophic zooids, degenerate simultaneously and are replaced by buds that have been developing in the atrial walls. Normally two buds are produced in each zooid, but this does not necessarily mean that the number of zooids will double at each cycle. Some buds move out of their rosette and establish new systems. Under crowded conditions in a large colony, many buds do not develop, and usually only one replaces the old zooid (Sabbadin, 1978).

During the breeding season, paired ovaries and testes appear in the atrial wall of each bud. The ovaries mature first whilst still in the bud, and are retained in the atrium when the bud becomes the functional feeding zooid (Fig. 8.4(c)). They are fertilized and commence development before the testes mature and shed their sperm. The tadpole larvae are released when the zooid degenerates.

B. schlosseri has proved an ideal animal on which to study the genetics of ascidians, since individual zooids or rosettes (all of which have been produced by asexual means and are genetically identical) are easily cut out and cultured separately or implanted in other colonies, whilst the parent colony remains as a reservoir of a particular genotype for further experiments (Milkman, 1967).

The internal structure is shown in Fig. 8.4(c) and (d). The oral siphon has a smooth rim and a ring of eight or more tentacles at its base which project into the lumen. The pharynx is relatively large and has eight to ten rows of

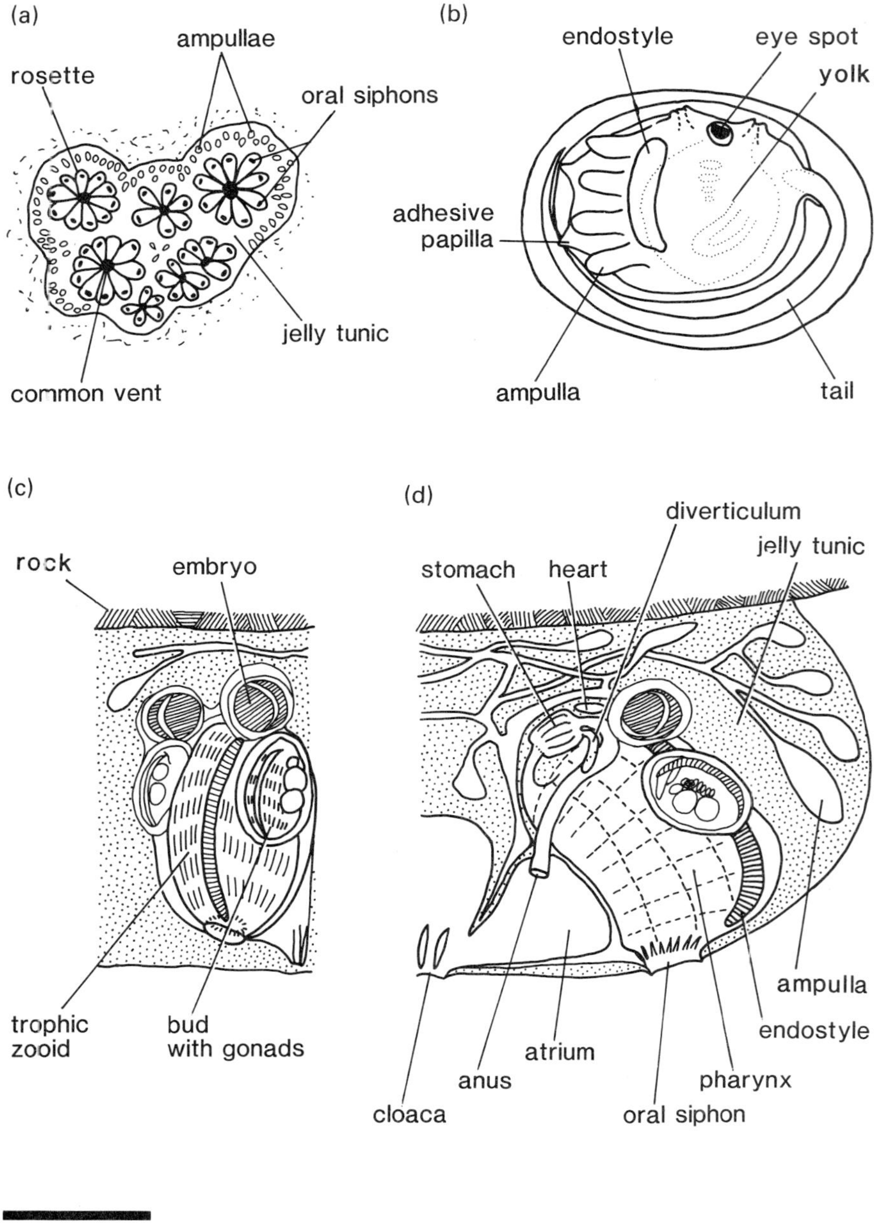

Fig. 8.4 Botryllus schlosseri. (a) Small colony × 5. (b) Tadpole larva. (c) and (d) Internal structure showing paired buds and embryos (c), and digestive system (d). Scale bar: (b) = 0.3 mm, (c) and (d) = 0.75 mm.

stigmata, a prominent endostyle, and three longitudinal bars on each side. There is no separate abdominal region, for the digestive tube lies to the left of the pharynx. The stomach is ridged lengthwise and gives off a blind outgrowth, the diverticulum, before passing into the short intestine. The atrial siphon is conical and discharges into the common cloaca at the centre of a system.

The botryllid tadpole (Fig. 8.4(b)) appears simpler than that of *Aplidium* because prospective adult organs have not started to develop. Three adhesive papillae and a ring of ampullae are present, but there are no torch-shaped adhesive organs. The sensory organ, called a photolith, appears as a single black dot.

Solitary ascidians

One of the best-adapted sea squirts for life in the intertidal zone is *Pyura stolonifera* (Fig. 1.1(b) g). Thousands of these large ascidians may cover rocky reefs that are just exposed at low tide and subjected to heavy surf. Individuals may grow to 20 cm in height and 9 cm in diameter, but much of this bulk is taken up by the tunic, and the zooid inside seldom exceeds 7 cm in diameter. The surface of the tunic is hard because of the incorporation of much sand, and is usually heavily encrusted with algae. The X-shaped apertures of the inhalent and exhalent siphons appear orange or red when open.

When a specimen is bisected (Fig. 8.5(a)), it will be seen that only the outer 1–2 mm of the tunic is hardened with sand; the very extensive central part is of a soft cartilaginous texture. Two blood vessels from the zooid can be traced into the tunic for some distance by the purple coloration of their walls. They do not join up with one another, and end in small blind swellings, the ampullae (Endean, 1955b). The tunic is firmly attached to the body wall or mantle round each of the siphons, but elsewhere it is easily separated. There is an extensive layer of muscle in the upper part of the body wall, consisting of circular fibres running concentric to the oral and atrial siphons and longitudinal fibres that radiate out from the siphons to insert on the tunic. The siphons appear to be closed by the longitudinal muscles, for when they contract they draw down the four lips and close the aperture. More vigorous contraction of both circular and longitudinal muscles pulls down the upper disc and siphons, causing water to be ejected from the branchial and atrial chambers. At the same time the cup-shaped region of the tunic is deformed. When the muscles relax, the elasticity of the deformed tunic opens the lips of the apertures, elevates the siphons to their normal position, and passively draws water in.

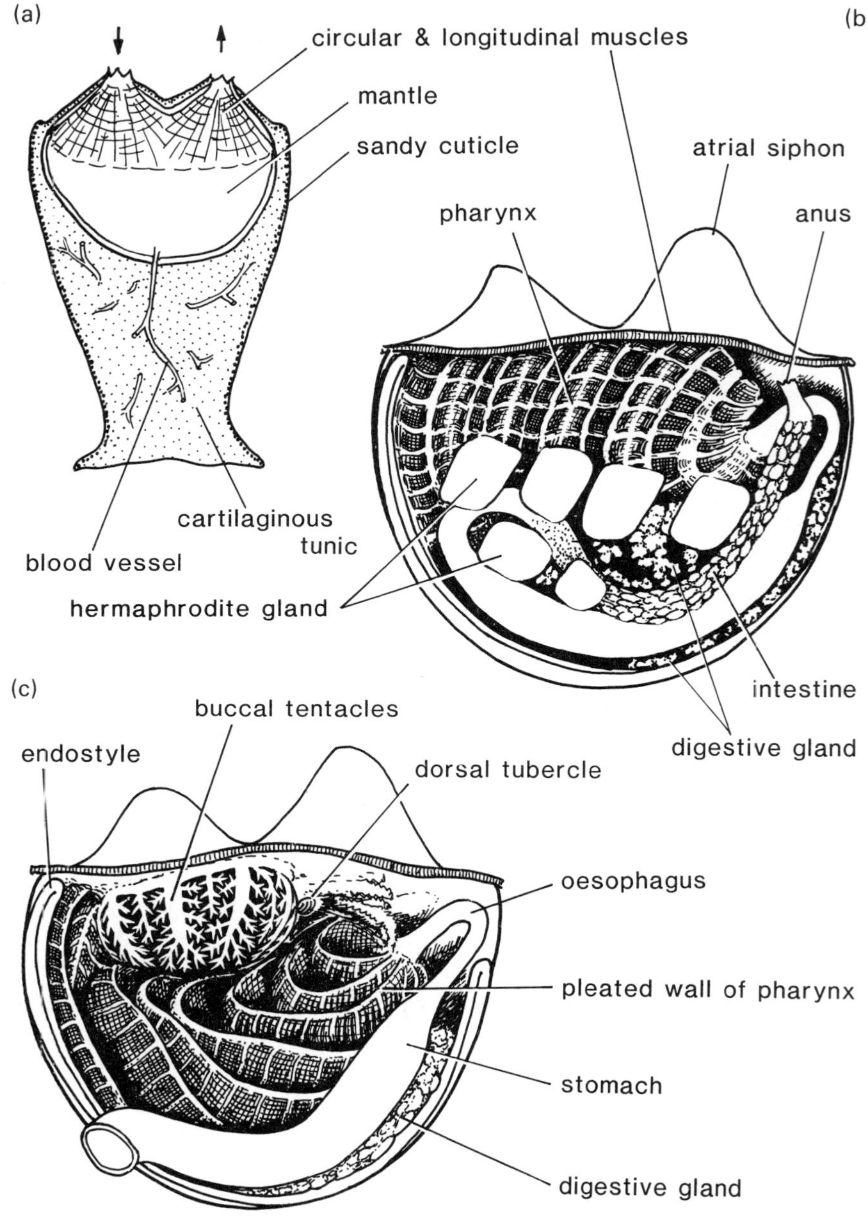

Fig. 8.5 Pyura stolonifera. (a) Tunic cut away to show zooid and blood supply to the tunic, ×0.5. Arrows indicate water flow. (b) Mantle removed to show gut. (c) Pharynx opened to show folded walls and buccal tentacles.

The internal structure of *Pyura stolonifera*, particularly the pharynx, is much more complex than that of the previous species (Fig. 8.5(b) and (c)). The wall of the pharynx is thrown into seven U-shaped longitudinal folds on each side of the endostyle, each fold has a number of longitudinal and transverse ridges on the inside (Fig. 8.5(c)). The walls are perforated by large numbers of rectangular stigmata. The remainder of the alimentary canal shows little regional differentiation. A short oesophagus passes to a wider tube, the stomach, lying on the left side of the branchial region. It then makes a sharp loop and runs back parallel to the stomach as the intestine, which terminates in an anus near the base of the atrial siphon. A large, lobulated digestive gland (sometimes called the liver) lies between the stomach and the left wall of the pharynx. Histological sections show that the branching ciliated tubes of this gland have lateral pockets lined by secretory cells, and the ducts open into the lumen of the stomach. A pyloric gland is present in the walls of the stomach and intestine. Buccal tentacles are pinnate and form a conspicuous basket guarding the entrance to the pharynx. Close behind this on the dorsal side is a button-like structure, the dorsal tubercle, which is associated with the dorsal ganglion or brain. In another solitary ascidian, *Phallusia mammillata*, Hoyle (1952) found that removal of the brain did not prevent contraction of the siphons in response to stimulation, but it did interfere with a cross-response in which stimulation of one siphon caused closure of the opposite siphon.

The gonads are hermaphrodite glands attached to the atrial wall of the mantle. Ovarian tissue packed with oogonia and developing oocytes occupies the centre of each lobe and is surrounded by testis. The separate oviduct and sperm duct open into the atrium. The sperm duct is used to store sperm and appears white when the animal is dissected. Vast numbers of eggs are shed into the sea, where they develop into tadpole larvae of relatively simple structure.

8.2 THE INTEGUMENT OF TUNICATES

The tunic is a structure unique to the sub-phylum Tunicata. The way in which it differs from the integument of all other animals calls for further discussion.

Tunicate species living on the lower shore tend to have a tough and very bulky tunic which is sufficiently rigid to maintain their characteristic shape out of water. This is in marked contrast to *Ciona intestinalis*, from deeper water, whose tunic is thin and incapable of preventing collapse if water is squeezed out of the body cavities. The gelatinous interior of the tunic in colonial species allows the siphons or whole zooid to be withdrawn deeply into the protection of the tunic; moreover, it incorporates a large quantity of water which, in addition to that in the body cavities, prevents desiccation of *Aplidium* and *Didemnum* species that are regularly emersed at low tide.

Littoral species are not easily dislodged from the substratum. Their low profile and smooth surface offer minimal resistance to flowing water, and the area of basal attachment is usually very large in relation to the body or colony volume. Encrusting species like *Didemnum moseleyi*, only a few millimetres thick, offer no lateral resistance to water flow. The common presence of sand grains or calcareous spicules incorporated in the surface layers not only increases rigidity against wave impact, but must increase surface hardness. All the above features strongly suggest that they have evolved as adaptations to violent wave action, abrasion and occasional exposure to the atmosphere.

The role of the tunic has been inferred from its structure rather than by experimental analysis, but it undoubtedly protects the delicate internal organs from physical damage and small predators. For species like *Pyura stolonifera*, the rock-hard exterior gives ample protection from wave action and desiccation during emersion. It probably prevents attack by predatory fish, and would be difficult for carnivorous gastropods to penetrate.

The elastic properties of the tunic enable it to function as an elastic skeleton which antagonizes the mantle musculature and so restores the length of muscles after they have contracted. It may play an excretory role because, in *Botryllus schlosseri*, blood cells (called nephrocytes) containing uric acid pass into the tunic and accumulate there. The fact that few species of ascidians have other sessile organisms living on their surface suggests that the tunic inhibits settlement of larvae or has antibiotic properties. It is possible that iodine bound to the tyrosine residues in the protein of cuticle, reported for *Ciona* by Barrington and Barron (1960), is responsible for keeping the tunic free of other organisms.

Structure and origin of the tunic

The structure and development of the tunic has been investigated from two angles, its histological appearance and its biochemical properties. Typically there is a thin protective outer cuticle to the main bulk of tunic, which consists of fibres in a watery matrix. In species with a jelly-like tunic, water may constitute more than 90% but in tough, leathery tunics, fibrous elements predominate. An early discovery was the presence in the tunic of *Ciona intestinalis* of a polysaccharide that was given the name tunicin. Subsequent analysis and X–ray diffraction studies showed that it was identical to plant cellulose (Barrington, 1974). The presence of tunicin in other species has led to a widespread belief that the tunic of all ascidians is composed of cellulose. However, this is not so: it is absent in *Pyura stolonifera* (Endean, 1955b) and *Halocynthia aurantium* (Smith and Dehnel, 1970). Their tunic is composed of other substances, such as proteins, mucopolysaccharides, glycoproteins and polysaccharides other than cellulose. The histological appearance of the body wall and tunic of *Halocynthia papillosa* is shown in Fig. 8.6(a) and (b). Externally it is covered in small (0.1 mm) cone-shaped spines, between which there is a thin cuticle. The cuticle and outer part of the spines appears to be produced by a particular type of blood cell, the dispersed vesicular cell, that collect in the outer part of the tunic (Smith, 1970). The cuticular spines are composed of protein which may be tanned, since there is evidence of quinones in the tunic (Stievenart, 1971).

The bulk of the tunic is fibrous. Under the polarizing microscope it is seen to consist of many concentric layers of fibres oriented in alternate circular and longitudinal directions, except towards the spines, where the fibres lose their regular orientation (Fig. 8.6(b)). Below the tunic, there is the epidermis of tall columnar cells. Glandular cells are absent, but in favourable sections it can be seen that the epidermal cells have fibrous tails which pass outwards into the tunic, pointing in the same direction as local fibres. In some cells it can be seen that these fibres originate in a zone of vesicles and there can be little doubt that they have been produced by the epidermal cells (Fig. 8.6(c) and (d)). The epidermal cells themselves are unusual in that they are attached to a basement membrane but otherwise not closely bound to each other. A similar condition has been noted in *Pyura pachydermatina*. Such an arrangement may allow the cells to flex from side to side and orientate the fibres they secrete, first in one direction, then in the other, to give the characteristic banding.

Barrington and Thorpe (1968) have produced histochemical evidence for the secretion of fibres (presumed to be cellulose) by the epidermal cells in *Dendrodoa*. They found two types of blood cell that appeared to contribute to the tunic: morula cells (rich in glycoprotein) that come to lie under the epidermis of the mantle and appear to pass proteinaceous materials to the epidermal cells, and polyphenol cells (containing phenolic substances) which pass through the epidermis into the base of the tunic. Barrington and Thorpe also demonstrated the presence of the enzyme polyphenol oxidase, which suggests that the abundant protein present in the tunic of *Dendrodoa* is hardened by tanning.

Biochemical studies on *Halocynthia aurantium* by Smith and Dehnel (1970) have shown that cellulose is absent and that the fibres consist of a complex of protein with a polysaccharide that is chemically different from cellulose in its properties and substitutions. There is evidence that mature morula cells, derived from the blood, tend to collect in the tunic just above the epidermis and may contribute the protein component of the fibrous tunic (Smith, 1970).

A radically different method of tunic formation has been put forward on the basis of histochemical studies. Endean (1955b) has shown that cellulose and protein are absent from the tunic of *Pyura stolonifera* and that the fibres are composed of another polysaccharide which is different in its chemical

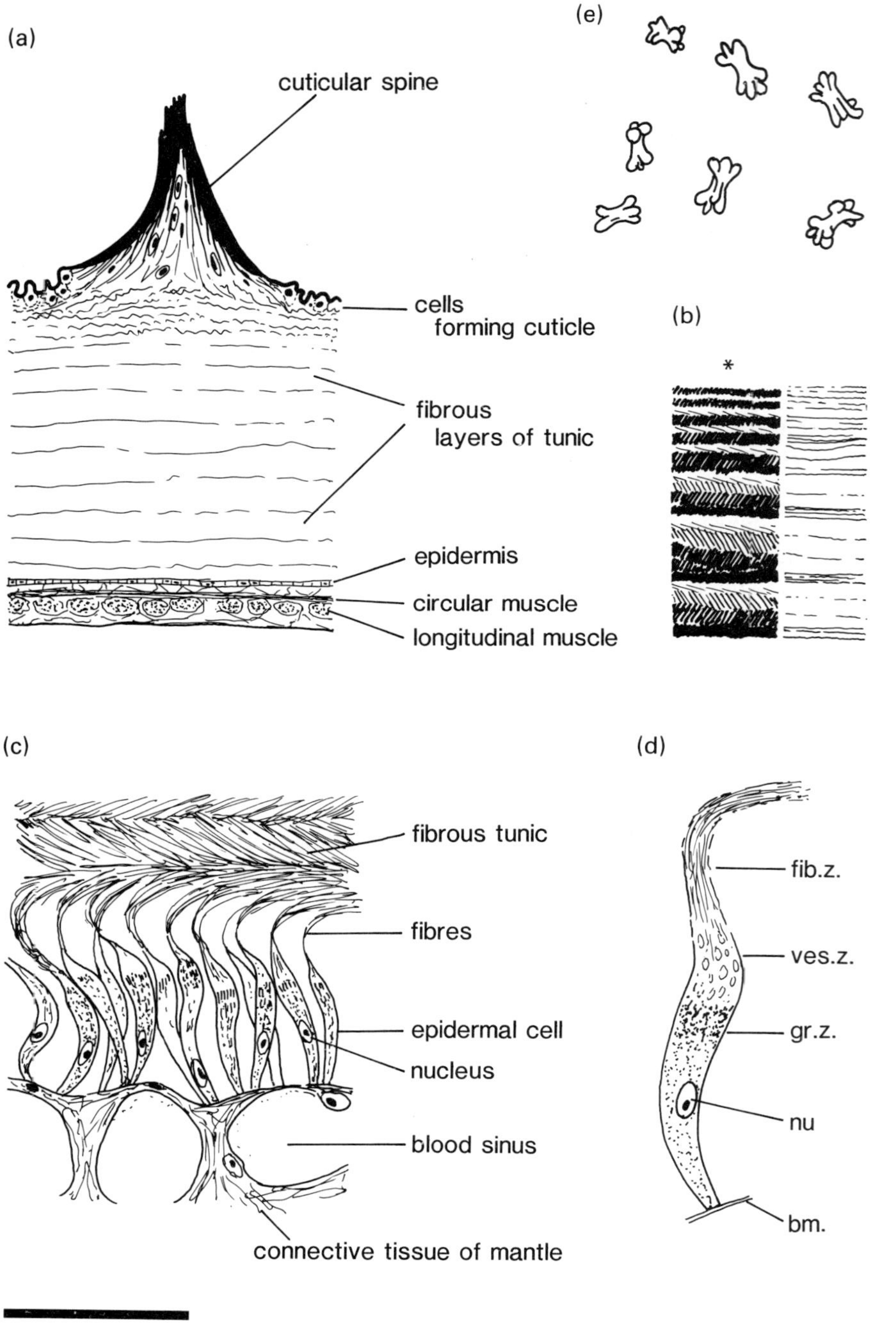
(a)
cuticular spine
cells forming cuticle
fibrous layers of tunic
epidermis
circular muscle
longitudinal muscle
(e)
(b)
*
(c)
fibrous tunic
fibres
epidermal cell
nucleus
blood sinus
connective tissue of mantle
(d)
fib.z.
ves.z.
gr.z.
nu
bm.

properties. The fibres appear to be bound together by a mucopolysaccharide. Blood cells (called ferrocytes) containing an iron chromogen were found to migrate into the tunic from the blood vessels, and then produce fibres in a manner reminiscent of fibroblasts (Endean, 1955c). Endean considered that the tunic of *Pyura stolonifera* was produced from ferrocytes and not epidermal cells. In *Phallusia mammillata*, the tunic contains cellulose and mucopolysaccharide, but no protein (Endean, 1961). The celulose was thought to be produced by blood cells called vanadocytes (corresponding to the ferrocytes but containing vanadium chromogen), which collect in the tunic. Blood vessels invade the tunic, and various blood cells pass from the vessels into the matrix, where they contribute to its substance. In this respect the tunic is partly of mesenchymal origin, and Brien (1948) regarded it as an ecto-mesenchyme.

The tunic of *Diplosoma listerianum* (a member of the Didemnidae) is remarkable in being contractile. Mackie and Singla (1987) have shown that a network of cells containing contractile microfilaments (myocytes), especially round the cloacal apertures, can cause contraction of the tunic and closure of the exhalent apertures. This regulates the flow of water through the colony. The myocytes also provide a diffuse conducting system in the tunic through which very slow impulses (< 0.015 m s^{-1}) are propagated, but this conducting system is not connected to the nervous system of individual zooids in the colony. Thus the contractile behaviour of the tunic is independent of the zooid colony.

It is evident that the tunic of ascidians shows a wide variety of structure among the various families. As an integument it is unique among all other animals in having blood vessels outside the epidermis.

8.3 BLOOD AND CIRCULATION

The blood of ascidians is remarkable for the variety of blood corpuscles it contains and their properties.

Fig. 8.6 Halocynthia papillosa. (a) Section through tunic, showing cuticular spines and muscles of the mantle wall. (b) Fibrous layers of tunic as they appear under polarized light (* to left) and ordinary light (right). (c) Epidermal cells and fibres of tunic. (d) A single epidermal cell. (e) Calcareous spicules from tunic of *Pyura pachydermatina*. b. m., basement membrane; fib. z., fibrous zone; gr. z., granular zone; nu., nucleus; ves. z., vesicular zone. Scale bar: (a) and (b) = 0.25 mm, (c) = 0.03 mm, (d) = 0.015 mm, (e) = 0.5 mm.

The plasma is nearly isotonic with the surrounding seawater, and has a similar ionic composition (except for the concentration of sulphate ion, which is much lower).

Wright (1981) has classified the nucleated blood corpuscles into six categories – haemoblasts, lymphocytes, leucocytes, pigment cells, nephrocytes and vacuolated cells – and has summarized the known distribution of types within the Ascidiacea. Haemoblasts may be found in the blood, but are more numerous in blood-forming tissue. They are thought to give rise to all the other types of blood corpuscles, probably changing first to lymphocytes and then, by formation of various granules or vacuoles in their cytoplasm, become leucocytes, pigment cells, nephrocytes or vacuolated cells.

Lymphocytes show amoeboid activity and will phagocytose foreign particles like India ink or polystyrene beads injected into the blood. Lymphocytes with clear cytoplasm give rise to the vacuolated cells.

Leucocytes with granular cytoplasm may contain proteins, glycogen and lipids. They are thought to be responsible for carrying nutrients to different parts of the body. Large numbers may accumulate as trophocytes during asexual reproduction, and constitute a food reserve for future growth.

Nephrocytes contain granules of purines or uric acid in their cytoplasm and are connected with nitrogen excretion. The bright orange, red or purple colour of tissues in many ascidians, like *Pyura stolonifera*, is caused by pigment cells in the blood containing carotenoid and melanin pigments; their function is not known.

Vacuolated cells are a conspicuous component of ascidian blood. Three types may be found: signet ring cells with a single large vacuole, compartment cells with several vacuoles, and larger morula or mulberry cells with numerous vacuoles; the last are the most numerous and in *Pyura stolonifera* may account for 70% of all blood cells. Vacuolated cells are of special interest because of the wide variety of substances found in their cytoplasm: proteins, polysaccharides, sulphuric acid, vanadium or iron chromogen (vanadocytes, ferrocytes), other heavy metals, and possibly polyphenols. Polysaccharides present in the vanadocytes of *Phallusia*, and an amino-polysaccharide in the ferrocytes of *Pyura stolonifera*, have been implicated as the source of cellulose or polysaccharide in the tunic (Endean, 1961, 1955b, 1955c). Vanadocytes are found in some species of the Ascidiidae and Perophoridae, and ferrocytes occur only in certain Styelidae and Pyuridae. Neither act as oxygen carriers. The use of pH indicator stains suggests that the vanadocytes are strongly acidic (pH 4.5–6.0), but Hawkins *et al.* (1983) found the cytoplasm of blood cells of many ascidians to be neutral. In *Pyura*, a pH in the range 6.0–6.4 was found which was attributed to phosphates within the cytoplasm. Upon cytolysis of the vanadocytes or damage to the tunic, sulphuric acid is released equal to a concentration of *c.* 2N in *Phallusia* (Webb, 1939) or 0.4N in *Pyura* (Endean, 1955a). Stoecker (1980) has shown that the released acid and the relatively high vanadium concentration in ascidian tissues can act as a deterrent to predatory crustaceans and fish.

The blood corpuscles of ascidians are peculiar in that some (vanadocytes, ferrocytes, dispersed vesicular cells and nephrocytes) are destined to fulfil their functional role outside the body (i.e., in the tunic).

Neither haemoglobin nor haemocyanin are found in the blood or corpuscles of ascidians, but it is not clear that a respiratory pigment is necessary, for all tissues are in close contact with the water current passing through the body, and diffusion alone would provide sufficient oxygen for their low metabolic needs. During periods of emersion, *Pyura stolonifera* may squirt out much of the water from its atrial and pharyngeal cavities. If air enters to replace the water this could provide a source of oxygen until the animal is re-submerged by the rising tide.

The vascular system of ascidians is a haemocoel, but it exhibits a number of unusual features. The heart is a muscular tube along which slow peristaltic waves pass, pushing the blood into the vessels and sinuses of the body. There are no valves restricting flow to a particular direction, and periodically the peristaltic waves reverse, forcing blood back in the opposite direction. There is no capillary network; blood ebbs and flows between the sinuses. In one direction, it flows via the large sub-endostylar vessel into the network of small blood vessels in the wall of the pharynx and the dorsal sinus (Fig. 8.9(a)). In the opposite direction, blood passes into sinuses round the stomach and gonads in the abdominal region, as well as to the mantle and tunic. Vessels in the tunic end in blind swellings, the ampullae, which are themselves contractile and push the blood back towards the heart; presumably elasticity of the tunic will cause their expansion and draw blood into them. Contraction of the ampullae is unusual for, unlike the heart, muscle fibres are not present in the wall. However, electron micrographs of *Botryllus schlosseri* reveal microfilaments in the cytoplasm of the epithelial cells lining the ampullae which are believed to be contractile (DeSanto and Dudley, 1969). In this respect they resemble the epithelio-muscular cells of cnidarians. Mackie and Singla (1983) have shown that there are no nerves in the vascular system, and that contraction of the ampullae is co-ordinated by impulses propagated from cell to cell in the vascular epithelium. The rhythmic pulsations of the ampullae cannot originate from the cerebral ganglion, for the rhythm persists in isolated ampullae. In *Botryllus schlosseri*, buds swell rapidly at the time they replace the old trophic zooids, and contractions of the ampullae appear to be responsible for this sudden increase in size.

The peristaltic contractions of the heart are a property of the muscle cells (myogenic) and not regulated by the cerebral ganglion. The latter can be

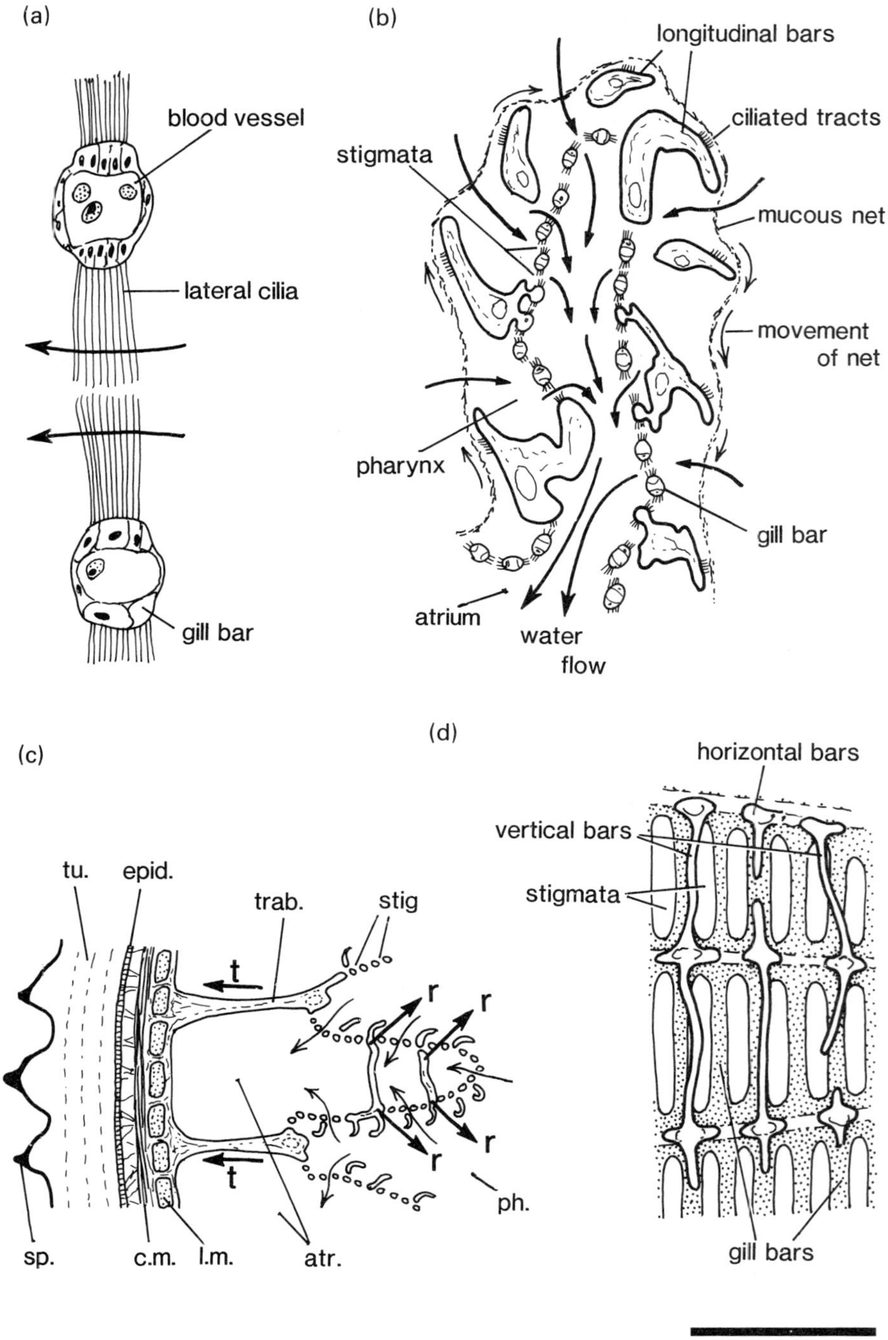
(a)
blood vessel
lateral cilia
gill bar
(b)
longitudinal bars
ciliated tracts
stigmata
mucous net
movement
of net
pharynx
gill bar
atrium
water
flow
(c)
tu.
epid.
trab.
stig
t
r
ph.
sp.
c.m.
l.m.
atr.
(d)
horizontal bars
vertical bars
stigmata
gill bars

removed without effect upon blood flow; however, nerve cells at either end of the heart are probably responsible for regulating reversal of the beat (Berrill, 1950).

8.4 GENERATION OF THE FEEDING CURRENT

Ascidians generate a current of water through the body by means of lateral cilia on the gill bars (Fig. 8.7(a)). This is analogous to the method used by bivalves, but the mechanism of feeding in ascidians differs in many important respects from that of mussels (see page 98).

The way in which the gill bars and stigmata develop can be seen in the tadpole (Fig. 8.8(a) and (b)). The atrium arises in the embryo of species like *Eudistoma* as two ingrowing pockets of ectoderm that come to lie on either side of the anterior part of the enteron or future pharynx. Eventually the pockets join to form one cavity, and their apertures fuse to form a single atrial opening. Where the atrial and pharyngeal epithelia come together, openings called protostigmata are formed, and the wall left between them becomes the gill bars. Later, as the young ascidian grows, the protostigmata are divided up into numerous stigmata by ingrowth of the gill bars. In *Didemnum* there are *c.* 50 stigmata, but in *Pyura* there are many thousands. The lateral cilia are borne on the gill bars and project into the stigmata, where their beating creates a water current from the pharynx into the atrium (Fig. 8.7(a) and (b)). Latero-frontal cirri and frontal cilia are absent, indicating that the method of particle capture is very different from that of bivalves.

The internal aspect of the pharynx, particularly in solitary ascidians, appears complex owing to the development of papillae, horizontal membranes, bars, and longitudinal ridges, some of which contain blood vessels. The way in which these structures develop can be seen in *Perophora multistigmata*, where some of the papillae are T-shaped and join to form longitudinal bars (Figs 8.7(d), 8.11).

A remarkable feature of ascidians is the complete absence of skeletal support to strengthen the pharynx in the way that skeletal rods strengthen the gill lamellae of bivalves like *Crassostrea*. Even in large sea squirts like *Pyura*

Fig. 8.7 Structure of pharynx. (a) Transverse section through gill bars, showing lateral cilia. Arrows, direction of water flow. (b) TS of fold in pharynx wall, showing longitudinal bars, water flow (heavy arrows) and movement of mucous sheet (light arrows). (c) Attachment of pharynx to mantle and tunic by trabeculae. Light arrows, direction of water flow, heavy arrows indicate direction of forces acting on pharynx (see text for explanation). (d) *Perophora*, internal appearance of pharynx. Scale bar: (b) = 0.3 mm, (d) = 0.1 mm.

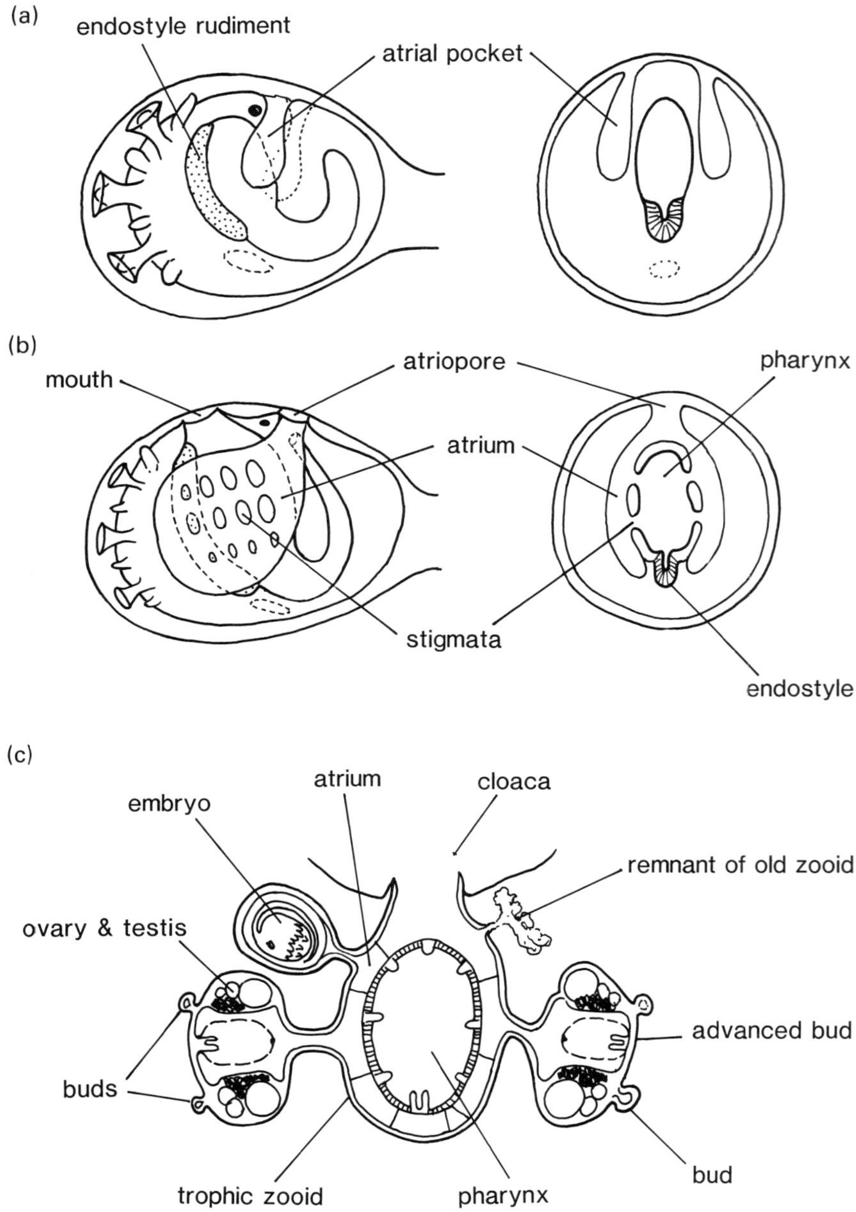

Fig. 8.8 (a) and (b) Stages in atrium formation of *Eudistoma* in the tadpole larva of *Eudistoma* (transverse section at right, see text). (c) Diagrammatic section through *Botryllus*, showing replacement buds and embryos (see text); an example of atrial budding.

stolonifera, the pleated pharyngeal wall remains soft and can easily be deformed, both by the longitudinal and circular muscles of the mantle and by intrinsic muscle fibres of the pharynx. Indeed, the presence of the latter is a clear indication that contraction or collapse of the pharynx is, for some reason, a necessary activity.

It is not obvious that a soft-walled pharynx should create a water current, for water forced through the stigmata by cilia will produce an equal and opposite thrust on the pharynx wall, causing it to collapse inwards (Fig. 8.7(c) **r**). However, this is prevented, not by stiff gill bars, but by anchoring the pharynx to the body wall (and hence the tunic) with fibrous trabeculae that run through the atrium and take the tension of the pharynx wall pulling inwards (Fig. 8.7(c) **t**).

Lying in the ventral mid-line of the pharynx, there is an elaborate mucus-producing organ called the endostyle, whose structure and mode of operation is best understood from transverse sections (Fig. 8.9). It is a deep gutter formed by tall columnar cells of various types. At the bottom of the gutter, the cells bear exceptionally long flagella which project into the lumen of the pharynx. The side walls have alternate longitudinal bands of secretory and ciliated cells that give different histochemical staining reactions, indicating several secretory products (Fig. 8.9(c)). The nature of these secretory substances varies widely between different species. Acid and neutral mucopolysaccharides as well as proteins have been identified; moreover, cells along the lip of the gutter concentrate iodine (Barrington, 1957, 1974). The function of the very long flagella is uncertain, but it is possible that they cause the secretion of the walls to remain as two separate sheets, which pass out to the right and left of the gutter. The reason for several components in the secretion is not known, but proteins, especially if fibrous, might give some strength to the mucous sheet. It seems likely that the mucous sheet is held away from the stigmata by the longitudinal and horizontal bars, but just how it is drawn across to the dorsal side of the pharynx is hard to determine. In some species (*Halocynthia*, for example), the horizontal and vertical bars are ciliated and help to draw the sheet across (Fig. 8.7(b)). There is also a ciliated band that runs from the endostyle round to the dorsal lamina just behind the tentacles (the peripharyngeal band). A ciliated dorsal lamina, sometimes bearing a row of J-shaped ciliated languettes which roll up the mucous sheet into a cord, passes food back to the oesophagus (Fig. 8.9(a)).

Regular periodic contractions of the branchial region may occur in undisturbed animals whilst feeding. It is possible that the mucous sheet is initially established by these contractions, for they would bring the endostyle and the longitudinal ridges or folds close to the dorsal lamina. Secretions of the endostyle might be spread out as a sheet when the pharynx is enlarged through elasticity of the tunic.

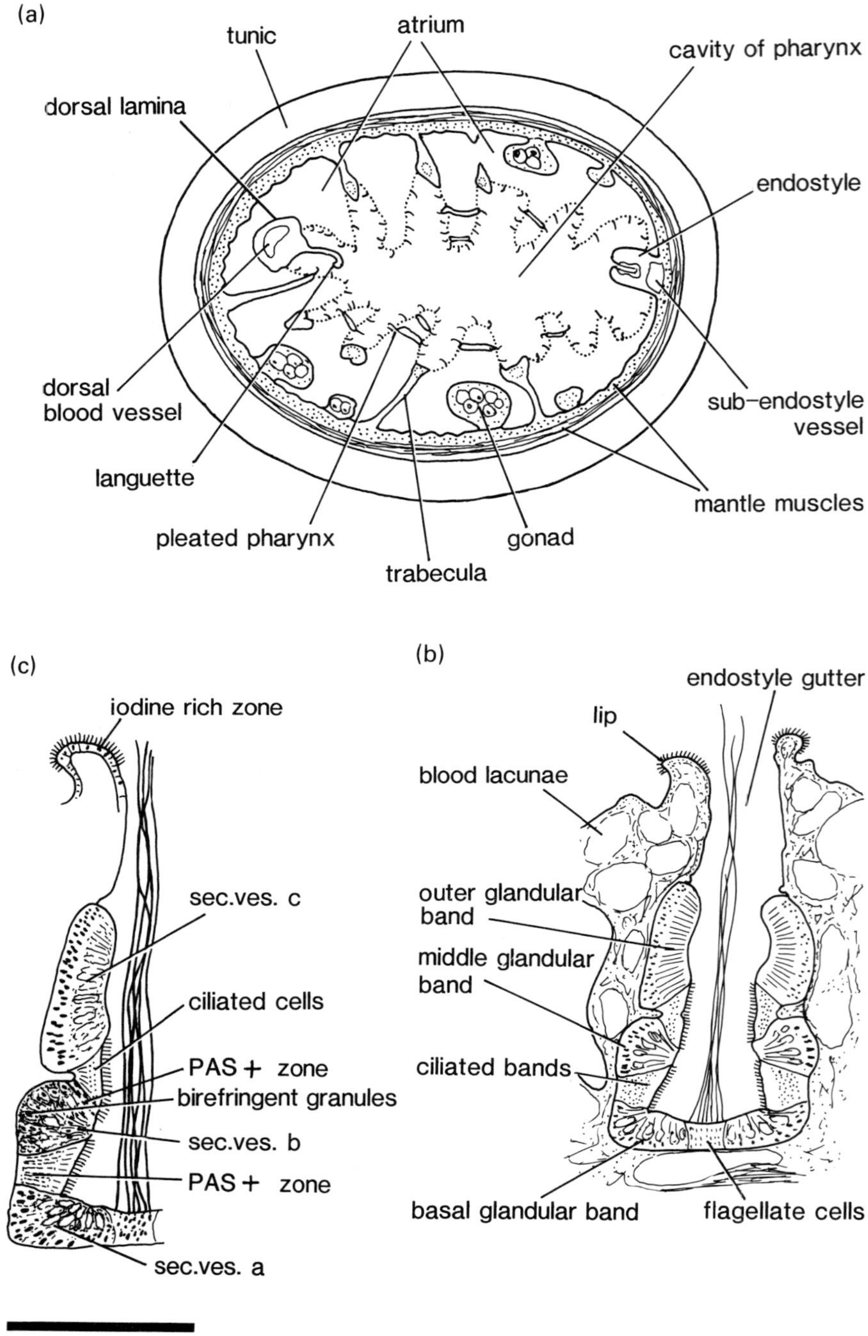
(a)
tunic
atrium
cavity of pharynx
dorsal lamina
endostyle
dorsal blood vessel
languette
pleated pharynx
trabecula
gonad
mantle muscles
sub-endostyle vessel
(c)
iodine rich zone
sec.ves. c
ciliated cells
PAS + zone
birefringent granules
sec.ves. b
PAS + zone
sec.ves. a
(b)
endostyle gutter
lip
blood lacunae
outer glandular band
middle glandular band
ciliated bands
basal glandular band
flagellate cells

Feeding

Ascidians can extract extremely small particles from the current of water passing through the branchial region. Jørgensen and Goldberg (1953) found that *Ciona intestinalis* could remove graphite particles 1–2 μm in diameter, so it is clear that the stigmata (diameter 30–50 μm) could not be acting as a sieve. The mucous sheet must act as a fine sieve to collect the graphite particles. They also found that some large molecules, such as crab haemocyanin (which would be expected to pass through the mucous sieve) were retained. Burighel (1979) has shown that protein substances such as horseradish peroxidase are rapidly taken up and passed to the stomach. This suggests that certain macromolecules (proteins), if present in the seawater, might bind directly to the mucopolysaccharides of the mucous sheet.

The picture that emerges is that the two functions – water-current generation and food collection – are kept separate. The mucous net, capable of collecting fine food particles, is kept away from the stigmata by the longitudinal and horizontal ridges of the pharynx, and the ciliated stigmata are only concerned with pumping water.

Although some attempt has been made to measure the volume of water passed through the pharynx, little attention has been paid to the rate at which water is drawn through the mucous sieve. It might be anticipated that this would be slow, to prevent the mucous sheet being disrupted or forced through the stigmata. Hoyle (1953) measured the amount of water pumped by *Phallusia mammillata* and arrived at an average flow of $0.06\ l\ h^{-1}$. Personal observations on *Pyura pachydermatina* in the laboratory, using milk to trace water movements, indicate that water is pumped slowly at about the same rate as Hoyle's animals. Hoyle did not give the weight or dimensions of his experimental animals beyond estimating their branchial capacity to be 17 ml. This would indicate a surface area of the pharynx in the order of 25–30 cm^2. If it is assumed that the stigmata occupy 30% of the pharynx wall, the total area through which water can flow will be *c.* 8–9 cm^2. At a pumping rate of $0.06\ l\ h^{-1}$, the velocity of water passing though the stigmata will be in the order of $0.02\ mm\ s^{-1}$, and the velocity through the mucous sheet about half this (i.e., $0.01\ mm\ s^{-1}$). This figure is similar to the velocity of water in the flagellated chambers of sponges (Table 12.1) and is consistent with the idea that the mucus forms a complete net within the pharynx.

Fig. 8.9 Structure of pharynx. (a) Transverse section of *Halocynthia*, showing folds of pharynx and position of gonads in the atrium. (b) T.S. endostyle of *Pyura pachydermatina*. (c) Endostyle showing different secretory zones. sec. ves. a,b,c, secretory cells with different staining properties; PAS + zone, regions staining positive to periodic acid Schiff reagent. Scale bar: (a) = 3 mm, (b) = 0.2 mm.

Many species of solitary ascidian have been observed to eject water vigorously from their siphons at regular intervals. Although such squirting has clear utility in washing away large particles that might clog the stigmata, or animals that might try to gain access to the interior of the sea squirt, its regularity, frequency and apparent spontaneity all point to some other function. Hoyle (1953) has shown that the frequency at which squirting takes place in *Phallusia mammillata* is to a large extent dependent upon whether food is abundant or not. He found that when food was plentiful, squirting occurred at intervals of 6–9 min, whereas in filtered seawater the rate rose to once every 3–4 min. Hoyle estimated that squirting draws in water at a greater rate ($0.09\ l\ h^{-1}$) than ciliary pumping ($0.06\ l\ h^{-1}$), and postulated that rhythmic squirting may be an adaptation to the intertidal environment, where food supply and suspended sand or mud fluctuate widely. Squirting offers an energy-efficient way of extending the range of flow rates beyond those produced by ciliary mechanisms.

Other hypotheses to account for periodic squirting can be suggested. It has already been pointed out that contraction of the pharynx might help to spread the mucous net. During normal ciliary pumping, parts of the mucous sheet might be pushed against the stigmata by the water flow and interfere with current generation. This could be rectified by closure of the atrial siphon, followed by contraction of the muscles in the mantle. Water would be forced back from the atrium through the stigmata, thus unblocking the system.

The high degree of structural specialization in the ascidian pharynx strongly suggests that its primary function is gathering food, but a possible respiratory function must not be overlooked. In the large pyurids, the major part of the body consists of gonads and digestive gland: both are likely to have high metabolic demands, so the feeding current must play an important role in supplying them with oxygen. When the siphons are closed and the current is stopped, that source of oxygen is cut off, but blood vessels and ampullae in the tunic could be important as alternative respiratory organs. It is not unusual for *Pyura stolonifera* to squirt water when exposed at low tide. This would draw air into the branchial and atrial cavities, but it is not known whether atmospheric oxygen can be used for respiration in the same manner as *Semibalanus balanoides* or *Mytilus edulis*.

Although the presence of food has an effect upon squirting, the rate of food intake can be regulated in other ways. Water flow is stopped when the siphons are closed, but even when they are open, there may not be a current, indicating that ciliary activity has been stopped. It is commonly observed that when an animal has been disturbed, some time elapses after opening of the siphons before the water current starts. Ciliated cells of the stigmata are innervated by nerves in the pharynx wall and are under the control of the cerebral ganglion. Stimulation of sensory receptors in the siphon causes the lateral cilia to stop beating (Mackie *et al.*, 1974). Secretion of the mucous net may not be continuous. Differences in the appearance of secretory cells of the endostyle suggest that their secretion may take place in cycles. Species that experience regular emersion (and hence cessation of feeding) could benefit from an endogenous rhythm controlling feeding and net secretion.

The nature of the food collected from the natural habitat is only known in general terms. Examination of histological sections through the stomach and intestine of various species suggests that feeding is indiscriminate, and that any particle, whether food or inert matter, is picked up. Diatoms, radiolarians and dinoflagellates are easily identified by their skeletons, but considerable quantities of cellular fragments of algae are usually present which are more difficult to identify. Fine particles of mineral matter may make up a significant proportion of the material in the gut of intertidal species like *Aplidium* and *Pyura*. Other organisms, such as bacteria and flagellates, are likely to be rapidly destroyed in the stomach and overlooked, although they may contribute significantly to the ascidian's diet. Planktonic crustaceans are rarely found in the food entering the stomach of *Pyura stolonifera*, possibly because copepods generally are negatively rheotactic and can swim away from the current entering the oral siphon. However, copepods belonging to the Notodelphidae (Dudley, 1966) and the amphipod, *Leucothoë*, are very common commensals in the pharynx of solitary ascidians.

The gut of ascidians is relatively simple. The oesophagus is lined with a ciliated epithelium which passes food to the stomach. The epithelium of the stomach and intestine consists of mucous, zymogen and vacuolar absorptive cells, all of which may be ciliated (Thomas, 1970). The lining of the digestive tract is not static. Embryonic or stem cells pass through the following developmental cycle: proliferation, differentiation into functional types, a synthetic phase, and finally a secretory phase in which the secretory products are released and the cell lost or broken down by autolysis before being replaced by a new generation of cells (Ermak, 1982).

Zymogen and mucous cells release their enzymes and mucus respectively into the lumen, but there is difference of opinion about the fate of the vacuoles in the absorptive cells. Thomas (1970) considered that they were released into the lumen in *Ciona*, but electron microscopic studies on *Botryllus* (Burighel, 1979) and *Styela* (Ermak, 1982) show that the vesicles are lysosomes (membrane-bound structures containing hydrolytic enzymes). It has been shown that particles of molecular dimensions, such as the enzymes horseradish peroxidase (5 nm: Burighel, 1979) and ferritin (11 nm: Thorndyke, 1977), are captured by the mucous net, passed to the stomach and rapidly taken up into the absorptive cells by pinocytosis. This strongly suggests that ascidians can absorb complex molecules such as proteins and carry out further digestion within the cell by means of the lysosomes.

Digestion in ascidians has always been regarded as exclusively extracellular (Morton, 1960), on the grounds that enzymes from the zymogen cells are released into the lumen and phagocytosis has not been observed.

The fact that macromolecules can be taken up by pinocytosis indicates that intracellular digestion does take place, and this opens the possibility that protein molecules released by normal extracellular digestion may be degraded intracellularly.

Members of the Pyuridae differ in having the zymogen cells in the alveoli of the digestive gland or liver, and not in the wall of the stomach. A strong amylytic, but weak proteolytic and lipolytic, activity was found in the secretions of the digestive gland of *Pyura* (Berrill, 1929); however, *Halocynthia aurantium* was found to have a strong chitinase activity (Elyakova, 1972). This suggests that some ascidians feed principally on zooplankton. The pyloric gland, found on the wall of the intestine in most species, does not produce digestive enzymes, and its function is debated. An excretory function has been ascribed to it (Brien, 1948), but this is doubtful as blood cells called nephrocytes accumulate uric acid and are clearly involved in nitrogenous excretion. More recently Ermak (1977) has shown that glycogen is stored in the pyloric gland of *Styela*. Burighel and Milanesi (1973) found that considerable quantities of lipids were stored in the gastric epithelium of *Botryllus schlosseri* as an energy reserve. They also suggested that both ingestion and the digestive process are regulated by some neural or hormonal mechanism, which is dependent upon environmental factors and a biological rhythm.

An interesting case of an ascidian deriving nutritional benefit from symbiotic organisms is found in species of *Diplosoma* and *Lissoclinum* (Didemnidae). The prokaryote, *Prochloron*, which is capable of carrying out photosynthesis, occurs in the cloaca and tunic of *Diplosoma*. Pardy and Lewin (1981) have shown that products derived from the symbiont reach the host, possibly by the activity of amoebocytes, for Cox (1983) found that they phagocytose *Prochloron* in *Lissoclinum*. Growth of *Didemnum molle* was increased by 40% in sunlight, presumably due to the nutrients from the symbiont, but, although there was a marked decrease in chlorophyll in complete darkness, the ascidian did not suffer weight loss and remained healthy (Olson, 1986). *Prochloron* has aroused considerable interest as the possible origin of the chloroplasts of green algae and higher plants. The role of symbiotic algae will be discussed later (page 219).

8.5 SEXUAL REPRODUCTION

The hermaphroditic condition of ascidians has already been noted (see page 124). Breeding seasons are influenced by the temperature of the sea. For example, *Ciona intestinalis* and *Botryllus schlosseri*, both with a wide geographical distribution, have a short summer breeding season in Sweden but breed almost all the year in the Mediterranean Sea (Millar, 1971).

Ascidian gametes show a number of unusual features. Each developing oocyte in the wall of the ovarian tubules is surrounded by follicle cells that supply it with nutrients and, in solitary sea squirts like *Ciona* and *Pyura*, form a complex shell or chorion round the ovum which gives buoyancy to the egg. Follicle cells also secrete the outer cuticular layer of the tadpole's tunic (Cloney, 1982). Spermatozoa of ascidians are considered primitive in lacking a middle piece: they consist of a head with a single mitochondrion and a flagellum (Miller, 1982).

Synchronization of spawning (egg and sperm release) may be achieved by response to environmental and internal factors. Spawning does not occur in mature *C. intestinalis* kept in constant darkness, but gametes are released within a few minutes of the animals' being illuminated (Whittingham, 1967). Constant light also inhibits spawning, and a period of darkness at least 1 h in duration is necessary before animals will respond to light by spawning. Mature *Ciona* kept in darkness will respond to a light pulse of duration as short as 1 min, and the most effective wavelength is 415 nm. Delay in spawning of *c.* 27 min after the light pulse was attributed to secretion of a hormone, possibly from the neural gland, which causes release of gametes. However, control of spawning is much more complex and probably involves an endogenous biological rhythm.

On the Brittany coast during late summer, *C. intestinalis* spawn shortly before dawn (9 h after the onset of darkness), but more individuals spawn on neap tides than spring tides. Georges (1978) found that histological changes take place in the neural gland of *C. intestinalis*, suggesting periodical secretion into the blood of a hormone which is known to inhibit spawning. The secretory period occurred during a falling tide and followed the tidal cycle. High water of both morning and afternoon spring tides occurs before the onset of darkness (giving two secretory periods), whereas during neap tides, only the morning tide precedes darkness (resulting in one secretory period). Georges considered that on spring tides, spawning would be inhibited by a higher concentration of hormone. During neap tides it would be insufficient for suppression and spawning would occur.

A more likely explanation of these observations is that during spring tides, a secretory period coincides with the day/night transition. It has been shown in many animals that this transition is a period of sensitivity during which internal rhythms may be entrained or readjusted by light changes (Hinde, 1970; Saunders, 1977). With *Ciona intestinalis*, the sensitive period will occur during the secretory phase on spring tides, but not on neap tides. The absence of hormone secretion during the sensitive day/night transition would allow spawning to take place at the predicted time.

A vascular system common to all zooids in a colony of *Botryllus schlosseri* raises the possibility that intracolonial synchronization of the reproductive cycle is mediated by some factor circulating in the blood. Milkman (1967) has shown that when a colony is divided, the two halves may drift out of step, but synchronization is restored when they fuse again. Compound ascidians like *Aplidium* that lack a common vascular system do not show synchronization of reproduction within the colony.

Fertilization

When the solitary ascidian, *Halocynthia*, spawns it sheds both sperm and eggs, but Fuke (1980) found that only a small percentage (1–5%) of eggs were fertilized by autologous spermatozoa; the remainder were fertilized by other individuals. Self-fertilization may be prevented by gamete incompatibility. It does not occur in *Ciona intestinalis* because the follicle cells or chorion block entry of spermatozoa from the same individual. However, if the chorion is removed experimentally, many spermatozoa may penetrate the ovum, including autologous spermatozoa and even spermatozoa from different genera (Fuke, 1980; Lambert, 1982). The penetration of the chorion takes a relatively long time (*c.* 2 min) and appears to involve production of an enzyme by the spermatozoon (Lambert, 1982).

The deleterious effects of inbreeding in *Botryllus schlosseri* have been studied by Sabbadin (1971). Self-fertilization cannot occur naturally in this species because ova ripen several days before autologous sperm are shed, and, as there is a strict synchronization throughout the colony, fertilization can only be effected by spermatozoa from another colony. Sabbadin was able to desynchronize parts of the colony so that fertilization by autologous spermatozoa was possible (all zooids in a colony are produced amictically and have identical genomes). The progeny showed a marked inbreeding depression which was expressed in a lower percentage of normal larvae, fewer successful metamorphoses, reduced growth and lower survival of new colonies.

When the eggs of *Ciona*, *Halocynthia*, *Pyura* and many other solitary ascidians are shed, they release a substance that acts as a sperm chemoattractant (Miller, 1975, 1982). It is not known whether it is the ovum or the follicle cells forming the chorion that produce the attractant, but its effect upon spermatozoa is clear. It stimulates swimming and initiates chemotactic turning movements which lead spermatozoa towards the egg (Lambert, 1982). Miller (1982) has demonstrated that the chemoattractant shows a fairly high degree of species specificity.

Fertilization of the eggs in compound ascidians must take place in the atrial cavity because the eggs are not shed into the sea. Possibly a chemoattractant guides spermatozoa to the egg, but this aspect does not appear to have been investigated. It would seem unlikely that spermatozoa enter through the inhalent siphon, as they would be captured by the feeding mechanism. Entry through the exhalent siphon would imply that they are able to swim against the water current produced by the zooid, unless the current stops at the time of egg laying.

Behaviour of ascidian tadpoles

In compound ascidians, development takes place within the parent's body, and release into the surrounding sea appears to be triggered by environmental changes. For many species it has been found that larval release occurs in response to exposure to light after a period of darkness (Watanabe and Lambert, 1973).

Tadpoles of *Aplidium* and many other compound ascidians can be obtained in all stages of development in summer by squeezing or tearing open a piece of the parent colony. Whilst still in the egg capsule, the tail fin lies flat against the body, as shown in Fig. 8.2(c), but it undergoes an axial rotation of 90° when the larva hatches so that the fin is in a vertical plane. Larvae swim in a similar way to amphibian tadpoles. When first released from their capsule they may not be very active, but they soon swim upwards towards light. The free-swimming stage is short, lasting a few minutes to 1–2 d depending upon the species. Most larvae become photonegative and seek shaded places prior to settlement, although the tadpoles of *Pyura stolonifera* are indifferent to light and settle on horizontal rock surfaces exposed to direct sunlight. Preferential settlement on or near adults of the same species (Young and Braithwaite, 1980) or on vertical rock faces (Gotelli, 1987) have been reported. These and other observations emphasize the fact that the eye–statolith organ and chemoreceptor cells round the rim of adhesive papillae (Cloney, 1982) are primarily larval sense organs specialized for finding a suitable site for adult life. However, exploratory behaviour similar to that of barnacle and oyster larvae has not been reported for ascidian tadpoles: indeed, they may attach to the first solid object with which they come in contact.

The tadpole larva is clearly a dispersive phase, which relies upon its yolk reserve to supply energy and material for growth to the adult form.

Attachment to the substratum and the onset of metamorphosis occur rapidly, although the more complex changes in development of adult organs require several days for completion. Initial attachment to the substratum is effected by eversion of the three adhesive organs, the central part of which contains cells with an adhesive substance. This process may take only 30 s, and appears to be due to contraction of myo-epithelial cells in the organ (Cloney, 1977). Shortly after this, the ampullae (if present) undergo regular expansion and contraction, increasing the area of attachment and producing secretions from their gland cells. Within 2 min, resorption of the tail commences. In compound ascidians like *Aplidium*, this is brought about by contraction of the epidermal cells of the tail, which pushes the muscles and notochord into the body, where they are gradually absorbed by phagocytes (Cloney, 1977, 1982). Presumably nutrients from resorption of the tail are made available for growth of the young sea squirt. The adhesive organs, sensory structures and part of the larval nervous system (visceral ganglion) are also removed by phagocytosis.

Rotation of the viscera now takes place, so that the oral siphon points away from the substratum. Over the next few days, adult structures grow and differentiate as functional organs. Nakauchi (1986), studying *Parascidia*, a close relative of *Aplidium*, found that the heart started to beat 20 h after settlement and feeding began 4–16 h later, even though yolk was still present after 5 d. The first instance of asexual budding took place 19 d after metamorphosis.

Organs of the larva can be grouped into three categories: larval structures lost at metamorphosis, rudiments of adult organs which will develop after metamorphosis, and a few structures which function in both larva and adult. In the first category, the tail, sense organs, adhesive papillae and yolk are the most obvious, but the outer layer of the larval tunic is also a transient structure. Prospective adult organs include the pharynx, gut, atrium, epicardia, cerebral ganglion and neural gland. The epidermis, tunic, heart and blood cells function in both phases.

8.6 ADVANTAGES OF ASEXUAL REPRODUCTION TO SHORE LIFE

Sexual reproduction invariably results in a tadpole larva which settles and metamorphoses into a new sea squirt. *Pyura stolonifera* has a simple life cycle of this type.

An asexual phase may be incorporated into the life cycle so that multiplication takes place after metamorphosis. When this occurs, there will be an alternation of generations (or metagenesis) comparable to that of hydroids and bryozoans. The fertilized egg eventually develops into a feeding zooid, called an oozooid, which is incapable of sexual reproduction. It reproduces by budding (blastogenesis), giving rise to further zooids called blastozooids. There may be little or no difference in appearance between oozooid and blastozooid, although they will have developed by very different processes. As blastogenesis does not involve meiosis or fertilization, all blastozooids derived from the same individual will be genetically identical and form a clone.

The oozooid of *Aplidium* has only four rows of stigmata, and its post-abdomen accumulates nutritive cells or trophocytes. About 3 weeks after metamorphosis, the first instance of asexual budding takes place, and a number of daughter blastozooids are produced, each with six to ten rows of stigmata. Periodically these blastozooids undergo budding (strobilation), producing a colony which grows rapidly in size. During the reproductive season the blastozooids produce sperm and ova.

The life cycle of *Botryllus schlosseri* is more complicated and shows an extreme case of reproductive synchronization within the colony. The founding oozooid produces two buds, but only the one on the right-hand side develops into the first blastozooid. This in turn produces two buds, which develop into blastozooids that move apart and establish a common cloaca and blood system. Second and third generations, each producing two buds, succeed one another to form a rosette of zooids with a common cloaca and common peripheral vascular system (Herdman, 1924). Later budding may not be so regular, and sometimes only one of the buds develops or a bud may be pushed out of the system to start its own rosette. In the summer, gametes are produced in the buds and sexual reproduction occurs (Fig. 8.10(d)). Two or four ova mature at about the time that the bud takes over as the feeding zooid. The fertilized eggs are retained in the atrium, where they develop into tadpoles, which are released when the zooid regresses and is replaced by the next bud. At 18 °C, replacement takes place every 7 d (Sabbadin, 1978).

This life cycle shows several points of interest. Three generations are present at any one time (Fig. 8.8(c)). The trophic or feeding phase is short (*c.* 6 d) and presumably very efficient, since it has to supply nutrients for the growth of up to four large eggs as well as the new buds. Because the ova develop in the bud before it becomes the trophic zooid, reproduction can be considered as a case of paedogenesis or precocious sexual development. Spermatozoa are produced during the trophic phase. In the lagoons of Venice, sexual reproduction takes place from April to November, but in colder climates a shorter breeding season occurs during the summer months (Sabbadin, 1978). Zooids may regress during the winter and regenerate the following spring by asexual budding to form a new colony.

Patterns of budding in compound ascidians

Solitary ascidians show considerable ability to regenerate lost or damaged parts of the body, but they rarely reproduce by asexual budding. In contrast, the large colonies of compound ascidians are the direct consequence of repeated budding, and asexual reproduction has become an integral part of the life cycle.

A wide diversity in patterns of budding is to be found in these colonial forms. Nakauchi (1982) has reviewed the subject and defined nine modes of bud formation, five of which are found in the species described here.

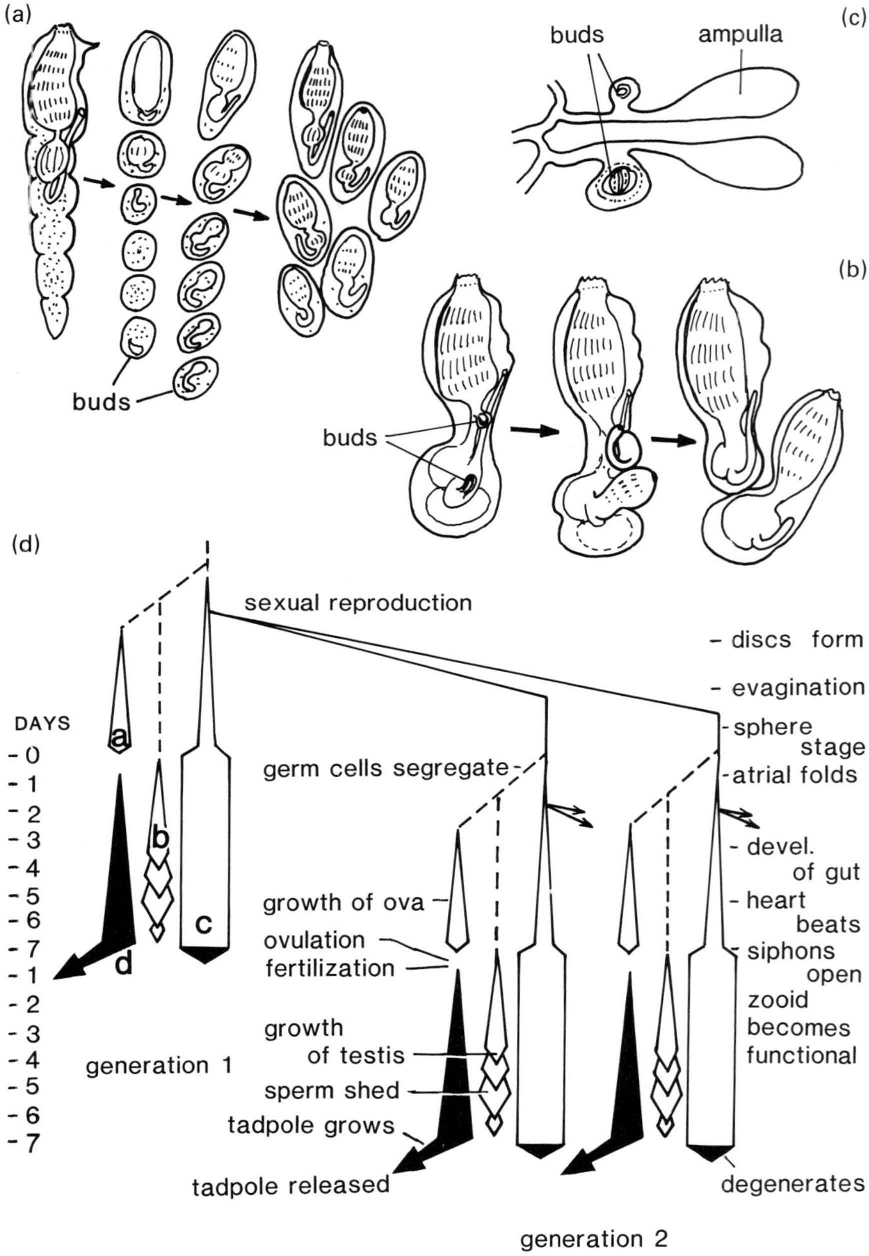

Fig. 8.10 (a) Strobilation of *Aplidium*. (b) Pyloric budding of *Didemnum*. (c) Vascular budding. (d) Time sequence in zooid replacement in *Botryllus schlosseri*.

Strobilation

When *Aplidium* is about to undergo asexual reproduction, the pharynx, stomach and intestine undergo partial regression, and large numbers of cells rich in nutrients (trophocytes) appear, particularly in the post-abdominal region. Transverse constrictions now take place to divide the abdomen and post-abdomen into a chain or strobilus of six or seven buds, each capable of developing into a new zooid. The anterior bud, derived from the old branchial region, develops into an entire new zooid, as do each of the other buds. Those from the post-abdomen do not receive any remnant of the digestive system; the only tissues present are epidermis, mesenchyme and epicardium (the last bud will have remnants of the heart), yet they all develop into normal zooids (Fig. 8.10(a)). An interesting aspect of strobilation is that all the new zooids reorientate themselves so that they form a group with a communal cloaca in the common test (Nakauchi and Kawamura, 1978). How the developing zooids move and reorientate is not understood, but Nakauchi and Kawamura (1978), and Nakauchi (1986) have shown that they respond to the presence of an older zooid by migrating towards it. Presumably a chemical substance is involved, but whether it is some metabolic by-product or a pheromone has not been established.

Pyloric budding

In the family Didemnidae, an unusual form of budding takes place, in which two buds arise simultaneously from the oesophagus and intestine respectively. The oesophageal bud forms a new abdominal region, and the intestinal bud differentiates into a new branchial region. Depending on the species, two things can happen: either the two buds join to form a new zooid, or the old pharynx joins with the new abdomen, and the old abdomen with the new pharynx, to form two new zooids (Fig. 8.10(b)). Each bud receives epicardial tissue as well as mesenchyme and part of the digestive tube.

Atrial or peribranchial budding

Atrial budding is typical of *Botryllus schlosseri*. Here two buds arise, one on either side of the body, as outpushings from the atrial epithelium of the mantle. The origin of the buds can be traced back to a disc of cells in the atrial epithelium of an older bud, which is developing in the atrium of a feeding zooid (Fig. 8.8(c)). Berrill (1941) has shown that this disc of cells, covered externally with epidermis, evaginates to form a hollow sphere. The epidermis forms a stalk connecting the bud at an early stage to the ampullary blood system of the colony. The primordia of the gonads separate at a very early stage from the inner sphere, which then grows and differentiates into all the organs of the adult body. Of particular interest is the fact that the atrium is not formed by an invagination of the epidermis as in the tadpole, but is produced by infoldings that divide the central cavity into three regions: a central pharynx and two lateral cavities which form the atrium. The future digestive tube arises as an outgrowth from the central chamber (Berrill, 1961).

Vascular budding

In *Botryllus primigenus*, buds may arise from the vascular ampullae (Fig. 8.10(c)). A spherical vesicle, similar to that described for atrial budding, forms at the base of an ampulla, but Oka and Watanabe (1959) have shown that in the related *Botrylloides*, the inner layer of cells is formed by aggregation of lymphocytes circulating in the blood, and not from the epidermal wall of the ampulla. Its future development, however, is similar to that of atrial buds. *Botryllus schlosseri* can be induced to produce vascular buds under experimental conditions by removing all the zooids and buds from a colony (Milkman, 1967).

Stolonic budding

In contrast to hydroids, the formation of buds from a branching stolon is rare among ascidians. The branched stolon of *Perophora multistigmata* produces buds at regular intervals, which develop into zooids that remain in organic contact through the stolon (Fig. 8.11(a)).

From the account of budding given above, it is evident that a new individual may develop from a variety of tissue or cell types; for instance, individuals can differentiate from endoderm, epicardium, atrial epithelium, and even blood corpuscles. The cells of these tissues must be multipotent and have the capacity to differentiate into cells of other functional types. Even more remarkable is the fact that similar structures can be formed by very different developmental processes. The atrium, for example, develops in an entirely different way during embryogenesis to that during blastogenesis, and yet its basic structure (apart from differences in the number of stigmata) is the same in both instances.

SUMMARY

Ascidians do not colonize the middle shore, but their tough outer tunic, smooth form and firm attachment to the substratum suit conditions on the

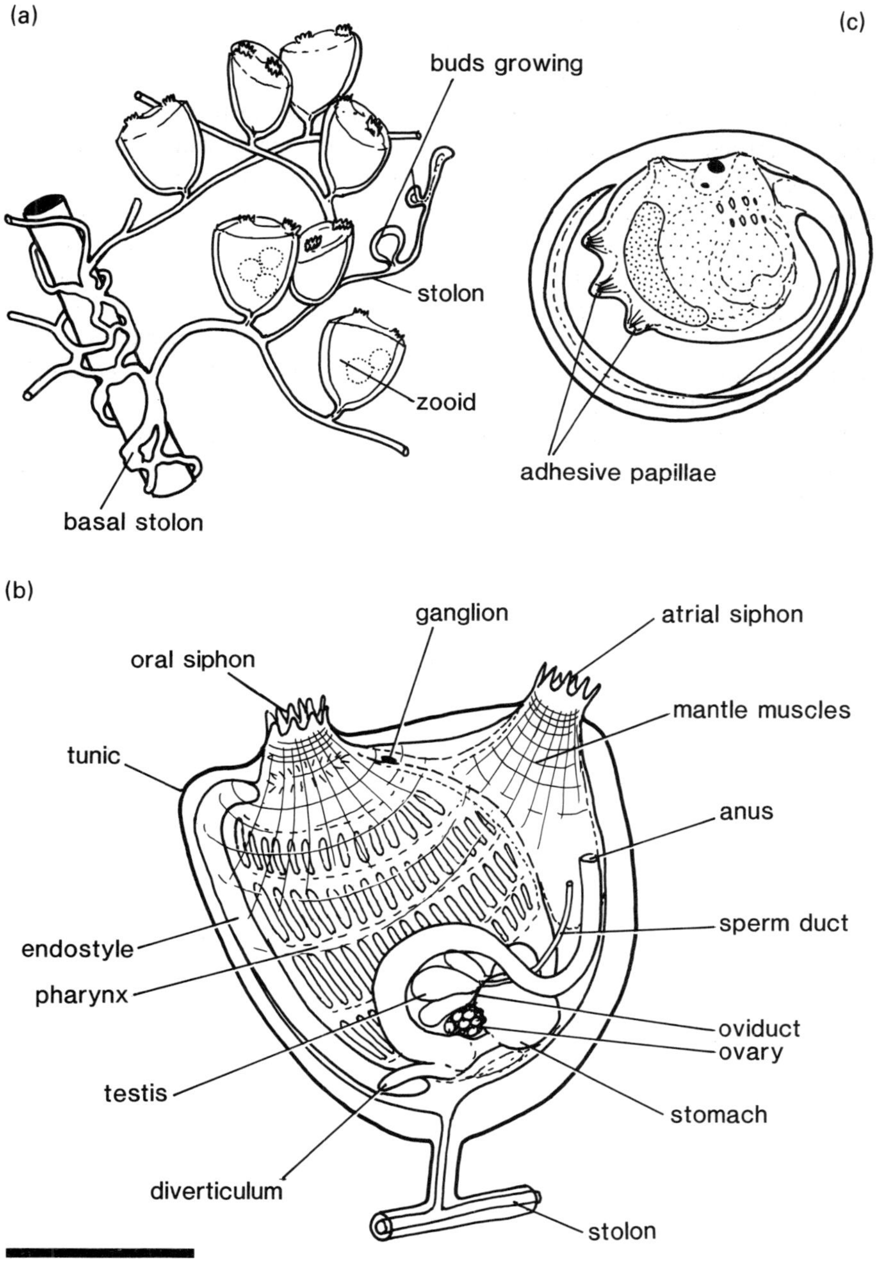

Fig. 8.11 Perophora multistigmata. (a) Small piece of a colony. (b) Single zooid, showing structure. (c) Tadpole larva in its follicular capsule. Scale bar: (a) = 2 mm, (b) = 0.15 mm, (c) = 0.5 mm.

lower shore. Several species live where they are subject to regular periods of emersion, but most species live under boulders or rock ledges away from direct sunlight.

The tunic has an outer cuticle (of tanned protein in certain species) which may be hardened with sand grains or calcareous spicules. The thicker inner layer, composed mainly of polysaccharide, is fibrous or gelatinous. It contains blood vessels and numerous cells derived from the blood. The tunic is elastic and restores the original shape after muscle contraction. The tunic provides effective protection against water loss during emersion, wave impact and abrasion. A large volume of water is retained in the body cavities whilst out of the water. Vanadium compounds, sulphuric acid and possibly iodine compounds in the tunic provide protection against predators and close settlement of competitive species.

Feeding is by a mucus net and is not selective. Detritus forms the major component of food, but the mucus may absorb macromolecules from the seawater. There are no special reproductive adaptations for shore life.

9

Bryozoans (*Polyzoa: Ectoprocta*)

Bryozoans are well represented on the lower shore, although the small size, drab coloration and plant-like growth form of most species render them inconspicuous. Few aspects of their structure can be attributed directly to specialization for life in the littoral zone. Nevertheless, certain features of bryozoan organization enable many species to survive strong wave action and short periods of emersion.

They are found on the shore in similar places to compound ascidians: on the underside of rocks, on the shells of other animals, on seaweeds, on wharf piles and even on moored boats. They take various growth forms; some are flat incrustations, others are sponge-like growths, broad flat fronds like seaweed, or tufted growths resembling plants, but in all cases these growths are colonies composed of numerous minute zooids less than 1 mm in size. Each zooid is housed in a box-like exoskeleton, and when left undisturbed it protrudes a crown of feeding tentacles (Fig. 9.1(a)). The colony originates from a single founding individual which, by sequential budding, gives rise to a colony of zooids that are genetically identical, in organic contact and dependent upon one another to varying degrees. Where the exoskeleton is heavily calcified, colonies may be rigid structures like the lace coral, *Sertella jacksonensis* (Fig. 9.1(b)), found on Australian shores, but species with non-calcified exoskeletons, such as *Alcyonidium gelatinosum*, are pliable and have a gelatinous appearance (Fig. 9.1(d) c).

The skeletal box, together with the tissues lining it, is called the cystid, while the crown of feeding tentacles, or lophophore, plus the U-shaped gut consitute the polypide (Figs 9.1(a), 9.2(d)).

9.1 SOME COMMON BRYOZOANS OF THE LITTORAL ZONE

Three species are common on the seaweed *Fucus serratus* growing on the lower shore in Britain. *Alcyonidium hirsutum, Flustrellidra hispida* and less

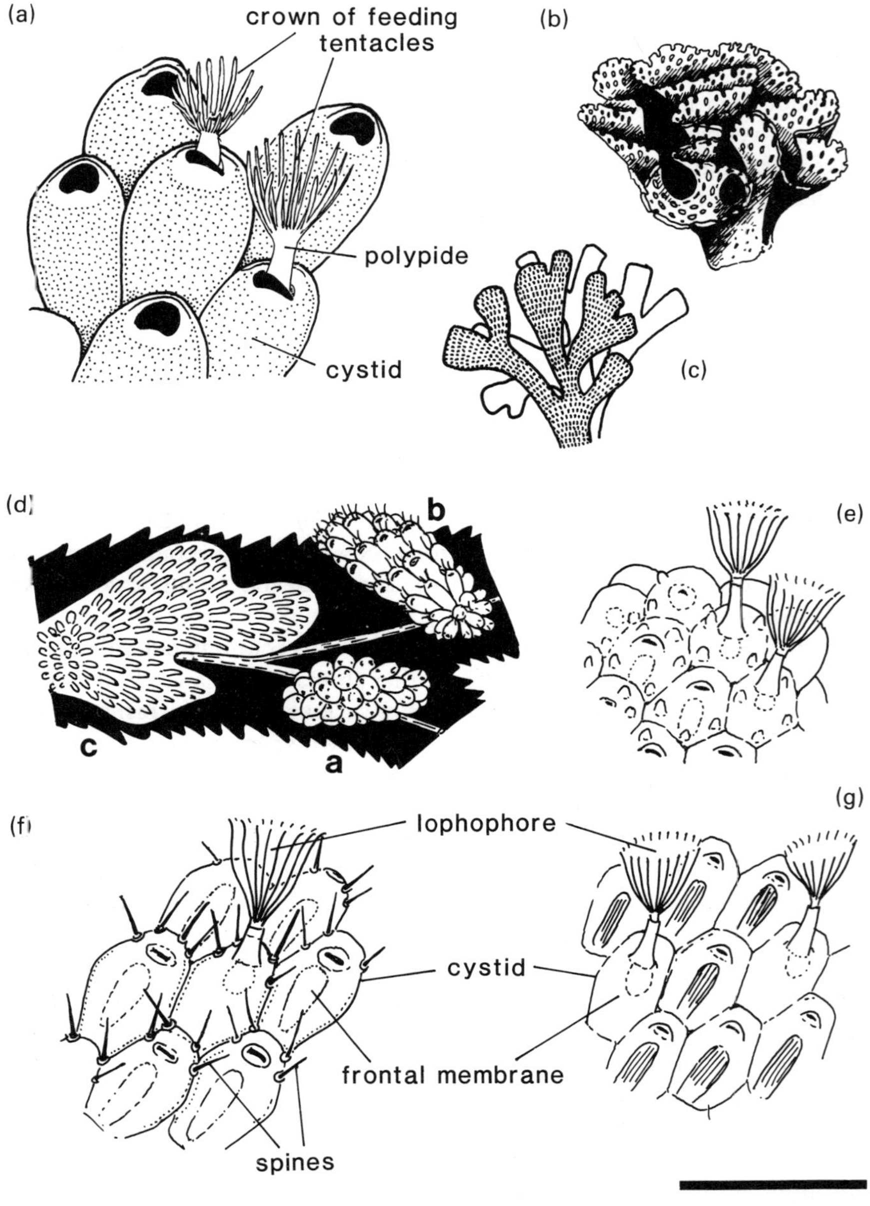

Fig. 9.1 Growth form of bryozoans. (a) *Watersipora arcuata*. (b) *Sertella jacksonensis*, a lace coral, about natural size. (c) *Flustra*, about natural size. (d) to (g) Bryozoans growing on *Fucus serratus*: (d) a and (e) *Alcyonidium hirsutum*; (d) b and (f) *Flustrellidra hispida*. (d) c and (g) *Alcyonidium gelatinosum*. Scale bar: (a), (e), (f), (g) = 1.5 mm.

commonly *Alcyonidium gelatinosum*, form transparent colonies with a gelatinous appearance that spread out over the surface of the seaweed. The zooids are clearly visible (Fig. 9.1(d)a,b,c), and if left undisturbed may protude their lophophore through an aperture towards the distal end of the frontal (ventral) surface. When the lophophore is withdrawn into the pocket-like tentacular sheath, the aperture is closed by contraction of its circular lips. A feature of these and related species is a collar below the tentacles, which may appear as a comb-like structure when the lophophore is withdrawn. Species possessing this collar are placed in the order Ctenostomata*.

The majority of marine Bryozoa have a calcified wall to the cystid, and the posterior lip of the aperture to the tentacular sheath is formed into a plate, the operculum, which completely closes the aperture when the lophophore is withdrawn. Operculate bryozoans are placed in the order Cheilostomata and show a wide variety of structure, illustrated by the following examples.

Membranipora membranacea (Fig. 9.2(a) and (b)) is a species that occurs abundantly on *Laminaria digitata* and is often found on *Fucus serratus*. A related species occurs on *Ecklonia radiata* in Australian waters. It forms a conspicuous white lace-like mat of numerous rectangular boxes arranged in parallel rows like bricks. Their side walls and base (dorsal) are calcified, but the frontal membrane is a transparent flexible cuticle through which the polypide can be seen. A D-shaped sclerotized plate, the operculum, is hinged to the frontal membrane.

From a section through a zooid (Fig. 9.5(f), it will be seen that the polypide is suspended in a large perivisceral coelom. A separate tentacular coelom round the base of the tentacles sends a branch up each tentacle and is connected to the perivisceral coelom by a pore. A ciliated tube, the inter-tentacular organ, puts the coelom in contact with the outside seawater (Figs 9.2(d), 9.6(c)) and allows eggs to escape from the coelom. The interior of the cystid is lined with mesenchyme, from which strands of tissue, the funicular cords, are derived. These pass through fine pores in the pore plates or septula from one zooid to another. The coelomic cavities of adjacent zooids are not confluent, therefore pressure differences cannot be transmitted from one zooid to another. This is a functional necessity for eversion of the lophophore. Nevertheless, all zooids are in communication through nerves that pass through the pore plates.

A ring of tentacles round the mouth forms a funnel-like lophophore. Lateral cilia on the sides of the tentacles (Fig. 9.5(e)) create a water current from which food particles are collected and passed to the mouth. Water flows into the funnel in the direction of the mouth (Fig. 9.3(c)) and, as the anus is located outside the lophophore (Fig. 9.2(d)), faeces will be washed away by the feeding current.

* See Appendix for classification of bryozoans.

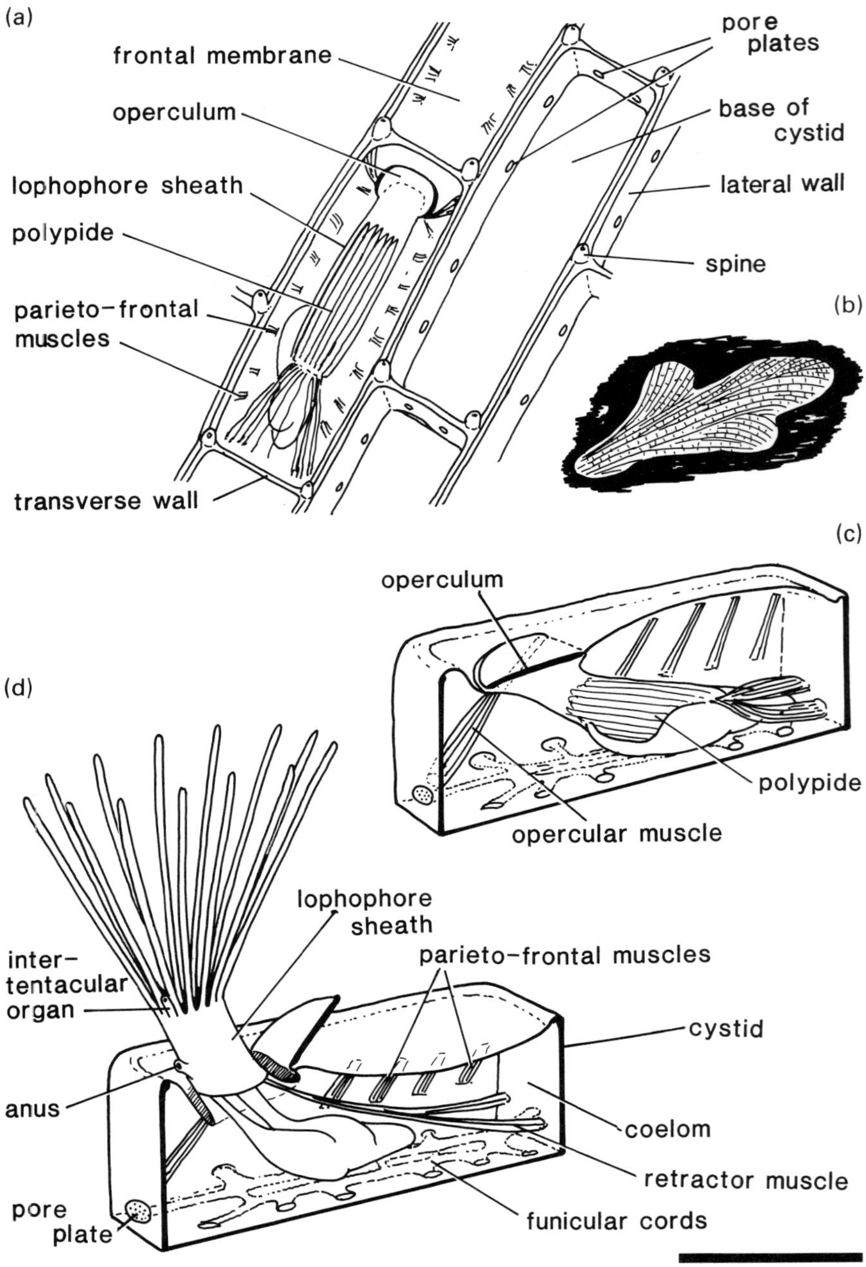

Fig. 9.2 (a) Structure of *Membranipora membranacea*. (b) Appearance on seaweed. (c) and (d) Generalized zooid, showing polypide everted and withdrawn. Scale bar: (a) = 0.2 mm.

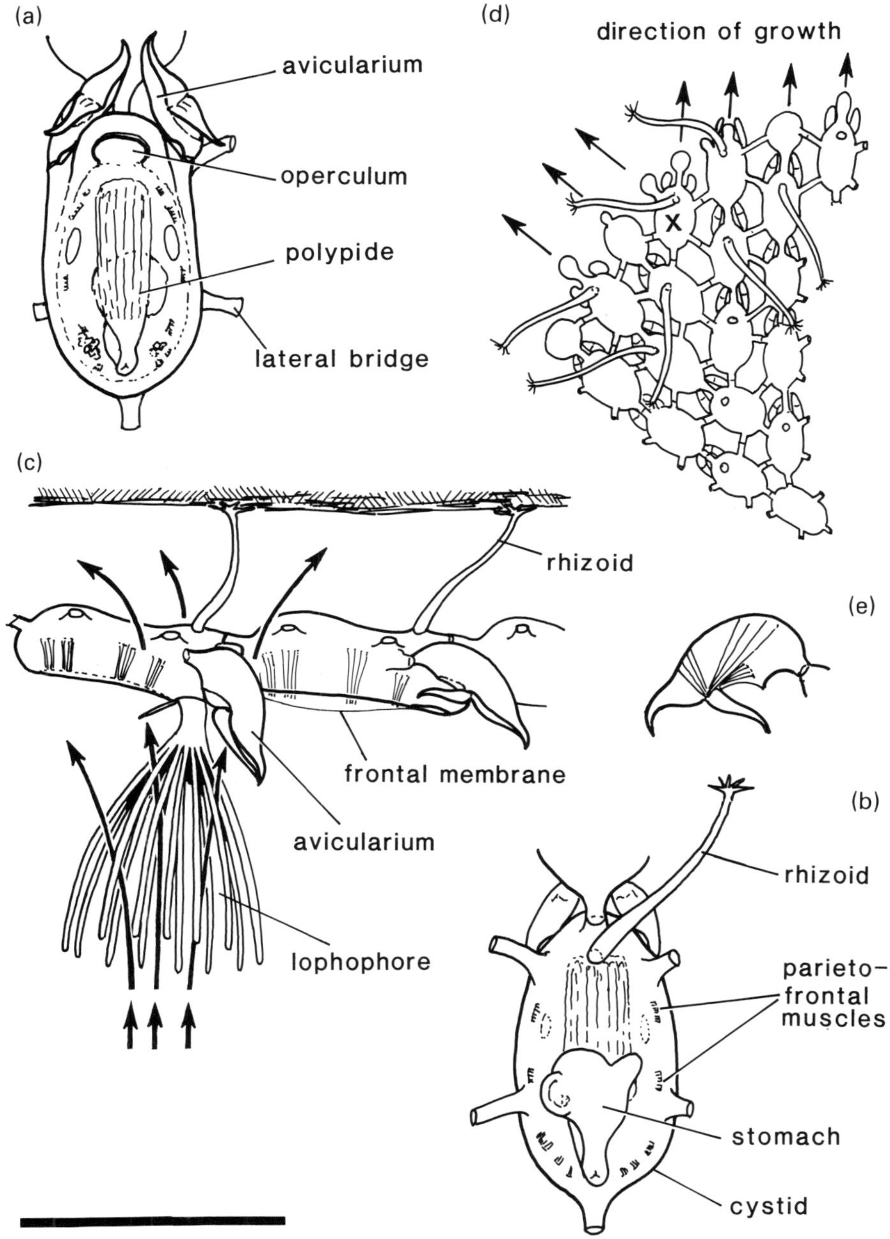

Fig. 9.3 Beania magellanica. (a) Zooid viewed from frontal aspect. (b) Zooid viewed from abfrontal side. (c) Side view, showing natural orientation and direction of water currents (arrows). (d) Growing edge of colony. Zooid marked **X** will give rise to a frontal and a lateral bud; arrows show direction of growth. For explanation see page 178. (e) An avicularium. Scale bar: (a) and (b) = 0.5 mm.

Membranipora membranacea is a relatively simple cheilostome. Other members have evolved a variety of protective devices to safeguard the delicate frontal membrane and the zooid inside.

Beania magellanica (Fig. 9.3(a), to (c)) is a common species living on the underside of boulder rocks below low water of spring tides on Australasian shores. It is easily recognized by its yellow or brown colour and lace-like sheet of zooids, which are suspended about 1 mm off the rock surface by numerous vertical rhizoids. Each zooid is ovoid and is connected to the zooid in front and behind as well as the four lateral neighbours by bridges that form an open hexagonal mat (Fig. 9.3(d)). From the basal surface the rhizoids anchor the mat-like colony to the substratum. The frontal membrane is thin, with a D-shaped operculum, and the polypide can be seen inside. The lophophore has 18 or 20 tentacles, and on either side of the operculum there is a defensive structure, called an avicularium, that looks remarkably like a bird's head and beak (Fig. 9.3(e)). The avicularia are fascinating to watch on live specimens, for at the slightest disturbance, such as a jet of water from a pipette, the beak will snap shut repeatedly and the whole structure will move from side to side. They are regarded as highly modified zooids that have lost their polypide. In other species of *Beania*, numerous spines border the frontal area and help to protect the delicate frontal membrane. Eggs are retained within the parent zooid during development, but without the formation of a special brood chamber.

On account of its relatively large zooids (1 mm), *Beania magellanica* is an ideal species in which to watch the movements of the polypide and passage of food through the alimentary canal. The mat of zooids can be removed without damage by cutting through the rhizoids and, by turning this upside down, the internal organs and rotation of the food in the stomach can be seen through the transparent floor of the cystid. Rotation is not continuous. Periodically it may stop, indicating that the cilia can be controlled, presumably by the nervous system. From time to time peristaltic contractions of the stomach occur. If the colony is turned the right way up, eversion and movement of the lophophore can be followed. The parietal and retractor muscles can be seen through the frontal membrane. If there are small particles suspended in the water, the feeding currents generated by the lateral cilia will be revealed (Fig. 9.3(c)).

Bicellariella ciliata, (Fig. 9.4(a)) like *Membranipora membranacea,* is another species of nearly world-wide distribution. It looks like a small piece of down feather 5–10 mm high, growing on the underside of rocks in the same situation as *Beania.* The small zooids (0.4 mm) alternate to form a double row, which branches frequently to give a bush-like growth. Long spines round the edge of the frontal membrane give the branches their characteristic feathery appearance. Many of the zooids carry a single lateral

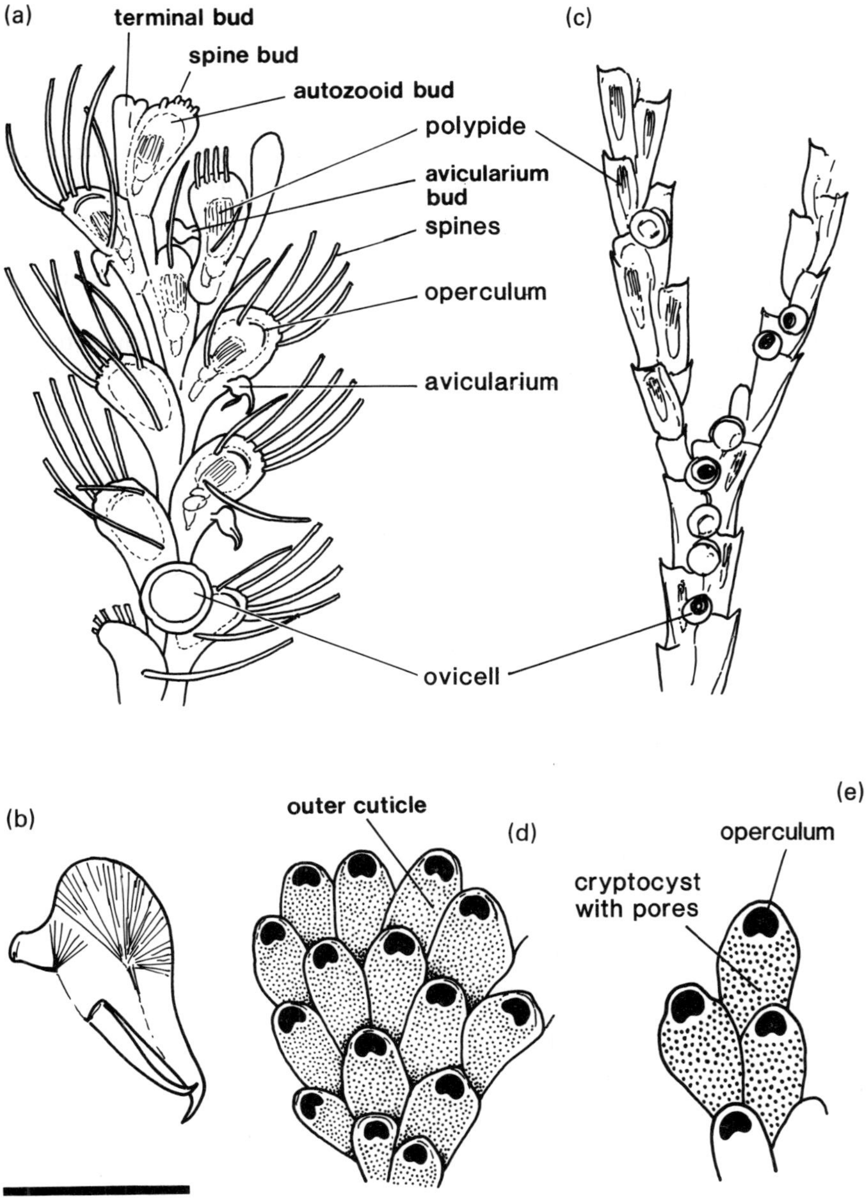

Fig. 9.4 (a) and (b) *Bicellariella ciliata*: (a) portion of the colony, showing spines and ovicells. (b) Avicularium of (a). (c) Portion of *Bugula neritina* colony, showing ovicells. (d) *Watersipora arcuata*. (e) Same as (d), with outer cuticle removed to show cryptocyst. Scale bar: (a) and (c) = 1.5 mm, (d) = 2.5 mm.

avicularium (Fig. 9.4(b)). Older zooids have a nearly spherical brood chamber (or ovicell), attached medially to the edge of the frontal surface, in which a single egg develops to a ciliated larva.

Watersipora arcuata (Fig. 9.4(d)) is an Australian species that is often found on the upper surface of rocks at lower tide levels, where it might be mistaken for a patch of tar. The upper surface of each zooid is dark brown or grey with a black operculum, and in some colonies the growing edge is orange-brown. Zooids are oval in outline, but their proximal border is indented by the zooid behind. Spines, avicularia and brood chambers are absent in this species.

The thin outer cuticle can easily be peeled off to reveal a horizontal calcareous plate, called the cryptocyst, which is pierced by numerous pores and forms a secondary roof to the cystid (Fig. 9.4(e)). Below this plate there is a sac-like compensation chamber called an ascus. This arrangement is characteristic of a large number of marine bryozoans which are grouped together in the suborder Ascophora. *Beania, Membranipora* and *Bicellariella* do not possess an ascus and are placed in the suborder Anasca. Fig. 9.9 illustrates the structural difference between anascan and ascophoran cheilostomes and the way in which eversion of the lophophore is effected (see also pages 175–77).

A very different type of bryozoan, belonging to the order Cyclostomata, may be common under rocks at low tide level. A species of *Crisia* (Fig. 9.5(a) and (b)) forms a colony of zooids that are fused together in groups or sections of five to 17 tubular zooids. The exoskeleton of the cystid is calcareous, with numerous lenticular markings, the pseudopores, and an external cuticle (Fig. 9.5(d)). A ball-and-socket joint between the sections gives flexibility to the colony (Fig. 9.5(c)). Each tubular zooid has a spine attached near the opening, but these are easily broken off and may be missing. One or two pear-shaped brood chambers may be present on certain sections (Fig. 9.5(a)) in which embryos undergo asexual reproduction or polyembryony.

Fig. 9.5 Crisia sp. (a) Portion of colony. (b) Enlargement, showing spines and joints. (c) Disarticulated joint. (d) Pseudopores in cystid wall seen in optical section. (e) Transverse section of bryozoan tentacle. (f) Diagrammatic cross-section of *Membranipora* cystid. Scale bar: (a) = 1.5 mm, (b) = 0.5 mm, (e) = 0.02 mm. Abfr. n., abfrontal nerve; bas. calc., basal calcification of cystid; calc., calcified wall of cystid; coll., collagenous skeleton of tentacle; ect., ectoderm; fr. cil., frontal cilia; front. cut., frontal cuticle; fun. cor., funicular cords; lat. cil., lateral cilia; lat. fr. cil., latero-frontal cilia; musc., muscle fibres of tentacle; n., nerve; perivisc. coel., perivisceral coelom; sept., septulum (pore plate); spl. perit., splanchnic peritoneum; tent. coel., tentacular coelom.

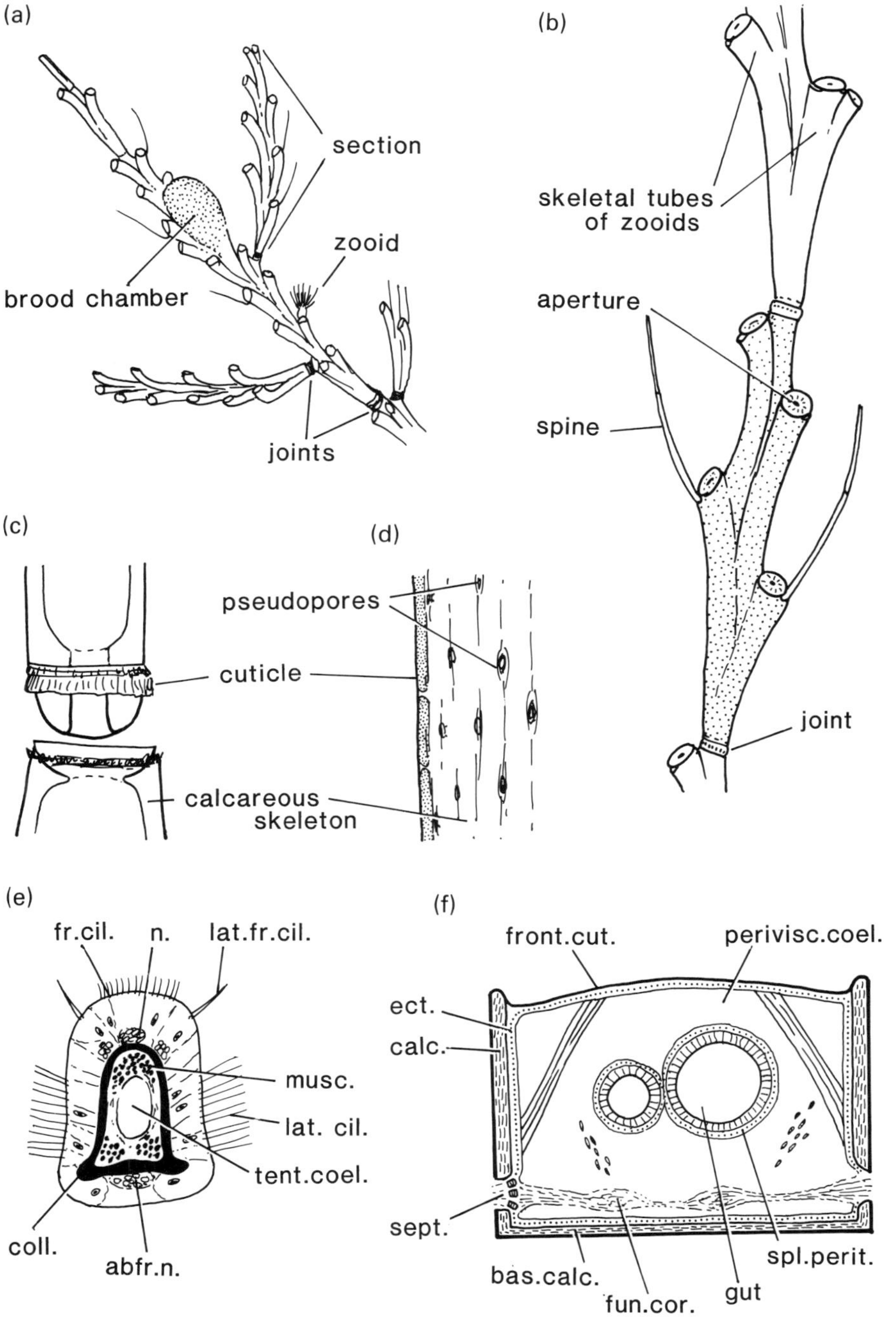
(a)
section
zooid
brood chamber
joints
(b)
skeletal tubes
of zooids
aperture
spine
joint
(c)
cuticle
calcareous
skeleton
(d)
pseudopores
cuticle
calcareous
skeleton
(e)
fr.cil.
n.
lat.fr.cil.
musc.
lat. cil.
tent.coel.
coll.
abfr.n.
(f)
front.cut.
perivisc.coel.
ect.
calc.
sept.
bas.calc.
fun.cor.
gut
spl.perit.

9.2 PROTECTION AND ATTACHMENT

The very low profile and smooth surface of species like *Alcyonidium gelatinosum, Membranipora membranacea*, and *Watersipora arcuata* virtually eliminate drag forces in fast-flowing current and heavy wave action. Many species are confined to the holdfast and lower branches of algal thalli, where water movements will be moderated by the main growth of seaweed above. In all encrusting species, the area of surface attachment is very large relative to the volume of the colony.

The nature of the exoskeleton is important for both protection and attachment. All species are covered externally with a cuticle composed of protein, chitin and some mucopolysaccharides (Ryland, 1970). The cuticle is secreted by epidermal cells in the growth zone, although materials elaborated in the mesenchyme may contribute to its structure. In encrusting species like *Membranipora membranacea*, the cuticle is soft when first secreted and conforms closely to the contour of the substratum. Later it hardens, but this does not appear to be due to quinone tanning. A thin layer of mucopolysaccharide between cuticle and substratum may improve adhesion in species growing on seaweeds. Cuticle is the only skeletal material forming the external covering of the zooid in *Zoobotryon* (Fig. 9.8(a)), *Alcyonidium* and *Flustrellidra*, but in the majority of bryozoans, calcium carbonate is laid down, either as calcite or calcite plus aragonite, between the cuticle and the internal epithelium, forming the typical rigid box-like exoskeleton or zooecium (Fig. 9.5(f)). The exoskeleton is a non-living structure and in *Flustra foliacea* (Fig. 9.1(c)) it is composed of 14% scleroprotein, 1% chitin and 85% calcium carbonate with small amounts of phosphate (Hunt, 1972).

The calcareous component of the exoskeleton will give rigidity, but the cuticle gives flexibility, tensile strength and probably some degree of protection against water loss from species emersed at low tide. All these properties are valuable to life in the littoral zone. For most species, the cryptic habit of living in shade protects them from harmful radiations. *Watersipora arcuata*, however, lives on the surface of rocks periodically exposed to direct rays of the sun. It is probable that its dark coloration is a protection against harmful ultraviolet radiation.

Rigidity of a colony may be an advantage against wave impact for species living on solid substrata, but it is undesirable for species that attach to the flexible surface of seaweeds. Here some flexibility is essential. The cuticle allows sufficient flexibility in species like *Alcyonidium gelatinosum* for the colony to bend with the seaweed. In *Membranipora membranacea*, where the exoskeleton is calcified, the lateral walls of individual cystids are not fused, and slight movement is possible between them. Flexibility is also essential for branched forms that grow into the water column, like *Crisia, Bicellariella ciliata* or *Bugula neritina*. These rely upon flexibility of the cuticle between zooids, except in *Crisia* where, as already noted, the cuticle allows some movement at the ball-and-socket joint between the calcified segments.

Many species with a branched, plant-like growth form strengthen their attachment to the substratum by tubular outgrowths. For example, in *Beania magellanica* each zooid produces a single dorsal (abfrontal) rhizoid, which attaches the mat-like colony to the rock. Upright forms like *Bicellariella ciliata*, *Bugula neritina* and *Zoobotryon* sp. produce tubular rhizoids or siphons from some of the lower zooids; then rhizoids run down to the substratum, where they are firmly attached. These tubes are joined to the zooid by pore plates and contain extensions of the funicular cords. They are regarded as kenozooids, or zooids that do not possess a polypide (Fig. 9.8(a)). A remarkably similar structure has been evolved independently in certain hydroids (see page 239).

Marine Bryozoa do not have stinging cells like those of hydroid polyps to protect them from predators, and defence rests upon the hard exoskeleton, spines and other structures. Spines are present in *Flustrellidra hispida* (Fig. 9.1(f)) and *Bicellariella ciliata* (Fig. 9.4(a)) towards the edge of the cystid. In *Beania discodermiae*, often found growing with *Beania magellanica*, spines at the edge fold over the frontal membrane to protect it. These spines are regarded as kenozooids.

From the typical operculate zooid, two important polymorphic zooids have evolved, which have abandoned their feeding role and taken on the function of defence. One is the avicularium, already described for *Beania magellanica* (see page 164), in which the mandible or movable 'beak' is homologous with the operculum, and the opercular occlusor muscle has become the massive striated adductor muscle that closes the 'beak' (Fig. 9.6(a) and (b)). Opening is caused indirectly by the abductor muscles, modified parietal muscles, which depress the membranous area behind the mandible (equivalent to the frontal membrane). Muscles at the base of the avicularium give limited movement of the whole structure. In many bryozoans, the avicularia are sessile and cannot be moved about. Another polymorph is the vibraculum, a large bristle-like structure that can sweep over the surrounding zooids, perhaps to dislodge foreign particles or prevent other animals from settling on the colony. The long seta of the vibraculum is homologous with the operculum and mandible of the avicularium.

9.3 THE LOPHOPHORE AND FEEDING

The delicate lophophore will achieve maximum efficiency collecting food in relatively calm water, thus specimens living on the lower shore will have their

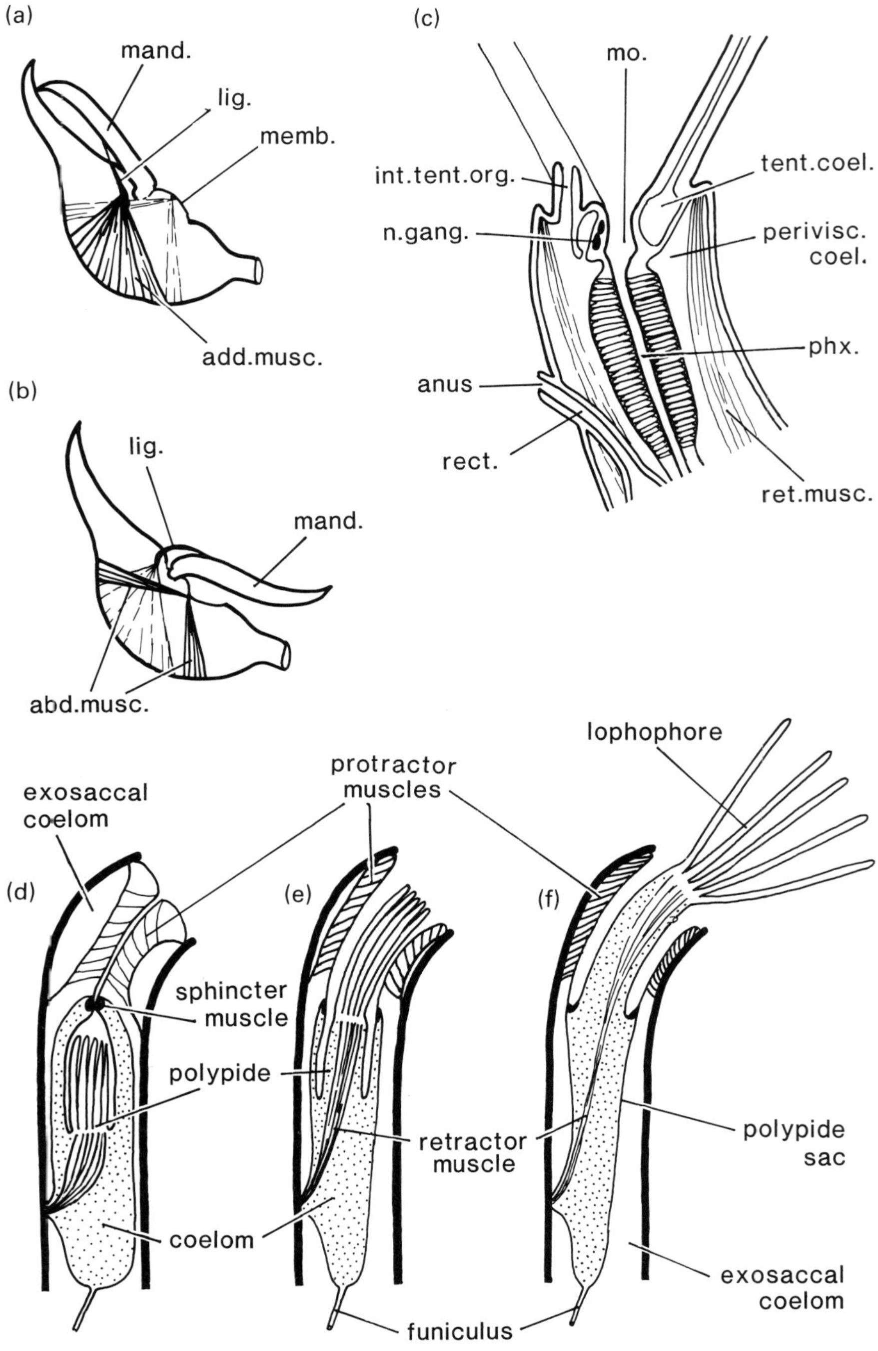
(a)
mand.
lig.
memb.
add.musc.
(b)
lig.
mand.
abd.musc.
(c)
mo.
int.tent.org.
tent.coel.
n.gang.
perivisc.
coel.
phx.
anus
rect.
ret.musc.
lophophore
protractor
muscles
exosaccal
coelom
(d)
(e)
(f)
sphincter
muscle
polypide
retractor
muscle
polypide
sac
coelom
exosaccal
coelom
funiculus

feeding time reduced by the period of wave action as well as actual emersion at low tide.

Individual tentacles have a central tentacular coelom, longitudinal muscle fibres enclosed in a collagenous sheath, and an external covering of ectodermal cells, some of which are ciliated (Fig. 9.5(e)). Individual tentacles may move independently by means of their intrinsic muscles. When these contract they shorten the tentacle and obliterate the tentacular coelom. Elongation is caused by increased coelomic pressure, which forces fluid into the tentacular coelom at the same time that the lophophore is everted. Each tentacle is provided with nerves that run down the frontal and abfrontal faces. On the side facing the mouth, tentacles bear a band of frontal cilia that carry particles towards the mouth. Longer lateral cilia which create the water current are present on the lateral faces, and at the latero-frontal edge there are stiff cilia which are believed to be sensory (Lutaud, 1973).

Bryozoans are not passive feeders: their cilia create the water current that brings food to the zooid. No mucus is produced by the tentacles or gut, so their method of collecting food must differ from that of tunicates. There are differences of opinion as to how the lophophore separates food from the water, but the most plausible hypothesis is that the lateral cilia locally reverse their beat and 'kick back' any food particle that starts to pass between the tentacles. Under the influence of the frontal cilia and the downward water flow, the particle would move closer to the mouth with each 'kick' (Strathmann, 1973; Ryland, 1976). Such a system would require a sensory receptor to detect the presence of the particle before it was swept past the tentacle in the water stream. The latero-frontal cilia and the extensive innervation of the tentacles probably provide this function and cause momentary reversal of the ciliary beat (Lutaud, 1973; Strathmann, 1973).

When all the zooids are expanded and creating their own downward water current, a route must be provided for the water to escape. In encrusting species like *Membranipora membranacea* regular spaces are found where there is no active zooid; this provides an exit or 'chimney' for the water (Banta

---

Fig. 9.6 (a) and (b) Opening and closing of an avicularium. (c) Relationship between perivisceral and tentacular coelom. (d) to (f) Eversion of lophophore in *Crisia*, a cyclostome. Contraction of the protractor muscles compresses the exosaccal coelom, which in turn compresses the perivisceral coelom ('coelom'), which is enclosed by the polypide sac (stippled). Gut of polypide not shown in diagrams. Abd. musc., abductor muscle; add. musc., adductor muscle; int. tent. org., inter-tentacular organ; lig., ligament; mand., mandible or modified operculum; memb., membrane equivalent to the frontal membrane; mo., mouth; n. gang., nerve ganglion or brain; perivisc. coel., perivisceral coelom; phx., pharynx; rect., rectum; ret. musc., retractor muscle; tent. coel., tentacular coelom.

et al. 1974; Ryland, 1976). In the case of *Beania magellanica*, the water current can pass between the widely separated zooids into the space under the colony formed by the rhizoids (Fig. 9.3(c)).

Little is known about the natural food of marine bryozoans, but small flagellates, diatoms, coccolithophorans and dinoflagellates appear to be taken from the plankton. From laboratory studies there is evidence that some species feed selectively. Food particles taken range from *c.* 8 μm to the maximum diameter of the mouth (45 μm for *Zoobotryon*, Bullivant, 1968b). Bullivant (1968a) has also shown that bryozoans are as efficient as other suspension-feeding animals in removing particles from seawater. Frontal cilia continue down into the mouth, which is normally kept open whilst feeding, and they convey small particles into the pharynx. In some species a tract of cilia in the mouth can reverse the direction of beat and carry unwanted material out of the mouth.

The gut is divided into functionally distinct regions. Whilst feeding, the pharynx (Fig. 9.6(c)) periodically increases in diameter, sucking in water and food particles that have accumulated in the mouth. As there are no direct muscles in this region to cause a rapid dilation, its mechanism has to be explained. Matricon (1973) has shown that the tall epithelial cells lining the pharynx each contain a large vacuole surrounded by radially-arranged contractile fibrils (Fig. 9.7(a)). When these fibrils contract, they reduce the height of the vacuole and consequently increase its diameter. This causes the pharynx to dilate. Circular muscles round the pharynx reduce its diameter and squeeze the vacuolated cells back to their tall columnar shape. It is evident that these cells have their own built-in hydrostatic skeleton.

The stomach varies in shape from species to species, but can usually be divided into a cardiac region with a sac-like caecum and a pyloric region leading to a short rectum and anus (Fig. 9.8(d)). In *Zoobotryon*, a conspicuous muscular gizzard at the base of the cardiac region consists of two opposing areas where the epithelium is covered with horny denticles (Fig. 9.8(c)).

The structure of the gizzard suggests that it is used to crush hard items of food, like diatoms. However, a gizzard is not essential for feeding on diatoms, for the gut of *Flustrellidra hispida*, which lacks this organ, is sometimes packed with diatoms. Markham and Ryland (1987) found that the frustrules of diatoms were broken up by the gizzard in species like *Amathia* and *Bowerbankia*, whereas in *Flustrellidra* the frustrules were separated intact, presumably releasing the cytoplasm for digestion. It is possible that some species of diatom are a poorer food source than naked flagellates, in which case bryozoans like *Flustrellidra* with 28–30 tentacles may be better equipped to feed on flagellates than *Bowerbankia* with only 8–10.

Enzymes secreted by the epithelium of the cardiac stomach carry out extracellular digestion of food, and particles passed by cilia into the caecum

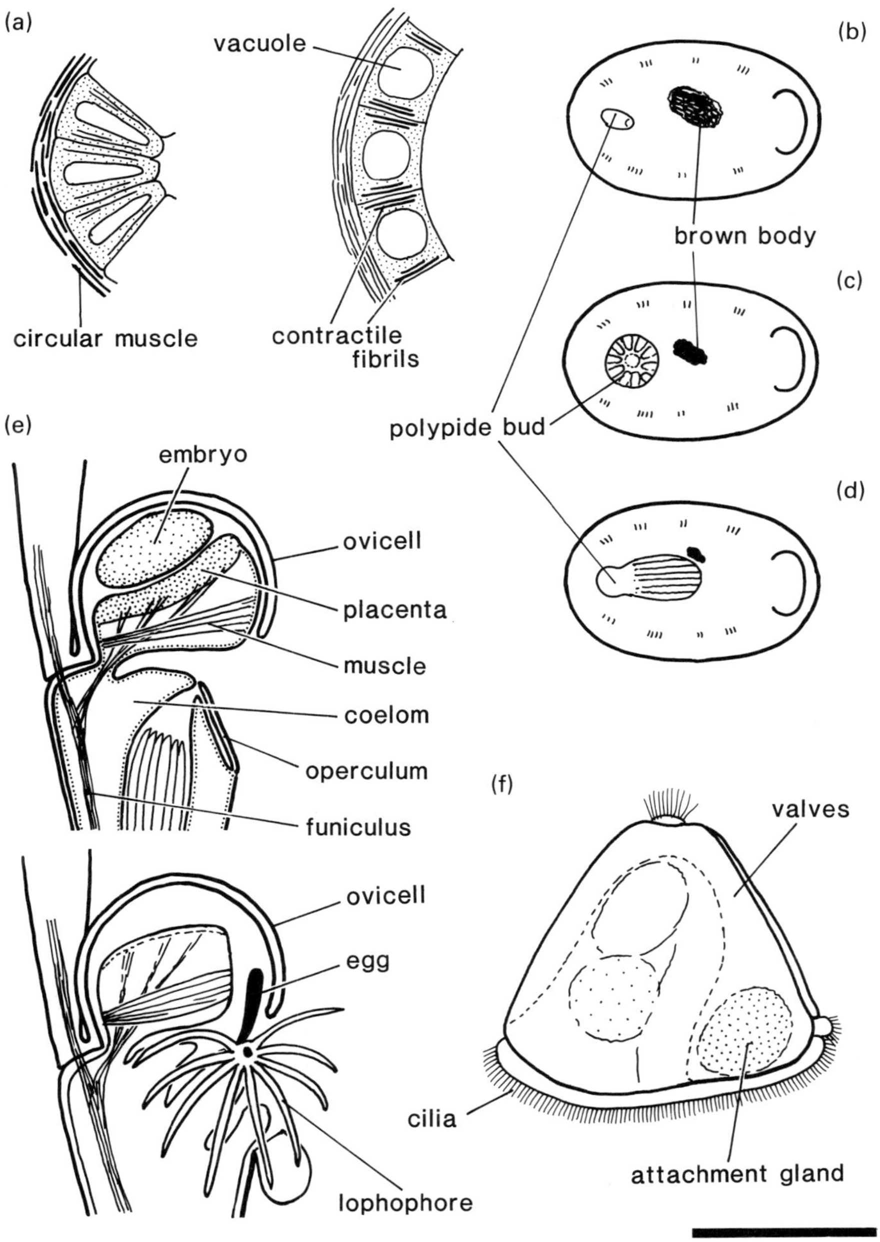

Fig. 9.7 (a) Dilation of the pharynx (see text). (b) to (d) Regeneration of polypide in *Beania*. (e) Diagram to show how egg is placed in ovicell (lower figure), and relationship between developing embryo and placenta (upper figure), in *Bugula*. (f) Cyphonautes larva. Scale bar: (f) = 0.5 mm.

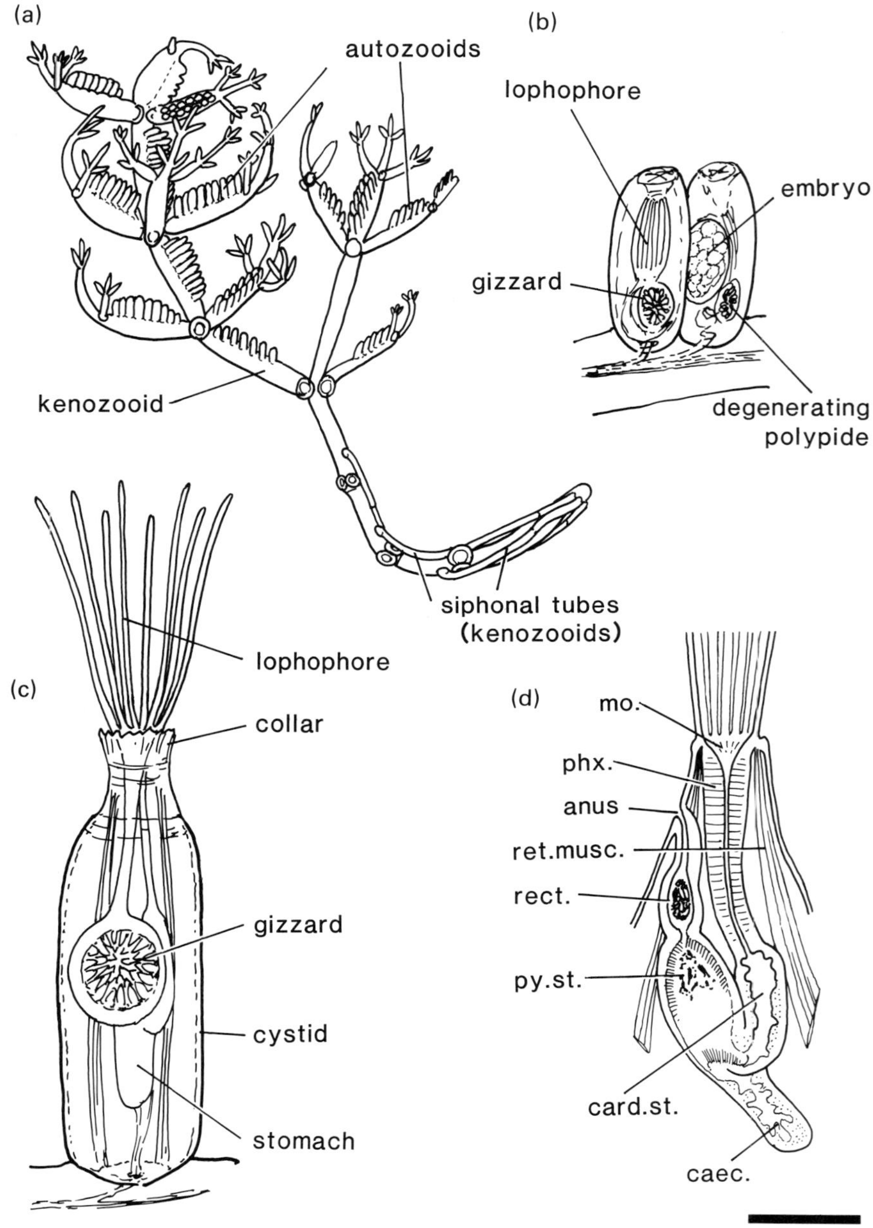

Fig. 9.8 Zoobotryon sp. (a) Portion of colony, showing siphons. (b) Autozooids contracted. (c) Autozooid with lophophore everted. (d) Optical section through polypide of *Flustrellidra*, showing structure of gut. Scale bar: (a) = 2 mm, (c) = 0.1 mm, (d) = 0.2 mm. Caec., caecum; card. st., cardiac stomach; mo., mouth; phx., pharynx; py. st., pyloric stomach; rect., rectum; ret. musc., retractor muscle.

are phagocytosed by epithelial cells and digested intracellularly. Undigested matter is collected as a bolus by cilia in the pyloric stomach and passed as a revolving mass to the rectum, where further absorption of the products of digestion takes place. The digestive process does not take long, judging from the time taken for food to pass through the gut. Residence times are influenced by temperature, but are in the order of 30–60 min (Menon, 1974).

Much of the undigested material left over from intracellular digestion accumulates in the cells of the stomach wall as brownish granules. These give the stomach of older polypides a brown colour. However, waste material cannot accumulate indefinitely, and periodically the polypide degenerates (Fig. 9.8(b)). The lophophore, pharynx and rectum are absorbed by phagocytes, which presumably recycle nutrients to the cystid. The stomach regresses to a dark brown mass which remains in the coelom of the cystid as the brown body. Brown bodies may accumulate in the cystid from previous cycles of regression and regeneration, or they may be incorporated into the gut of the regenerating polypide and eventually expelled to the exterior. This cycle of regression and regeneration must be regarded as a normal part of the process of digestion and excretion. It is not a form of asexual reproduction, for the original cystid continues intact. If a colony of *Beania magellanica* is examined it will be found that many zooids contain polypides, but others appear empty or have polypides in various stages of regression and regeneration. From these it is possible to follow the regeneration sequence (Fig. 9.7(b) to (d)). The first rudiment appears as a tubular ingrowth of the body wall in the proximal region of the frontal membrane. This bud gradually enlarges and differentiates tentacles, a U-shaped gut and retractor muscles, to become a functional trophic unit. The polypide is derived entirely from ectoderm and mesenchyme.

Bryozoans do not possess a circulatory system, but movement of the coelomic fluid occurs every time the lophophore is extended or retracted, and this may be adequate to distribute nutrients and metabolites to all parts of the body. Amoebocytes present in the coelomic fluid may transport nutrients as well as play a phagocytic role. Oxygen for respiration will reach the tissues by diffusion across the tentacle sheath when the lophophore is everted and across the thin frontal membrane (or ascus in ascophorans) when the tentacles are withdrawn. Although the lophophore offers a large surface area, it is likely that most of the oxygen diffusing into the tentacles will be used by the metabolic activity of the cilia, therefore it is unlikely that the lophophore is a respiratory surface for the rest of the zooid. In *Watersipora arcuata*, the ascus will not function as a respiratory surface when the polypide is withdrawn, but diffusion can take place through the outer membrane into the hypostegal coelom.

Eversion of the lophophore in *Alcyonidium, Flustrellidra* and anascan cheilostomes like *Beania* is caused by a rise in pressure when the parieto-frontal muscles depress the frontal membrane (Fig. 9.9(a) and (b)). In

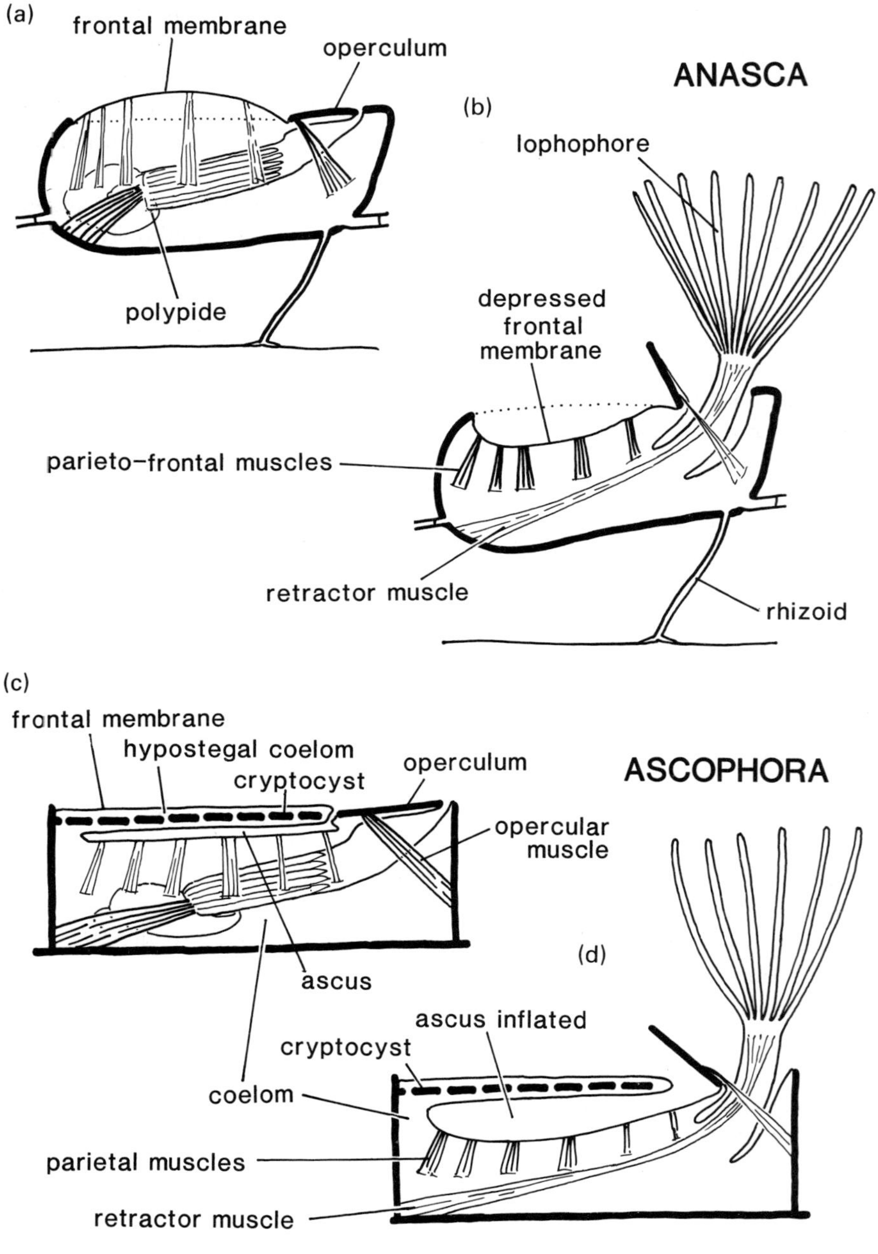

Fig. 9.9 Method of lophophore eversion in the Anasca and Ascophora. (a) and (b) *Beania*, Anasca. (c) and (d) *Watersipora*, Ascophora.

ascophoran cheilostomes like *Watersipora arcuata*, contraction of the parietal muscles inflates the ascus, and the increased pressure forces the lophophore out (Fig. 9.9(c) and (d)). In cyclostomes like *Crisia*, contraction of protractor muscles compresses the exosaccal coelom, which in turn compresses the perivisceral coelom and causes eversion of the lophophore (Fig. 9.6(d) to (f)).

Withdrawal of the tentacles is extremely rapid and is brought about by the retractor muscle fibres which are attached to the base of the lophophore and are under the control of the neural ganglion. Thorpe *et al.* (1975) found that it only took 60–80 ms for retraction of *Membranipora membranacea* and *Electra pilosa*. They suggested that the muscles are innervated by a giant axon from the ganglion.

When the frontal membrane is touched, adjacent polypides retract simultaneously. Lutaud (1969) found that a nerve runs round the walls of each cystid and connects with corresponding nerves in neighbouring zooids through the pore plates. Thus the nervous systems of the zooids of a colony are united, and nervous impulses can pass from zooid to zooid. Thorpe *et al.* (1975) found that stimulation of the frontal membrane of one individual resulted in a train of impulses passing across the colony through the nerves at *c.* 1 m s^{-1}. Initially this caused the almost simultaneous withdrawal of all expanded zooids, but the train of impulses may persist for several seconds thereafter. Instant co-ordinated contraction of the zooids in a colony would protect them from sudden wave impact and attack by predators.

9.4 REPRODUCTIVE STRATEGIES

Budding occurs at the periphery or terminal edge of a colony. In this region, zooids develop one or more bulbous outgrowths on their distal or terminal end which will grow into new zooids. A partition forms between the old and new zooid, but organic connection is maintained by the funicular cords and nerves which pass through pore plates in the partition.

The pattern of budding varies considerably from genus to genus. Buds may develop into feeding zooids (autozooids), kenozooids (such as spines and rhizoids), avicularia, vibracula or ovicells. In *Bicellariella ciliata* the terminal bud is divided into two by a longitudinal partition. Only the bud on the outside develops into an autozooid; the other remains as a kenozooid which develops a new terminal bud. The autozooid bud itself produces a number of buds on its frontal surface that become spines and the avicularium (Fig. 9.4(a)). Growth of *Beania magellanica* is more complicated. Eight buds develop in each peripheral zooid. The terminal or axial bud is the first to form. It will eventually give rise to a new autozooid, but before a polypide develops, two lateral outgrowths appear, each of which joins with a similar outgrowth from an older, neighbouring bud to form a bridge with a pore plate. Next a dorsal bud gives rise to the rhizoid that will anchor the new zooid to the substratum. The main bud has now grown to its full size and has developed five more buds: a new axial bud, two at 30° to the axis that will form the avicularia, and two at 60° to the axis which form bridges with younger adjacent axial buds. At the edge of the mat, one of the buds at 60° may assume axial status and give rise to a new autozooid instead of a bridge. In Fig. 9.3(d), the zooid marked X produces two axial buds that will develop into autozooids. This results in the mat growing in a new direction at 60° to the old axis, which still persists (see triplet of zooids at lower right of fig.).

It is evident that not all zooids are independent or equivalent. Only the autozooids can obtain and digest food; all the others derive their nourishment from the autozooids. Similarly only autozooids are capable of sexual reproduction, although not all may do so. In the case of *Crisia*, only one zooid in a section develops as a brood chamber (Fig. 9.5(a)).

It is perhaps unfortunate that the term zooid has been used to describe the cystid and its polypide, for it implies individuality. Bud formation becomes asexual reproduction, and the assemblage of zooids a colony. There is, however, another way to look at bryozoans. Budding is a process that leads to the replication of functional parts or units (i.e. multiplication of feeding, attachment or defensive units), and specific assemblages of different units form modules. For example, in *Beania magellanica* each module comprises four units – two avicularia, a rhizoid and a feeding autozooid – whilst in *Bicellariella ciliata* the module consists of one avicularium, six spinous kenozooids, a feeding autozooid and for some modules, an ovicell. The whole structure consist of many hundred modules, all derived from a single founding zooid, and must be considered as a single organism. The term modular organism is more appropriate than 'colony'.

The longevity of such modular organisms varies. *Flustrellidra hispida* is perennial, but *Bicellariella ciliata* produces two generations each year. Organisms that develop from larvae produced in late summer reproduce in the following spring and die. Their larvae give rise to the summer generation (Eggleston, 1972).

The dispersive phase is always produced by sexual reproduction. Bryozoans are hermaphrodite, but sperm appears to be produced before ova and this would prevent self-fertilization. In *Electra pilosa* and some other species, spermatozoa migrate through the coelom of the two dorsal tentacles and escape from a terminal pore (Silen, 1972). Ova are often large and yolk-laden. They are passed from the coelom to the outside through a coelomopore or inter-tentacular organ (Fig. 9.6(c)). Among littoral bryozoans, there is a marked tendency towards reduction of the number of eggs produced and an increase in their size. They are retained in the parent cystid or passed into a specialized zooid, called an ovicell, which is modified for the purpose of incubation. The consequences of such tendencies are that the planktonic phase is greatly shortened and larvae do not feed. By means of their cilia, newly released larvae of *Beania magellanica* glide for a short while over the surface of the substratum. This searching phase may last for only a matter of hours before they settle, thus it is unlikely that they will travel very far from their parents. For species living in a very narrow band parallel to the shore, wide dispersal in the plankton may be disadvantageous, and settlement close to the point of release a better strategy.

Membranipora membranacea produces numerous eggs which are shed into the sea, where they develop into a planktonic cyphonautes larva protected by two chitinous valves (Fig. 9.7(f)). This larva swims by means of cilia and has a gut, enabling it to feed on planktonic particles. In *Alcyonidium hirsutum* and *Flustrellidra hispida*, several eggs are retained in the cystid after the polypide has regressed, and are incubated there. They hatch out as flattened, round, ciliated larvae similar to those of *Beania magellanica*. A single egg is incubated in *Zoobotryon* after regression of the polypide (Fig. 9.8(b)). Retention of eggs by intertidal species is a protective measure and will enhance the probability of successfully completing larval life.

In *Bugula neritina* (Fig. 9.4(c), 9.7(e)), a single egg is extruded into a highly modified zooid, the ovicell, where it is nourished by the parent autozooid through a placenta (Woollacott and Zimmer, 1975). Nutrients could reach the placenta dissolved in the coelomic fluid, carried by coelomocytes or along the funicular cords which pass forward to the placenta (Fig. 9.7(e)). The embryo develops into a ciliated larva which is released from the ovicell by contraction of the muscles shown in Fig. 9.7(e), (upper figure).

Both the cyphonautes and the ciliated larva possess a large adhesive gland, which everts after settlement and secretes an adhesive which sticks the larva to the substratum prior to metamorphosis, in which there is a radical reorganization of the body to form the first zooid, the oozooid or ancestrula.

SUMMARY

Bryozoans are modular organisms; their constituent zooids are minute and show various degrees of inter-dependence. Their lack of appropriate physiological adaptations and small size prevent them from living on the middle shore, but they are well suited to life on the lower shore. Few live where they will be exposed at low tide. Encrusting forms have a low profile and may lie within the boundary layer where velocity of flow is minimal. Plantlike forms growing into the water column have a firm attachment to the substratum.

Zooids are protected by a tough cuticular covering which may be strengthened by calcification. When disturbed or taken out of the water, the feeding tentacles are withdrawn and the opening closed by the operculum. Feeding is by ciliary action, small particles down to 2 μm are taken, and there is evidence of selection. Mucus is not involved. Feeding may only be interrupted for relatively short periods during turbulence at low tide. There are no reproductive specializations specifically for shore life.

10

Sea anemones and related animals (*Cnidaria: Anthozoa*)

Despite their soft body and absence of a hard protective shell, several species of sea anemone are conspicuous members of the intertidal community. This they achieve by a number of behavioural and physiological adaptations. For example, the dark red waratah anemone, *Actinia tenebrosa* (named after the floral emblem of New South Wales), avoids the stresses of high temperature and desiccation by living in dark places where it is shielded from direct sunlight and where humidity remains high. A large volume of seawater is retained in the body and may assist in respiration, but certainly guards against water loss from the tissues. In the laboratory, this sea anemone can be kept out of water in a humid atmosphere for more than 3 weeks without suffering harm.

Most other members of the Anthozoa* (hard and soft corals, sea pens, gorgonians) live in deeper waters; they will not be considered here, except for an example of a soft coral and a cup coral that are found in the sublittoral of temperate shores.

10.1 GENERAL STRUCTURE OF SEA ANEMONES

The appearance of typical sea anemones in different states of expansion is shown in Fig. 10.1(a) to (f). Their internal structure is more easily seen in transparent juvenile specimens which have not developed large numbers of internal partitions and is represented in Fig. 10.1(g).

The body takes the form of a cylindrical polyp attached to the substratum by its base and, at the opposite end, bearing numerous hollow tentacles

* See Appendix for classification of sea anemones.

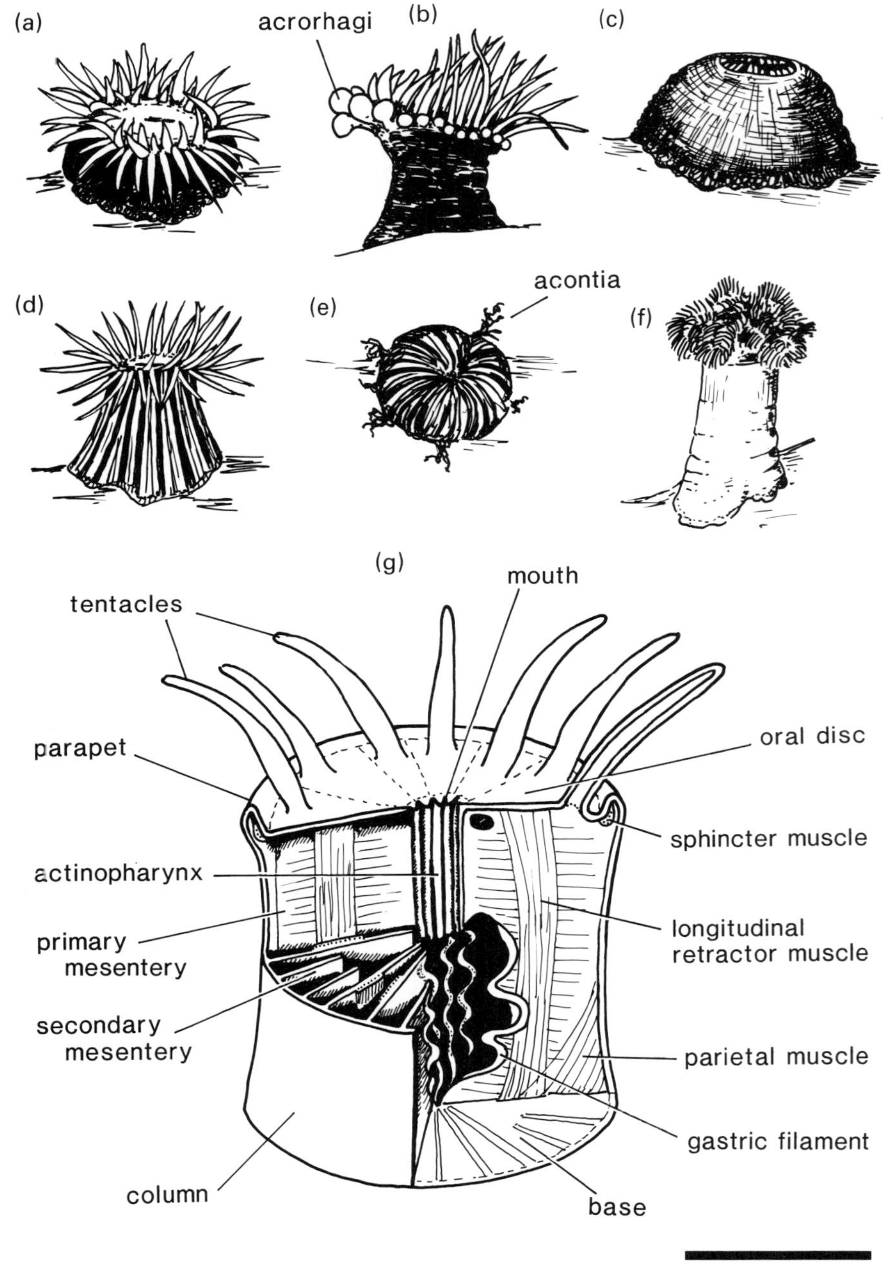

Fig. 10.1 Sea anemones. (a) to (c) *Actinia tenebrosa*. (d) and (e) *Anthothoë albocincta*. (f) *Metridium senile*. (g) Internal structure of a sea anemone. Scale bar: (a) to (c) = 35 mm, (d) = 12 mm, (f) = 80 mm.

arranged in concentric circles round a central opening that serves as a mouth and anus. The column of the anemone has a very large body cavity, the coelenteron, which is divided into many sectors by radial mesenteries. It is filled with seawater and acts as a hydrostatic skeleton during movement and as a digestive cavity. The mouth does not open directly into the coelenteron, but is connected with it by a tubular gullet (or actinopharynx), which possesses two flagellated grooves, the siphonoglyphs, that pump seawater into the body cavity (Fig. 10.2(c)).

The body wall and tentacles are relatively simple in structure. They consist of an outer ectoderm of epithelial and mucous cells, an inner endoderm of epithelio-muscular cells, some of which play a digestive role, and a middle fibrous layer, the mesogloea. The mesenteries arise as folds of the endoderm supported by mesogloea (Fig. 10.2(b)). Nematocysts (or stinging cells) occur in the ectoderm and certain parts of the endoderm; they are a characteristic feature of cnidarians and are disussed in Section 10.3. Contractile cells are present in both the ectoderm and endoderm of the tentacles, but in the column, muscles are confined to the endoderm. Of particular importance are the longitudinal retractor muscles on the mesenteries, which draw in the oral disc and shorten the column, and the sphincter muscle, which closes the top of the column over the tentacles to protect them (Fig. 10.1(g)).

The internal structure of an adult sea anemone is difficult to interpret unless a well-expanded specimen is examined, because numerous mesenteries pack the coelenteron. A section cut at the level of the actinopharynx (Fig. 10.2(c)) reveals that six pairs of mesenteries join this organ; these are called the primary mesenteries. Mesenteries occur in pairs. In the lower part of the column many more mesenteries of the second, third and higher orders may be present (Fig. 10.2(d)). The longitudinal retractor muscles run down one face of each primary and secondary mesentery. The cavity bounded by two mesenteries whose muscles face each other is called an endocoel, and the cavity between adjacent mesentery pairs is called an exocoel (Fig. 10.2(c)). Tentacles arise in the endocoelic position, one to each mesentery pair in *Actinia*, but in the jewel anemone, *Corynactis*, a single tentacle is found in each exocoelic position in addition to the normal endocoelic series (Fig. 10.3(c)). The mesenteries associated with the siphonoglyphs are called the directives; they establish a plane of bilateral symmetry in the body.

The free edge of each mesentery below the actinopharynx is highly convoluted and constitutes the gastric filament (Figs 10.1(g), 10.2(b), 10.3(e)). This has a central glandular region containing mucous cells, enzyme cells and nematocysts (called the cnido-glandular tract), and on either side there is a broad band of flagellated cells which move particles towards intermediate tracts of digestive or phagocytic cells, where most of the food is absorbed (Van-Praët, 1985). In many sea anemones the gastric filaments also contain cells with symbiotic algae. In very young, transparent sea anemones it

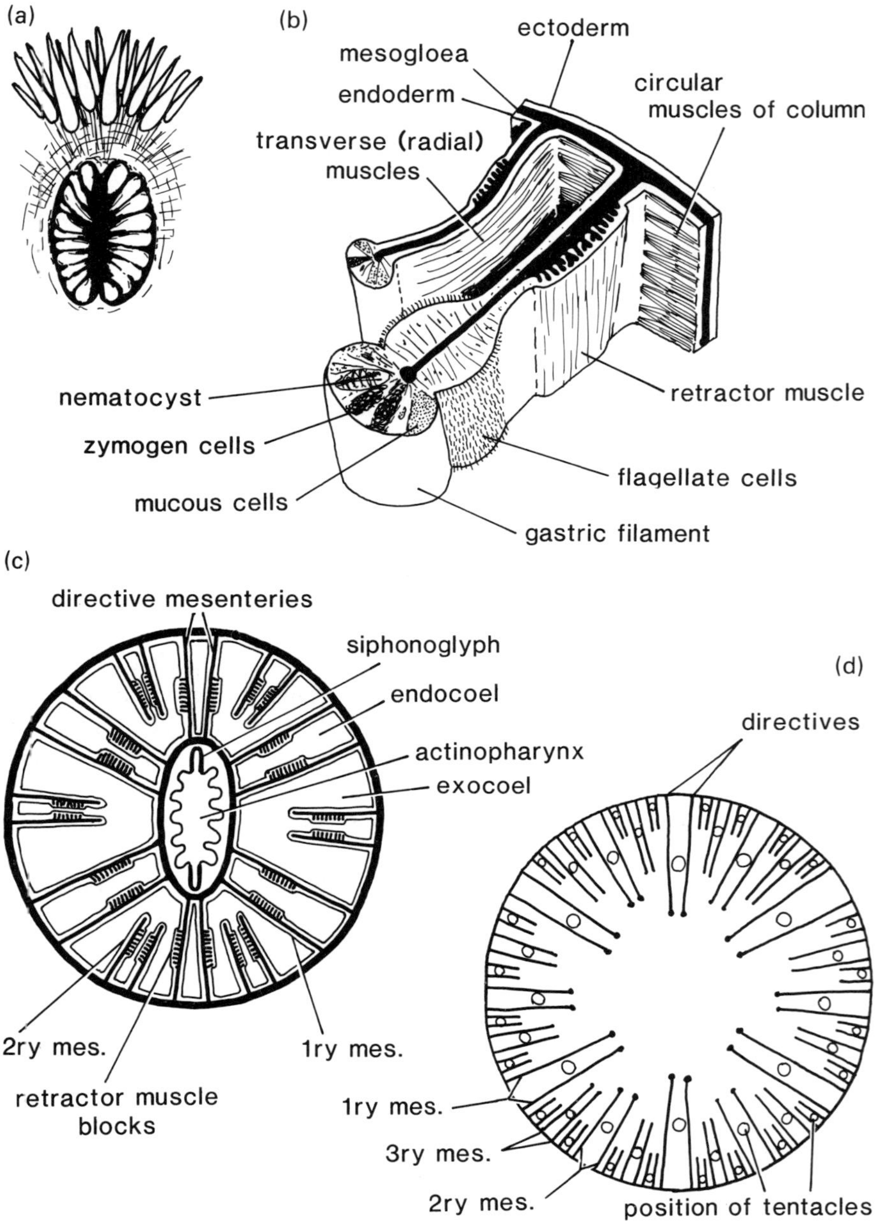

Fig. 10.2 Structure of sea anemones. (a) Partly open mouth, showing wall of actinopharynx. (b) Structure of mesentery (diagrammatic). (c) Arrangement of primary mesenteries in region of the pharynx. Directive mesenteries are primary mesenteries lying in the plane of bilateral symmetry. (d) Arrangement of mesenteries lower in column and position of tentacles. 1ry mes., primary mesenteries; 2ry mes., secondary mesenteries; 3ry mes., tertiary mesenteries.

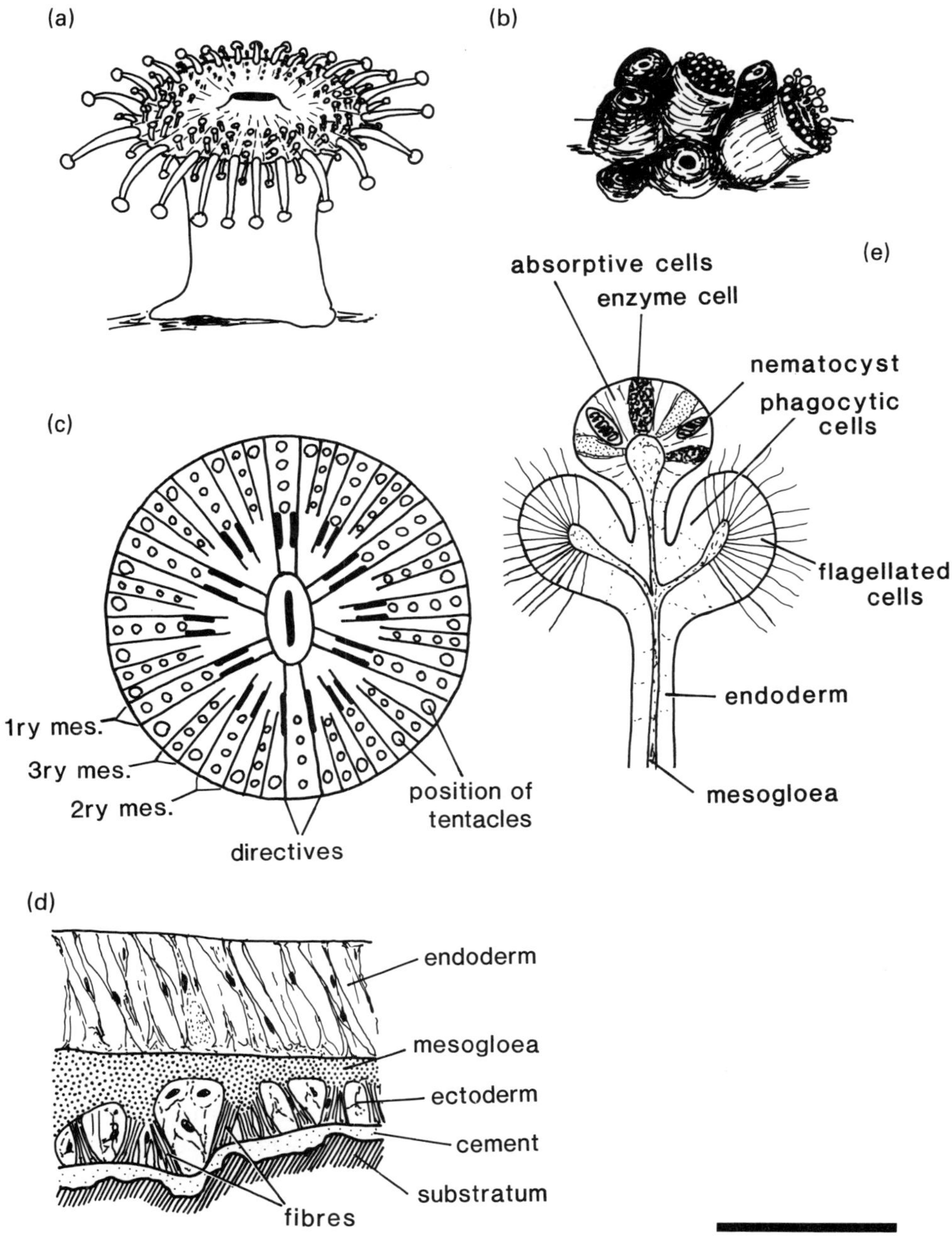

Fig. 10.3 Corynactis australis. (a) Expanded specimen, × 2. (b) Partly contracted specimens. (c) Arrangement of mesenteries and position of tentacles. (d) Histological section through base, showing attachment fibres. (e) Transverse section through upper part of gastric filament. Scale bar: (a) = 6 mm, (d) = 0.02 mm.

is possible to see that the water in the coelenteron is kept in constant circulation by the flagellated endodermal cells. At certain times of the year, gonads develop on the secondary and tertiary mesenteries of adults.

10.2 REPRESENTATIVE SEA ANEMONES

Among the numerous species of sea anemone found in the intertidal zone of rocky shores, the following species feature prominently in experimental studies on behaviour and physiology.

On British shores the beadlet anemone, *Actinia equina*, is one of the commonest and is found up to mean high water neap tide levels (MHWN). It is usually reddish brown, but may be green or yellowish brown. The closely related crimson waratah anemone, *Actinia tenebrosa*, of Australia and New Zealand is more constant in colour and occupies a similar position on the shore. Both species have below the tentacles a ring of pale blue swellings called acrorhagi (Fig. 10.1(b)). The North American Pacific Coast anemone, *Anthopleura elegantissima*, frequently forms clusters of genetically identical individuals (a clone). Most specimens of this species harbour symbiotic zooxanthellae in their tissues (see page 220), particularly the gastric filaments. The large parasitic anemone, *Calliactis parasitica*, which lives commensally with a hermit crab, *Eupagurus bernhardus*, and the dahlia anemone, *Urticina (= Tealia) felina*, have both been used extensively in neurophysiological studies. They are sublittoral species but are sometimes encountered on the lower shore. The plumose anemone, *Metridium senile* (Fig. 10.1(f)), may be found under rocks about low tide level on many British shores, although large specimens are sublittoral. It has a large number of fine, feather-like tentacles and varies in colour from white to pale pink, orange or brown. Many classical studies on the anatomy and physiology of sea anemones have been carried out on this species.

Another important group of sea anemones, represented by *Sagartia elegans* in Britain, and the green and white striped *Anthothoë albocincta* from Australia (Fig. 10.1(d)), are interesting because of their mechanism of defence against predators. When vigorously prodded with a glass rod, these anemones contract maximally and eject writhing white threads, the acontia, from the mouth and slits in the walls of the body. The acontia are extensions of the gastric filaments and are heavily charged with nematocysts.

The common jewel anemone, *Corynactis viridis*, of British shores is sufficiently different in structure from true (actiniarian) sea anemones to be placed in a separate order, the Corallimorpharia. The related Australian *Corynactis australis* (Fig. 10.3(a) and (b)) may smother exposed vertical rock faces with a carpet of bright orange polyps, 5–10 mm tall, that have been derived by asexual budding from the basal disc to form a clone. They are more firmly attached to the rocks than other anemones, and can withstand

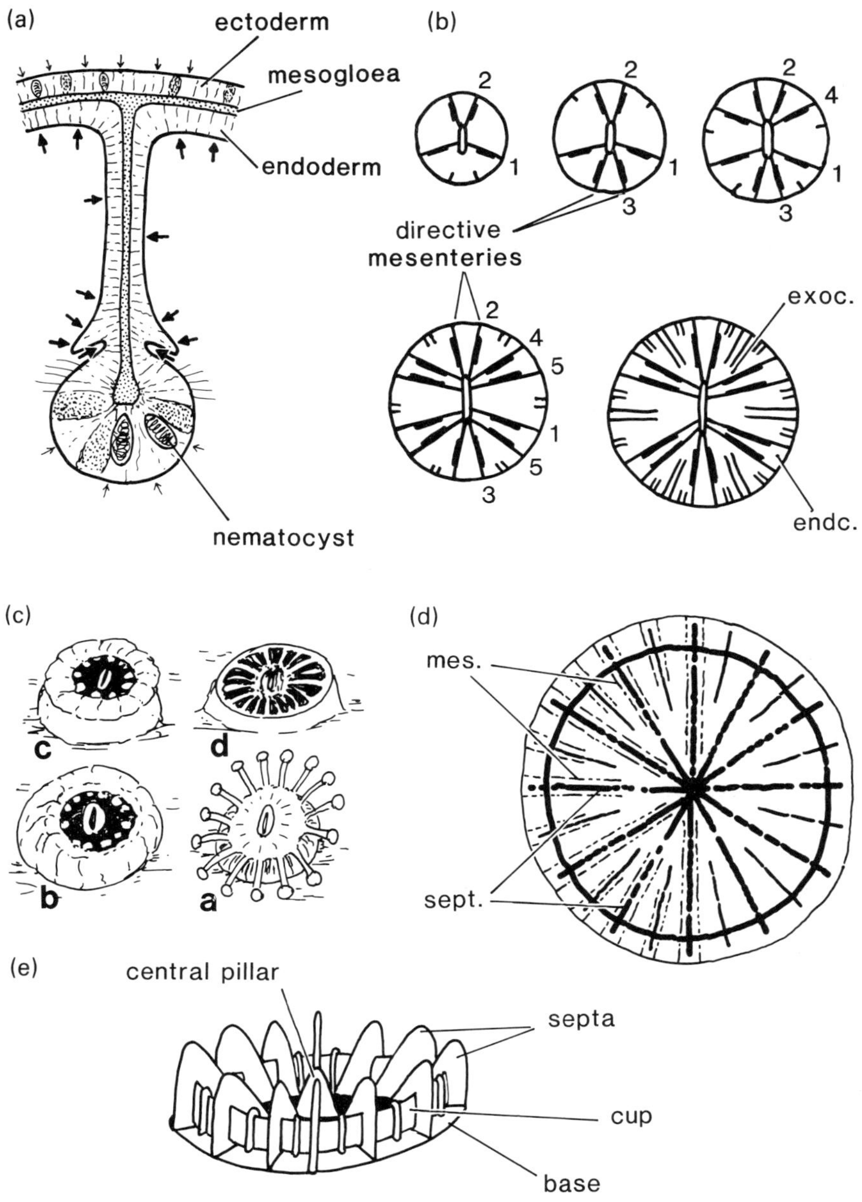

Fig. 10.4 (a) Section through body wall and mesentery to show region of absorption. Fine arrows, amino acids; bold arrows, macromolecules and particulate food. (b) Development of mesenteries. (c) *Culicia quinaria*, a cup coral: a, polyp expanded; d, polyp contracted. (d) Arrangement of calcareous septa relative to mesenteries. (e) Cup with polyp removed. Scale bar: (c) = 4 mm. endc., endocoel; exoc., exocoel; mes., mesentery; sept., septum.

the full force of wave action. The tentacles are arranged differently on the oral disc to actiniarian anemones (Fig. 10.3(c)) and terminate in a conspicuous white knob, the acrosphere, which is laden with spirocysts (see page 192) and large nematocysts.

Asexual reproduction occurs in many actiniarian sea anemones. *Anthopleura elegantissima*, for example, forms tightly packed clones of individuals that have been derived by vertical fission of the whole anemone, including the oral disc. Both the waratah and beadlet anemones brood their young in the coelenteron. It is not an uncommon experience, when prising an anemone from the rock, to observe juveniles only a few millimetres in diameter being ejected with water from the body cavity. This would not be unusual if it were not for the fact that male and non-sexual anemones have been found brooding young. In any one anemone all the young are of the same colour and genotype (as determined by electrophoretic studies of their enzymes), and this would not be the case had they been produced by cross-fertilization. Self-fertilization must be ruled out as they are not hermaphrodite (Ottway and Kirby, 1975; Gashout and Ormond, 1979; Quicke and Brace, 1983).

The origin of the brooded young is uncertain, but the studies by Black and Johnson (1979) on *Actinia tenebrosa* show that it must be the result of asexual reproduction. They found no genetic segregation among young of heterozygous adults and, although allelic frequencies and genotype varied between discrete populations, very little genetic variation occurred within a population. Apomictic parthenogenesis (the development of unfertilized eggs), as reported for the sea anemone *Cereus pedunculatus* living in the tidal zone (Chia, 1976), cannot explain brooding males or non-sexual individuals. Possibly a form of somatic embryogenesis, as suggested by Gashout and Ormond (1979), may account for this. Carter and Thorp (1979) found evidence of asexual reproduction of very young anemones brooded by *Actinia equina*, and suggested that a rare zygote that was histocompatible with the adult might divide within the parental coelenteron to form a clone of genetically identical individuals.

Reef-building corals (order Scleractinia) occurring on tropical and subtropical shores will not be dealt with here, but a species of solitary cup

---

Fig. 10.5 Clavularia novaezelandiae, a soft coral. (a) Natural appearance. (b) Expansion of polyp. (c) Section, showing internal structure. (d) Sclerites. (e) *Clavularia crassa*, showing branching growth. (f) and (g) Transverse sections of (c) at X and Y to show mesenteries and actinopharynx. (h) Retraction of polyp of (e). Scale bar: (b) and (e) = 5 mm, (d) = 0.2 mm. actph., actinopharynx; coel., coelenteron; cut., cuticle; mes., mesentery; spic., spicule.

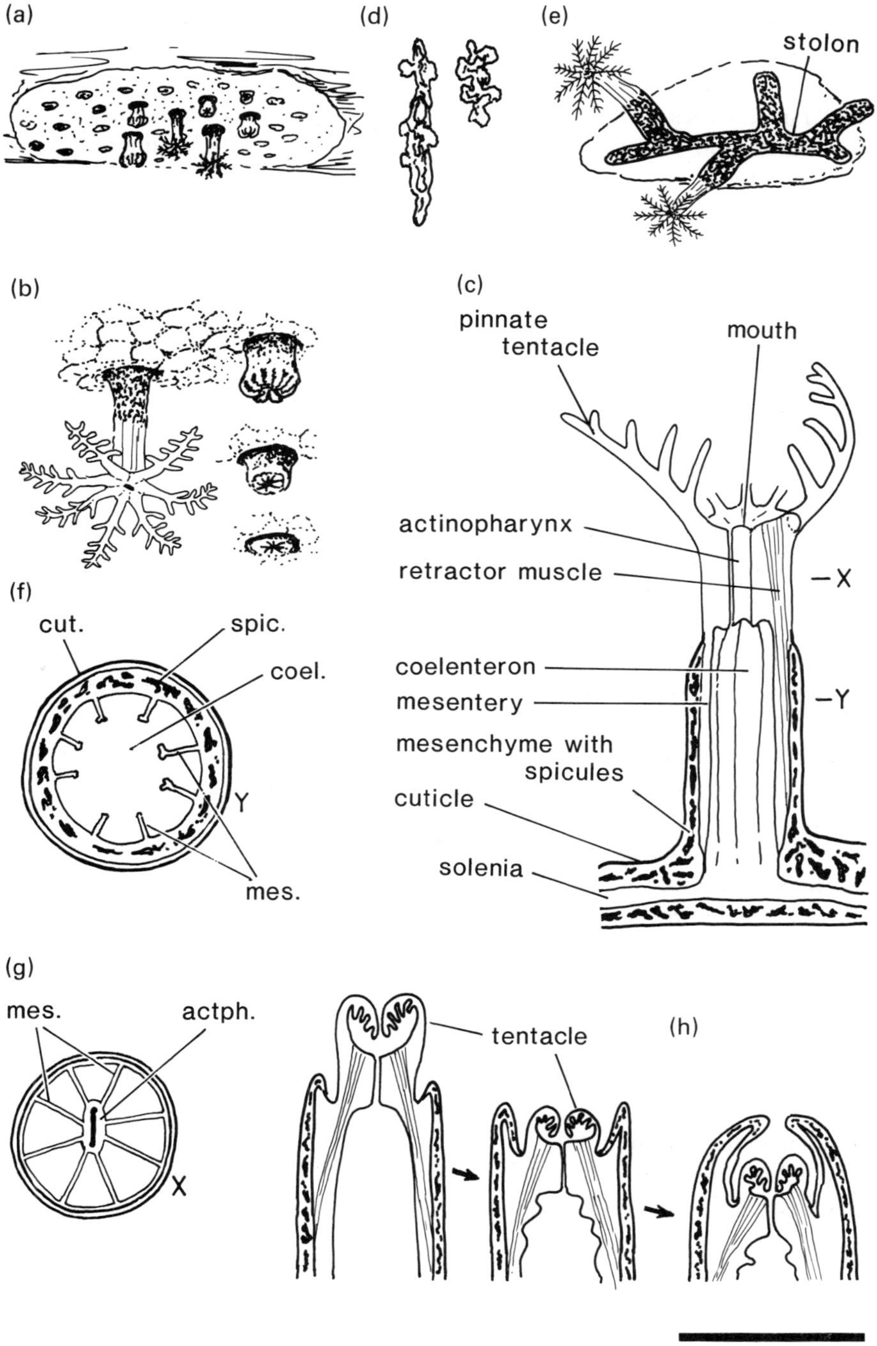
(a)
(d)
(e)
stolon
(b)
(c)
pinnate tentacle
mouth
actinopharynx
retractor muscle
–X
coelenteron
mesentery
–Y
mesenchyme with spicules
cuticle
solenia
(f)
cut.
spic.
coel.
Y
mes.
(g)
mes.
actph.
X
tentacle
(h)

coral found at low tide level and in the sublittoral of temperate Australian shores will serve to illustrate the structure of a hard coral.

Culicia quinaria (Fig. 10.4(c)) is a small orange cup coral common on the underside of rocks. The polyp is very similar to *Corynactis australis*, but the basal disc deposits calcium carbonate on the rock along radial lines between the mesenteries. As deposition proceeds, the lines build up into vertical plates, the septa, which form a cup or corallite. The first septa appear in the endocoels of the primary and secondary mesenteries, but later another series forms in the exocoels and a circular wall is laid down at the periphery, joining all the septa to form the corallite (Fig. 10.4(d) and (e)). The corallite is external to the polyp and is exposed when the polyp is fully retracted.

The perimeter of the basal disc spreads out beyond the corallite as a sheet with ectodermal and endodermal layers, called the coenosarc. More polyps develop from this sheet, each producing its own cup, but remaining in organic connection with its neighbours through the common coenosarc.

Soft corals and related members of the Octocorallia usually live in deeper water, but some members of the order Stolonifera may be found at low tide on the underside of boulders. Two examples from Australia and New Zealand, *Clavularia novaezelandiae* and another species believed to be *C. crassa*, are shown in Fig. 10.5. *Clavularia novaezelandiae* forms purple-brown encrustations *c.* 1 mm thick and up to 15 cm in diameter, with numerous polyps that can be retracted completely into the encrusting mass (Fig. 10.5(a) and (b)). The surface feels hard owing to numerous calcareous sclerites embedded in the mesogloea (Fig. 10.5(d)). The brown colour is the result of vast numbers of symbiotic zooxanthellae in the endoderm of the tentacles and, to a lesser extent, the network of endodermal tubes called solenia, which ramify throughout the thick mesogloea. The other species consists of several small transparent polyps (*c.* 1 mm high) bearing pinnate tentacles and joined to one another by a branched stolon, which is protected externally by a cuticle (Fig. 10.5(e)). The actinopharynx, eight mesenteries and internal circulation of water can easily be seen through the transparent wall (Fig. 10.5(c)). The polyp has the same three layers noted for sea anemones (ectoderm, mesogloea and endoderm), but in the stolon the mesogloea is very thick and contains numerous warty calcareous sclerites (Fig. 10.5(d)) as well as tubular branches of the coelenteron, the solenia. Small nematocysts are found on the tentacles, but symbiotic algae are absent. The delicate tentacles and column can be pulled down into the thick-walled basal part of the polyp for protection (Fig. 10.5(h)).

10.3 ANTHOZOAN CNIDAE

Three classes of cnidae or discharging capsules are found in the Anthozoa: spirocysts, ptychocysts and nematocysts (cnidocysts). All consist of a capsule

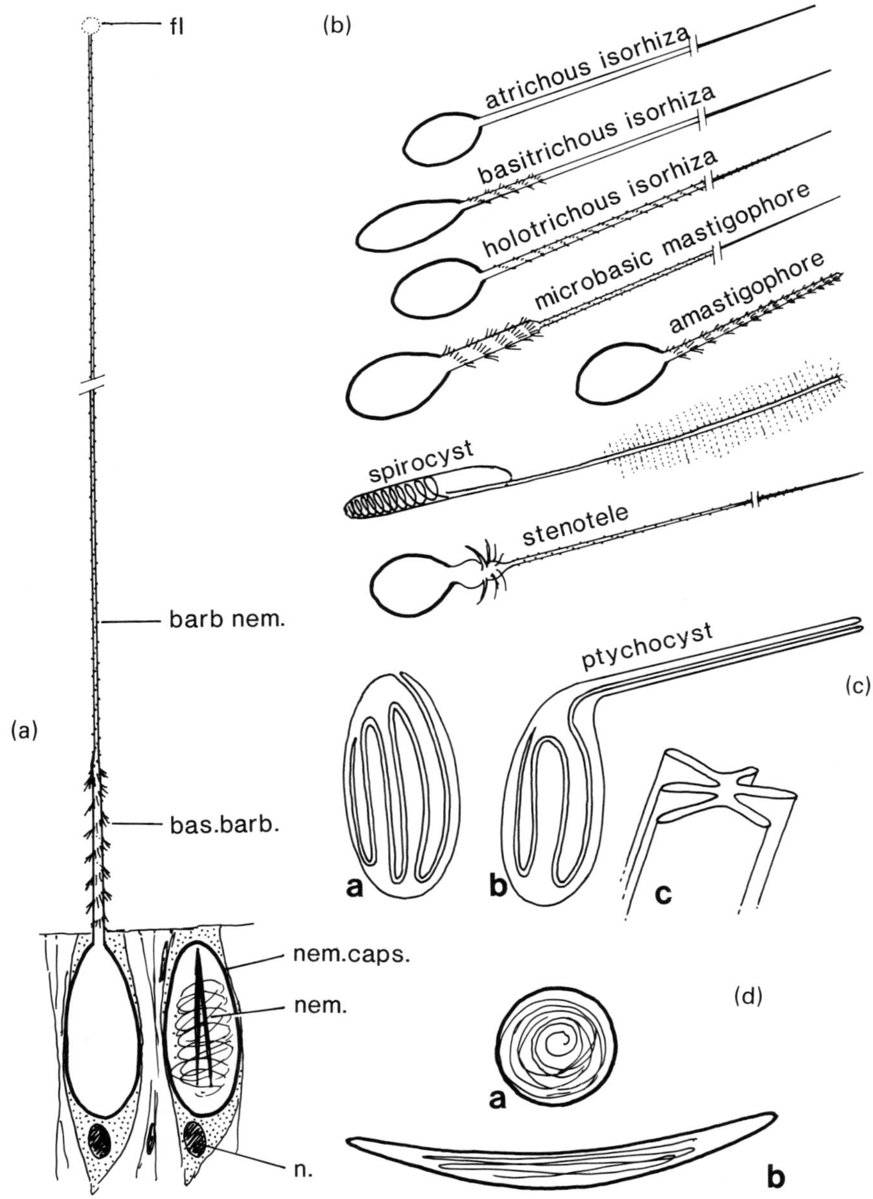

Fig. 10.6 Cnidae of Hydrozoa and sea anemones. (a) Discharged and undischarged nematocyst from *Actinia*. (b) Types of nematocysts and a spirocyst, discharged. (c) Ptychocyst from *Cerianthus*: a, undischarged; b, partly discharged; c, cross section of pleated tube. (d) Undischarged nematocysts: a, *Physalia*; b, *Aglaophenia*. fl., fluid issuing from tip of nematoneme; barb nem., barbs on nematoneme; bas. barb, basal barbs; n., nucleus of nematocyte; nem., undischarged nematoneme; nem. caps., capsule of nematocyst.

containing a hollow inverted tube which can be discharged to the exterior by a process of eversion.

Spirocysts (Figs 10.6(b), 10.7(a)) are by far the most common type found on the tentacles of sea anemones and scleractinian corals. When stained they look like a long, tightly coiled, cylindrical helix. The capsule is very thin and barely visible. It has been shown that on discharge the thread everts in much the same way as nematocysts (Mariscal and McLean, 1976). There are no barbs; their place is taken by microtubules, which swell to a tangled mass of sticky threads when they come in contact with water (Mariscal *et al.*, 1977b). Despite their abundance and common occurrence, little is known about their function. They cannot penetrate and do not appear to contain toxin; nevertheless, they seem to be connected with feeding, presumably trapping prey with their sticky surface. Eversion is extremely rapid (*c.* 10 ms) and spirocysts are discharged before any nematocysts (Picken and Skaer, 1966).

Ptychocysts (Fig. 10.6(c)) are found in the ectoderm of burrowing anemones belonging to the order Ceriantharia. The capsules contain a longitudinally pleated tube which is folded instead of spirally wound. It has no barbs or microtubules, but sand grains adhere to the pleated tube when everted. Vast numbers of ptychocysts are discharged from the ectoderm to form a felt-like lining to the tube in the sand where *Cerianthus* lives (Mariscal *et al.*, 1977a).

Nematocysts are present in nearly all members of the phylum Cnidaria and show wide diversity of form and structure. Typically each nematocyst is contained in a cell, the nematocyte (cnidocyte), which originally produced it, although in some instances several nematocysts may be deployed in a special battery cell.

A nematocyst consists of a thick-walled capsule containing a narrow coiled tube, the nematoneme (Fig. 10.7(c) and (d)). In the Hydrozoa some capsules have an operculum covering the point where the nematoneme will emerge, but this is absent in the Anthozoa. The nematoneme is everted (turned inside out) as the nematocyst discharges, and this causes the tube to rotate. Barbs on the inside of the tube perform a boring action as they flick over the lip of the rotating tube and come to lie on the outside (Fig. 10.7(b)): this gives the delicate nematoneme its remarkable power to penetrate the cuticle of crustaceans. Many nematocysts contain toxic substances in their intracapsular fluid, which exudes from the open tip of the nematoneme when it is fully discharged (Fig. 10.6(a)).

The nematocyst is a dead structure. It retains its capacity to discharge indefinitely, but once discharged it cannot be used again and is discarded. It is not surprising, therefore, to find that nematocysts are continually being produced to replace those that have been fired. However, when the number used in food capture is compared with the number present on the tentacles

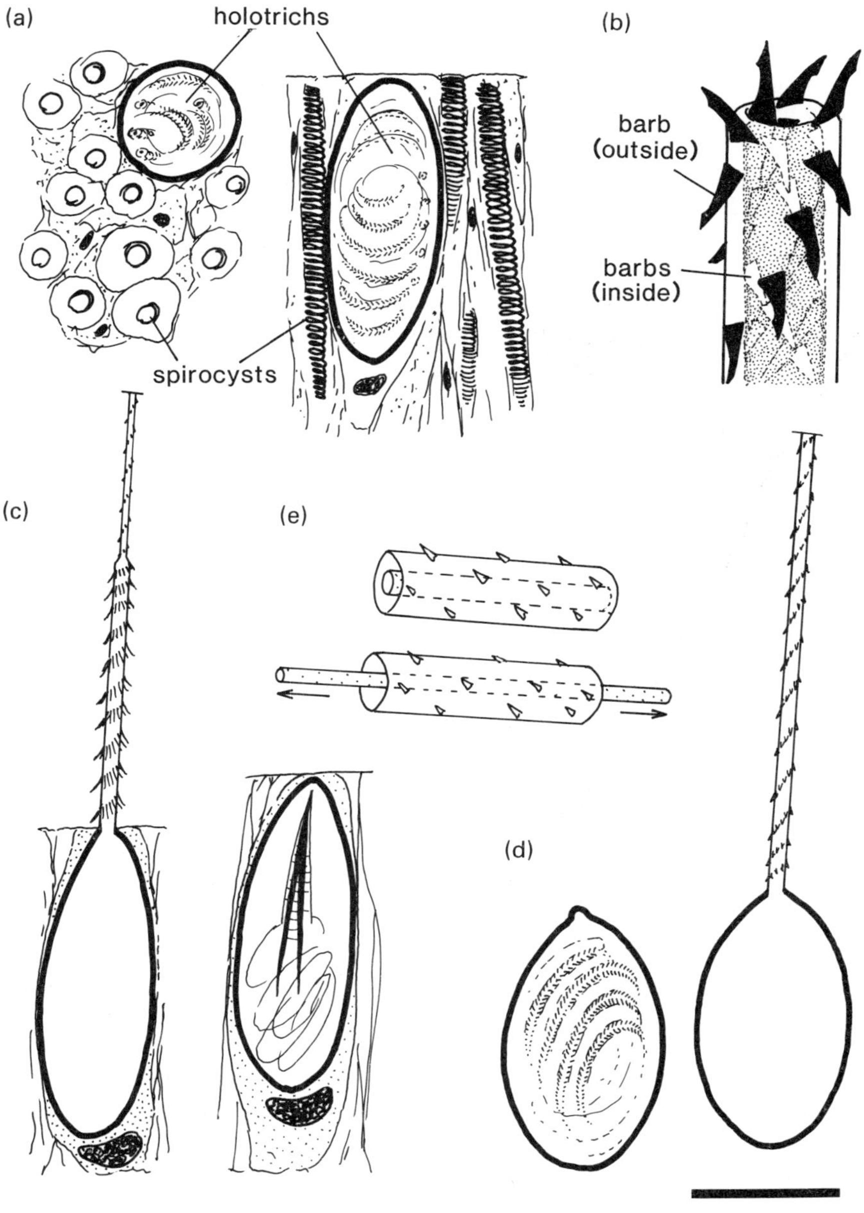

Fig. 10.7 (a) Spirocysts and holotrichous nematocyst in histological section. (b) Motion of barbs during eversion of the nematoneme. (c) p-Microbasic mastigophore discharged and undischarged. (d) Holotrichous isorhiza from *Corynactis*. (e) Robson's experiment (see text). Scale bar: (a) = 20 μm, (c) = 10 μm, (d) = 35 μm.

and the rate of addition to the stockpile, the reason for their continual production is not at all clear. Examination of brine shrimp nauplii caught by the hydroid *Thyroscyphus marginatus* (Fig. 11.5(c) and (d)) seldom shows more then a few detached nematocysts, sometimes only one or two, yet the tentacles of each polyp have *c*. 30 rings containing 30–40 nematocysts each, thus the tentacles have a total complement in the order of 25 000–30 000 per hydranth. Estimates for sea anemones are even more staggering. From sections of the jewelled anemone, *Corynactis australis* (8 mm high), it was estimated that each acrosphere contained over half a million nematocysts, and the stem of the tentacle an even higher number of spirocysts. Thus the armoury of nematocysts seems to be out of all proportion to the number used.

There are a number of reasons for this. Cnidarians do not seek food; they wait for prey to collide with their tentacles, so it could be argued that the greater the number of nematocysts, the more certain it is that food will be captured. Similarly in defence their deterrent effect is likely to be proportional to their abundance. There seems to be a wide variation in the threshold at which individual nematocysts discharge. Moreover, they may be specialized for different roles, thus requiring higher numbers than would be expected if they all performed the same function. Continued production is essential in species like the hydroid *Thyroscyphus marginatus*, where the entire cnidom of the hydranth is lost and has to be replaced at each hydranth cycle, or in the Hydrida, where continuous growth results in the loss of cells and unused nematocysts from the ends of the tentacles (Campbell, 1967).

Types of anthozoan nematocyst

There is a wide diversity of nematocyst structure, particularly in the Hydrozoa. An extensive survey by Weill (1934) distinguished 25 types on the structure of their discharged nematoneme; Mariscal (1974) has given a revised version of this classification.

The nematoneme of anthozoan nematocysts is open at the tip (stomocnidae), but in a number of hydrozoan nematocysts it is closed (astomocnidae). Further subdivision is based on the shape of the discharged nematoneme and the presence or absence of barbs. Heteronemes have a well-defined basal swelling or shaft; two common types are the microbasic mastigophores (with a short shaft and long nematoneme bearing barbs, of which two categories, p- and b-mastigophores are recognized) and amastigophores (with only a shaft covered in barbs) (Fig. 10.6(b)). Haplonemes do not have a basal shaft, and in isorhizas the nematoneme is about the same diameter throughout its length. Three common types are atrichous isorhizas (atrichs) without barbs, basitrichous isorhizas (basitrichs) with barbs at the base only, and holotrichous isorhizas (holotrichs) with barbs along the whole length of the nematoneme. Fig. 10.8 summarizes their relationship.

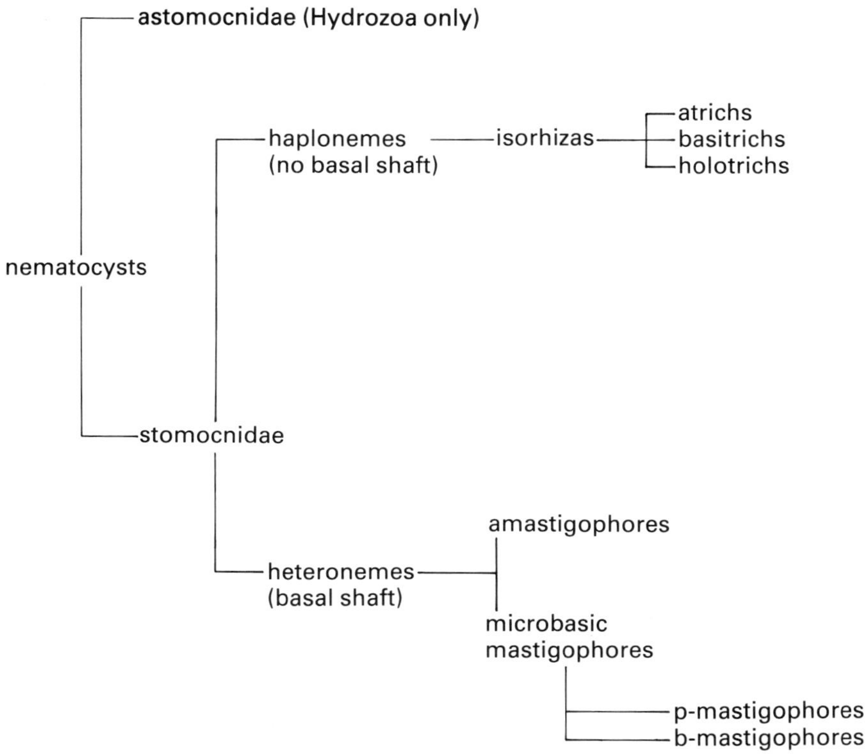

Fig. 10.8 Summary of relationships among anthozoan nematocysts.

Because this system of classification is based on discharged nematonemes, it is not helpful in identifying undischarged nematocysts in histological material, but the common types in sea anemones can usually be recognized on the appearance of the coiled nematoneme (Fig. 10.7(c) and (d)).

Distribution and function

Nematocysts and spirocysts are not uniformally distributed over the body. In most cnidarians more than one type can be found, and from their location in the body it seems evident that they play a number of different roles in the life of each species. Although much is known about differences in their shape and ultrastructure, we are largely ignorant of what these differences mean in terms of function. Three facts, however, are certain: there is a range in nematocyst size as well as type within a species, particular types are concentrated or restricted to specific parts of the body, and there is a wide variation in the ease with which nematocysts can be made to fire, indicating that there may be specific conditions for their discharge. Taken together these observations strongly suggest a wide range of functional specificity.

Nematocysts are rarely encountered in the endoderm of hydroids. The few instances where nematocysts have been seen among digestive cells in *Thyroscyphus marginatus* can be attributed to ingestion of free nematocysts after disintegration following regression. Therefore nematocysts are to be regarded as strictly ectodermal in the Hydrozoa. The picture is strikingly different in the Anthozoa, where nematocysts are as abundant on the gastric filaments inside the body as in the ectoderm outside. As nematocysts are not found in the endoderm of the tentacles, column or mesenteries, their presence in the gastric filaments suggests that the latter may have been derived from the ectoderm of the actinopharynx during development.

From histological material a comparison has been made of the distribution of the different types of nematocysts in *Corynactis australis, Anthothoë albocincta* and *Actinia tenebrosa* (Table 10.1). The density as well as the type varies from region to region. A problem arises with size. The holotrichous isorhizas of *C. australis* all look alike, except that some are much larger than others. Statistical analysis of capsule dimensions shows that holotrichs on the acrosphere belong to a different population from those on the gastric filaments. This suggests that size may indicate some difference of function.

At first sight this diversity is puzzling, for the only functions that have been ascribed to nematocysts are food capture, defence against predators and attachment to the substratum; however, each of these functions may cover a wide spectrum of circumstances and so some diversity might be expected. For example, it is unlikely that all potential prey can be captured successfully by a single mechanism, for a nematocyst toxin that is lethal to one species of prey may not have any effect on another type of animal. A large animal capable of struggling will require a method of capture different from that effective on a very small one.

The acrosphere and stem of the tentacle of *Corynactis australis* have vast numbers of spirocysts in the ectoderm. These are accompanied by a smaller number of holotrichous isorhizas, p-mastigophores and a few b-mastigophores. At least four other sizes of nematocyst are present elsewhere in the body making a total of at least eight different cnidae. Whether they are all concerned with feeding is not known, but a wide variety of animals are captured as food. In the laboratory, polychaete worms (*Syllis* sp. and *Nereis* sp.), copepods, amphipods, isopods, small shrimps and a crab megalopa have been consumed that ranged in size from *c.* 1 mm to over 10 mm. In the natural habitat the variety of food organisms would be much wider. It is of interest to note that Picken and Skaer (1966) found a definite order in which cnidae were discharged after an electric shock. Spirocysts were the first to discharge, followed by the mastigophores. The holotrichs did not start to discharge until after the mastigophores had stopped discharging.

Table 10.1 Distribution of nematocysts*

	Column wall	*Tentacles*	*Actino-pharynx*	*Gastric filament*	
Corynactis australis					
					Acrosphere
Holotrichous isorhizas					
Small *c* 32 μm	+	–	+	+	–
Large *c* 58 μm	–	+	–	–	+
V. large *c* 80 μm	–	–	–	+	–
p-Mastigophores					
Small *c* 30 μm	+	–	–	+	–
Large *c* 55 μm	–	+	–	+	++
b-Mastigophores					
Small *c* 20 μm	+	–	–	–	–
Large *c* 30 μm	–	+	–	–	+
Spirocysts	–	+++	–	–	+++
Anthothoë albocincta					
					Acontia
Microbasic amastigophores					
Small *c* 20 μm	+	+	–	–	–
Large *c* 55 μm	–	–	–	–	++
p-Mastigophores					
c 10 μm	–	–	–	+	–
c 20 μm	–	–	+	+	–
b-Mastigophores	–	+	–	+	+
Spirocysts	–	+++	–	–	–
Actinia tenebrosa					
					Acrorhagi
Atrichous isorhizas					
c 25 μm	+	–	–	–	–
c 50 μm	–	–	–	–	++
Basitrichous isorhizas					
c 15 μm	+	+	–	–	–
c 25 μm	–	–	+	+	–
p-Mastigophores					
c 20 μm	–	–	+	+	–
Spirocysts	–	++	–	–	+

* +, present; ++, abundant; +++, very abundant; –, absent.

It is hard to imagine how nematocysts in the column ectoderm could contribute to food capture, and so a defensive or deterrent role must be assigned to them. Both the column and tentacles have holotrichs and mastigophores that appear similar, thus it might be concluded that these have a dual role of food capture and defence, and that spirocysts are primarily concerned with feeding. Clearly this is an area in which more investigation is needed. Apart from nudibranch molluscs, large anemones seem to have few predators. This could well reflect the effectiveness of their nematocysts as deterrents, for one might imagine that the gonads of an anemone would provide an excellent meal for a predator. A predator might acquire immunity against a particular nematocyst toxin, in which case its effectiveness as a deterrent would be lost. A counter strategy would be to have a number of nematocysts with different toxic properties.

The acontia of *Anthothoë albocincta* are heavily laden with very large (60 μm) microbasic amastigophores and basitrichs (25 μm) and there can be little doubt that these nematocysts are connected with defence, for the acontia are only pushed out when the anemone is attacked. It is not so easy to understand why the gastric filaments of non-acontiate anemones, like *Actinia tenebrosa*, should contain large numbers of nematocysts (p-mastigophores and basitrichs). Gastric filaments are not pushed out like acontia, nor are they involved in food capture, so their presence inside the anemone is enigmatic. Nematocysts may anchor the gastric filaments to the food so that enzyme-secreting cells are held close to where extracellular digestion takes place. Alternatively, they may facilitate penetration of enzymes by puncturing the cuticle of prey. Yet another possibility is that their discharge may in some way induce release or activation of the enzymes. In any event, their function is connected with the digestive process rather than food capture.

Relatively few nematocysts are found on the tentacles of *Clavularia novaezelandiae*, suggesting that the symbiotic zooxanthellae play a major role in nutrition. However, the ectoderm covering the sheet-like stolon contains numerous small nematocysts which must act as a defence against predators.

Certain types of nematocyst are used in territorial behaviour (discussed on pages 318–322). It is well established that many species of anemone (and some corals) react to the presence of another member of the same species by aggressive behaviour in which the acrorhagi and fighting tentacles (containing atrichous isorhizas not found elsewhere on the anemone's body) are used as weapons of offence (Francis, 1973b; Purcell, 1977).

Cause of discharge

What causes a nematocyst to discharge is not fully understood. It is possible to isolate the nematocysts from their nematocytes, dry them and then cause them to discharge at a later date by putting them in distilled water, dilute acid ($pH < 4$), or trypsin solution, thus the mechanism of discharge (as distinct from the circumstances of discharge) is not dependent upon the nematocyte or living tissue.

The early experiments of Pantin (1942) showed that nematocysts on the tentacles of *Anemonia viridis* (= *sulcata*) were not discharged in response to touch or chemical stimulus (food extract) alone, although discharge did occur when the two stimuli were applied together. His interpretation, that chemical stimuli lower the threshold at which touch will elicit a response, implied two things: the mechanism of discharge is initiated or triggered by the nematocyte, and the timing of discharge is determined by either the nematocyte or the nervous system. Opinions differ as to whether the nervous system is involved.

Many nematocytes are not innervated, yet they can initiate discharge of their nematocyst. This has led to the widely held view that a nematocyte with its nematocyst is an independent effector. In the case of hydrozoan nematocytes, the cnidocil is regarded as the site of sensory perception; however, this view does not rule out the possibility of conduction through epidermal cells (neuroid conduction) or neurosecretions influencing the threshold of response of the nematocyte. On the other hand, evidence from electron microscopical and histological studies shows that many nematocytes are innervated (Lentz, 1968). A wide range of behavioural studies indicate that hydras and sea anemones have the ability to modify the threshold or circumstances of discharge to suit the environmental situation; this implies some form of neural, neuroid or neurosecretory integration and control.

Ewer (1947) found that discharge of the atrichous isorhizas of *Hydra*, which are normally concerned with attachment, was inhibited by the presence of food extracts. The opposite phenomenon has been noticed in the swimming anemone, *Stomphia*. Ross and Sutton (1964) found that extracts from the starfish *Dermasterias* caused *Stomphia* to detach and start swimming. It was noted that food did not adhere to the tentacles whilst the anemone was swimming, suggesting that nematocyst discharge was inhibited. The anemone *Calliactis parasitica* is normally found adhering to a whelk shell (*Buccinum*) inhabited by a hermit crab. Davenport *et al.* (1961) found that the tentacles of such anemones do not respond to being touched by a piece of whelk shell, although the tentacles of an anemone attached to a piece of plastic adhered strongly to an empty whelk shell, and attempts were made to transfer to the shell. Mariscal (1972) found that the tentacles adhere to the shell by the discharge of basitrichous isorhizas and spirocysts. This shell-specific response has been attributed to some form of information from the basal disc, but the important point to notice here is that the threshold of nematocyst discharge has been altered. This alteration might be due to impulses in the slow conducting system (SS1) (McFarlane, 1969b).

Suppression of feeding behaviour and of nematocyst discharge after feeding has been noted for several hydroids and sea anemones, but Clark and Cook (1986) found that the accumulation of substances released from discharging nematocysts (stenoteles) during feeding inhibited or inactivated further discharge of nematocysts in the hydroid *Halocordyle disticha*. An interesting adaptation to the short feeding period available to *Anthopleura elegantissima* living on the upper shore has been noted by Zamer (1986). High-level anemones had an increased receptivity to prey which compensated for long periods of emersion.

The debate as to whether or not nematocyst discharge is under nervous control is somewhat irrelevant, for the studies of Carré and Carré (1980) elegantly demonstrated that both types of control may be present in the same animal. They found that nematocytes in the pneumatophore, nectophores and bracts of physonect siphonophores (planktonic colonial hydrozoans) have no connection with the nervous system. Their nematocysts are used only for defence and act as independent effectors. In contrast, nematocytes on the fishing tentacles and gastrozooids are connected with the nervous system, therefore physiological condition might control nematocyst discharge by modulating the threshold of excitation of the cnidocil or nematocyte through the nervous system. This separation of function into defence nematocysts, which fire at a set threshold, and feeding nematocysts, whose discharge can be altered (enhanced or inhibited) by nervous intervention acting on the nematocyte, may be a widespread phenomenon in the Cnidaria.

Development of nematocysts

In the Hydrozoa, nematocysts are derived from ectodermal interstitial cells. These undergo nuclear division to form a syncytium of four, eight or more nematoblasts. Slutterbuck (1961) has described the initial stages of nematocyst formation in *Hydra*. The Golgi complex becomes active and one of its vesicles enlarges to form the nematocyst. In this vesicle the capsule wall differentiates from the central cytoplasm, except at one end where the vesicle membrane grows out as a long external tube with the Golgi complex at its tip. During this phase an extensive endoplasmic reticulum fills the cell cytoplasm, indicating intense synthetic activity, and more Golgi vesicles fuse with the external tube, presumably passing synthesized material to the interior (Fig. 10.9(a) to (d)).

Holstein (1980) has shown that at first this external tube is a single unit membrane, but two layers are added externally so that it becomes a double membrane with a thicker, electron-lucent layer between. This process starts from the capsule and progresses to the tip of the tube. A rapid process (lasting a matter of minutes) now takes place, in which the external tube regresses. During this process the inner membranes (i.e., the electron-lucent and the

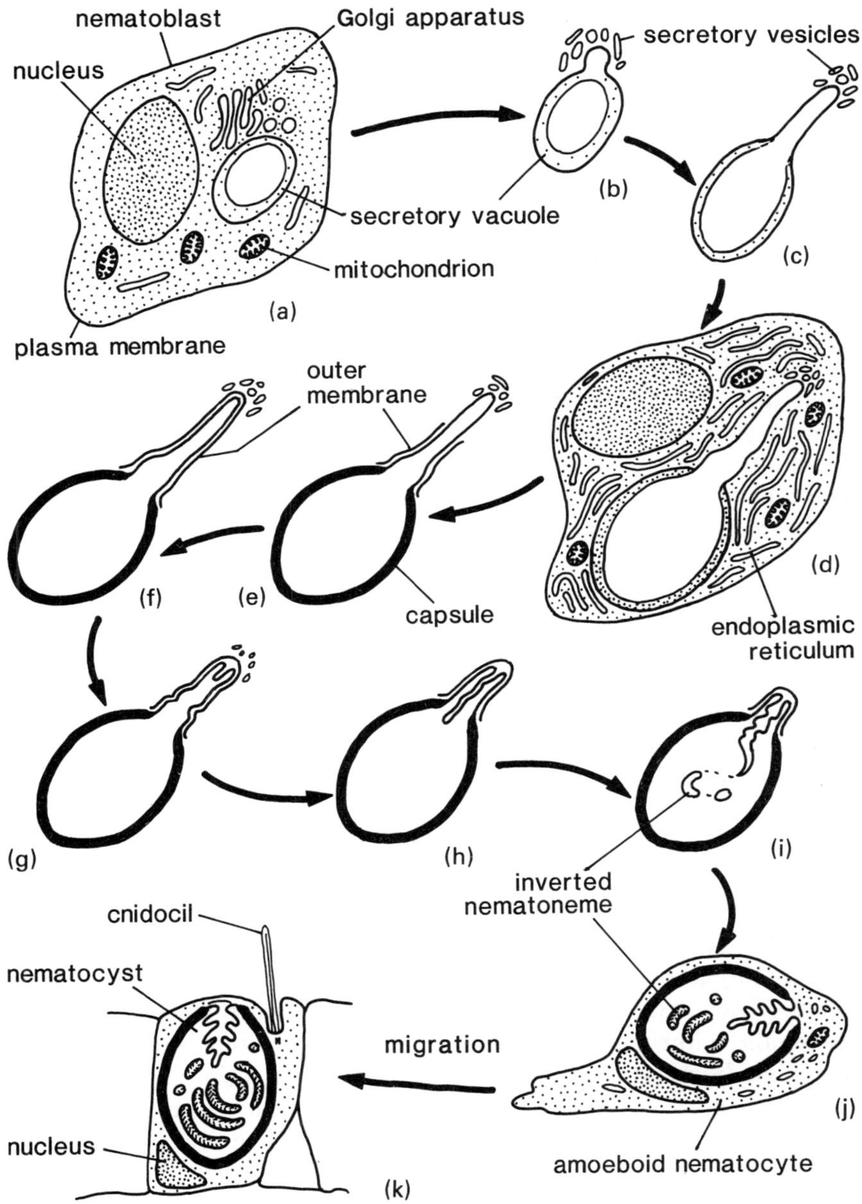

Fig. 10.9 Diagram showing sequence of events in the formation of a hydrozoan nematocyst (see text for explanation); a–d, formation of capsule; e–i, formation and inversion of the nematoneme; j–k, migration and orientation of nematocyte. Based on the findings of Slutterbuck (1961) and Holstein (1980). Nematocytes of sea anemones do not migrate, nor do they have a cnidocil.

original vesicular membrane, which forms a positive copy of the external tube) invert and pass back into the capsule to develop into the nematoneme (Fig. 10.9(e) to (h)). Barbs rapidly form within the tube from material lying in the tube. All trace of the external tube disappears. The endoplasmic reticulum regresses and an operculum may be formed over the pole of the capsule. Fig. 10.9 is a diagrammatic interpretation of the findings of Slutterbuck (1961) and Holstein (1980) to show how the invaginated nematoneme may be formed.

It is clear that the nematocyst is formed as an everted tube that then inverts, rather like the finger of a rubber glove that is pulled into the glove when the hand is withdrawn. How inversion comes about is not known. From Holstein's electron micrographs it would appear that the electron-lucent middle layer, which may be fluid, is important, for it separates the inner membrane from the outer layer of the external tube and appears to increase in volume. If this is so, reduction of the external tube would cause the inner membrane to invert and push back into the capsule. It is probable that other substances elaborated in the endoplasmic reticulum, such as toxins or material responsible for the internal osmotic pressure of the nematocyst, pass into the capsule during the external tube stage, for once the capsule is fully formed, it is extremely impermeable and seals off the contents from the outside world.

In hydroids the nematoblast now transforms into the mobile nematocyte, which will carry the nematocyst from the stolon to the tentacles of the hydranth or the nematophores. An instructive example of nematocyte migration has been noted in personal observations on the polyps of *Eleutheria* (Fig. 11.2(c)). Nematocysts are formed in the stolon and are carried up the column of the polyp by the nematocytes along definite pathways which lead to the four tentacles. Eventually they pass onto the acrospheres and are reorientated in the discharge position. A similar migration is seen on the tentacles of the medusae, but here the nematocysts are invariably carried to the acrosphere and not towards the sucker.

These observations strongly suggest that there is a physical separation between the epithelio-muscular cells of the ectoderm along these tracts, and that this guides the nematocytes to their destination. The problem of how the nematocytes move has not been explained, but a likely hypothesis is that they move in contact with the mesogloea by means of fine pseudopodia. The direction of amoeboid locomotion could be determined either by the orientation of the muscle tails or by a lack of adhesion between the epithelial cells along the pathway. When the nematocyte reaches its destination the nematocyst is orientated with its point of discharge towards the free surface. A stereocilium, the cnidocil, develops in the nematocyte and projects from the surface (Fig. 10.9(k)). In species like *Aglaophenia* (Fig. 11.8(a)), the cnidocils may be prominent and give the tentacles a prickly appearance. In some

Hydrozoa, nematocytes enter a battery cell, which contains many nematocysts (Chapman, 1974).

The anthozoan nematocyst probably develops in a similar way, but there is no migration through the body and a cnidocil is absent. Nematoblasts appear to form their nematocysts *in situ*, but their origin is uncertain, for interstitial cells are considered to be absent from the Anthozoa. Nematoblasts may originate from the numerous amoebocytes seen in the mesogloea and sub-epithelial cavity. This fact is reflected in the pattern of distribution of the different nematocyst types in the body. Nematocysts in the gastric filaments and acontia are presumed to be of endodermal origin, whereas those of the tentacles will be derived from ectoderm.

There is evidence that the composition of the cnidom (the sum of all types of cnidae in an animal) can be changed. Purcell (1977), studying the transformation of feeding tentacles to fighting tentacles in *Metridium senile*, found that over a period of 9 weeks the typical cnidom of the tentacles (80% spirocysts, 16% b-mastigophores and 2% amastigophores) gradually changed and was replaced by holotrichs (*c.* 60%) and atrichs (38%). This reflects a change in the role played by these special tentacles.

Mechanism of discharge

All cnidae (spirocysts, ptychocysts and nematocysts) discharge by eversion (Mariscal, 1974). During eversion of barbed nematocysts, the tubular nematoneme travels up the inside of the everted portion. As each of the barbs reaches the advancing lip it turns over and comes to lie on the outside of the everted tube (Fig. 10.7(b)). When the whole nematoneme has everted, fluid from the capsule slowly leaks from the open tip. The capsule does not alter shape, thus there must be an overall increase in the volume of fluid within the capsule during discharge.

Many theories have been put forward to account for the phenomenon of nematocyst discharge, but two observations are of particular importance to its interpretation. By an experiment remarkable for its simplicity, Robson (1953) has shown that the undischarged nematoneme elongates when it comes into contact with water, presumably by a process of hydration which causes reorientation of its molecular structure. A partly extruded holotrichous isorhiza of *Corynactis* was dried on a microscope slide. The capsule and tip were cut away, leaving a portion of thread with everted tube on the outside and inverted tube on the inside, both of the same length. When this was wetted with distilled water, the inner tube increased in length and protruded at both ends beyond the outer tube, which was unaffected (Fig. 10.7(e)). This showed conclusively that the nematoneme can elongate without pressure from the capsule.

In the intact nematocyst the inner tube, as it everts, only comes into contact with water at the advancing lip, and it is here that elongation must occur. Because the inner portion of the tube is confluent with the portion that has everted, it follows that the inner tube must be pulled up the outer tube as the thread discharges. Picken (1953) reasoned that the simultaneus elongation and dilation that takes place at the lip not only rearranges the barbs into three spiral rows, but also causes rotation of the inner tube as it is pulled up out of the capsule. It is this rotation that gives the barbs their cutting action (Fig. 10.7(b)).

Since the discharged volume is greater than the initial volume, water must have been taken into the capsule. Picken and Skaer (1966), by another noteworthy experiment, measured the osmotic concentration of the fluid in the undischarged and discharged nematocysts of *Corynactis*. They found it exceptionally high, corresponding to a notional osmotic pressure of 1.42×10^7 N m^{-2} (140 atmospheres). Water must enter the capsule under this osmotic gradient. This not only supplies the necessary increase in volume during discharge, but more importantly, the high pressure it generates is necessary to maintain the turgidity of the very narrow tube as it everts. Without this pressure in the capsule, elongation of the thread by hydration would not provide sufficient force to draw the tube out of the capsule.

Normally the capsule is extraordinarily impermeable to water. Chemical analysis of *Corynactis* nematocysts (Brown, 1950) suggests that the capsule may be a keratin-like protein with sulphide linkages. Yanagita and Wada (1954) found that trypsin and thioglycollate both caused nematocyst discharge, presumably by rendering the capsule permeable. It is possible that the nematocyte can cause a similar change in permeability in the capsule and so initiate discharge.

Toxicity

A wide variety of chemical substances (amino acids, proteins, various enzymes, phenolic substances and 5-hydroxy tryptamine) have been detected in nematocysts, and their toxic action (as neurotoxin, myotoxin or haemolytic agent) on such unlikely animals as mice and rabbits has aroused considerable interest (Mariscal, 1974), but surprisingly little attention has been paid to the physiological effects of nematocyst toxins on natural prey or predators.

The rapidity with which nauplii or amphipods are immobilized when they touch a tentacle strongly suggests that the toxin acts upon the neuromuscular system. If a substance in the toxin mimics the effects of the inhibitory neurotransmitter of crustaceans, muscles affected would not contract, and the animal would be temporarily paralysed. Polychaete worms, which have a different system of neural control, do not show the same paralysis, and may continue to struggle whilst being swallowed.

10.4 MOVEMENTS OF SEA ANEMONES

Sea anemones are capable of a wide range of movements of the column and tentacles, and some, such as *Actinia tenebrosa*, are capable of moving about over the substratum. Most movements and changes in shape are performed so slowly that they can seldom be detected by the unaided eye; devices such as kymographs, video equipment or time-lapse photography have been used to record and analyse them. To understand the movements and behaviour of anemones it is necessary to say more about their structure.

Nematocysts and gland cells secreting mucopolysaccharides are abundant in the ectoderm of sea anemones and perform a protective role, but the ordinary epithelial cells of the ectoderm do not possess contractile fibres, and all movements of the column are effected by muscles in the endoderm.

The basic cell type of the endoderm is the epithelio-muscular cell (Fig. 10.10(b); contraction is explained below). The main body of the cell containing the nucleus and usually bearing a flagellum, forms the epithelial lining to the body cavity. A narrow isthmus of cytoplasm connects the cell to a contractile 'muscle tail' which lies against the mesogloea and is attached to the latter by a number of protoplasmic processes (Fig. 11.7(a)). Epithelio-muscular cells are abundant in certain regions of the column, particularly the mesenteries, where folding of the mesogloea forms distinct contractile fields or sheets (Fig. 10.10(e)). For simplicity in the following account, these will be called muscles.

Fig. 10.10(a) shows the general arrangement of muscles in an anemone like *Actinia*. Longitudinal retractor muscles lie on the endocoelic side of mesenteries (except the directives) and run from oral to basal disc. The arrangement of these muscles round the vertical axis is important as it allows the oral disc and tentacles to be withdrawn symmetrically. However, bending of the column or withdrawal of the tentacles on one side of the mouth can be achieved by contraction of the longitudinal retractor or parieto–basal muscles on one side of the body. Parietal or parieto-basal muscles are attached to the body wall and basal disc on the exocoelic side of mesenteries. Basal muscles run radially at the base of mesenteries in some anemones. Circular muscles in the endoderm of the column run through the mesenteries at the point where mesentery and body wall meet. They also form the sphincter muscle (Fig. 10.10(d)) and the circular muscles of the actinopharynx and oral disc. Radial muscles lie in the mesenteries. Tentacles have endodermal circular muscles and ectodermal longitudinal muscles which continue on to the oral disc as radial muscles.

The 'muscle tails' containing contractile fibrillae are relatively short (< 1 mm, Batham and Pantin, 1951) and vast numbers are attached to the internal surface of the mesogloea lying parallel to one another. Each will cause only local shortening of the mesogloea when it contracts, and so there is nothing corresponding to the long muscle fibres running from origin to

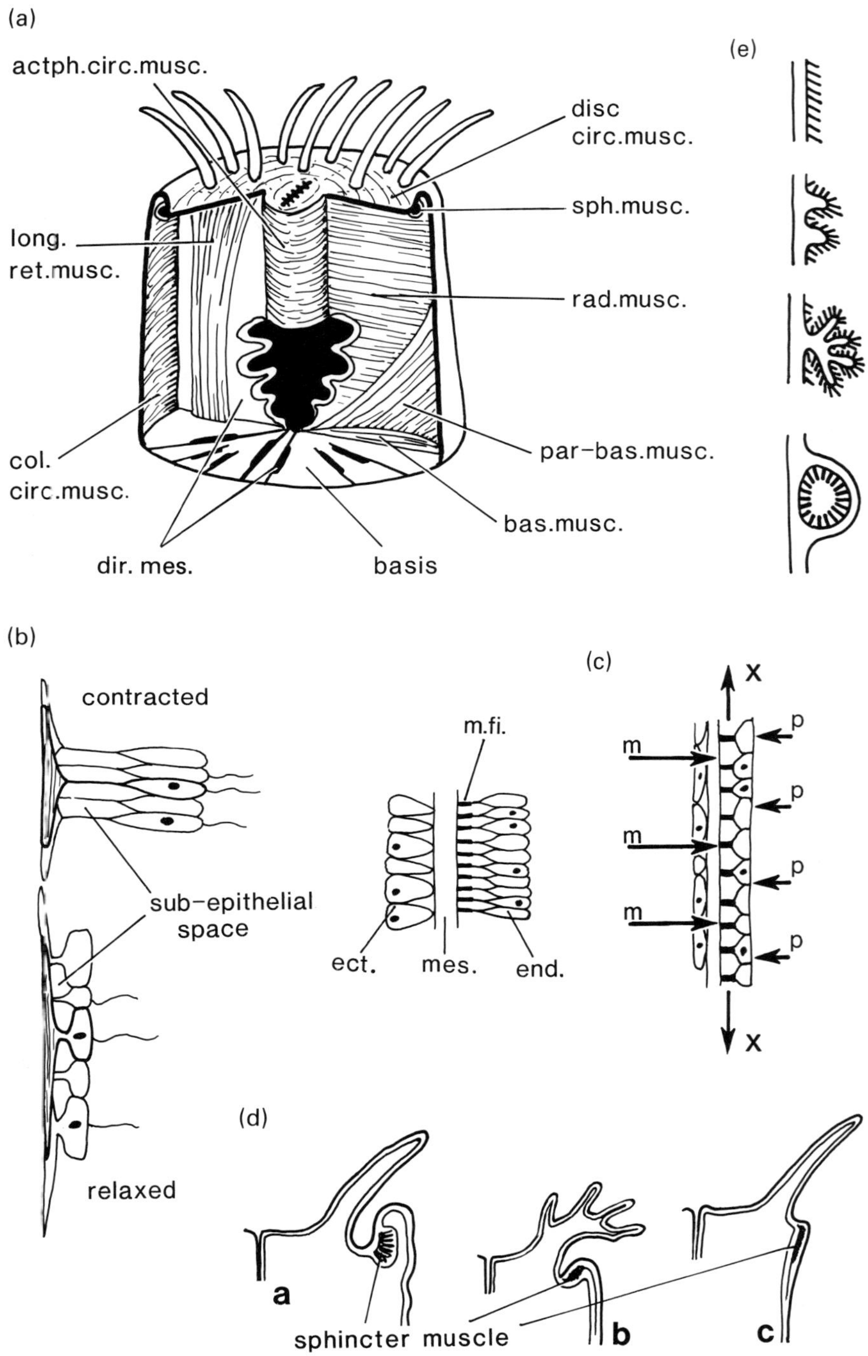
(a)
actph.circ.musc.
disc circ.musc.
sph.musc.
long. ret.musc.
rad.musc.
col. circ.musc.
par–bas.musc.
bas.musc.
dir. mes.
basis
(e)
(b)
contracted
sub–epithelial space
relaxed
m.fi.
ect.
mes.
end.
(c)
x
m
p
(d)
a
b
c
sphincter muscle

insertion which develop tension between these two points in other animals (arthropods, for example). When large numbers of 'muscle tails' belonging to the same muscle field contract, their collective effect upon the mesogloea will be to shorten it (exert tension) in the direction of the contractile fibres. The effect that this has on the sea anemone as a whole, depends upon many factors. Tension in the mesogloea and muscles will ultimately exert pressure on the fluid contained in the body cavity. Since pressure within a fluid is exerted equally in all directions, an inescapable consequence of any one muscle contracting is that it will affect every other part of the body: no part of the body is immune from the effects of activities performed elsewhere.

The coelenteron acts as a hydrostatic skeleton, thus the immediate effect of, say, the retractor muscles contracting, is a rise in coelenteric pressure which will be felt in all parts of the body. Any area where the muscles of the body wall do not exert an equal and opposite pressure will yield, and fluid will flow to this region, allowing the contracting muscle to shorten the column. If, for example, the circular muscles do not oppose the pressure increase, the diameter of the column will be increased and the circular muscle fibres will be stretched. In this case the retractor and circular muscles act as antagonists. For this to happen, muscles elsewhere in the body, such as the tentacles or oral disc, must exert a force by tonic contraction which will oppose the initial pressure rise and stop them from distending; they are acting as synergists to the parietal muscles. Batham and Pantin (1954) have shown that mutual inhibition of the parietal and circular muscles occurs in *Metridium senile*. In other animals, some form of reflex inhibition of antagonists and co-ordination of synergists is built into the central nervous system, but little is

---

Fig. 10.10 Muscle systems of cnidarians. (a) Sea anemone muscle fields. (b) Change in shape of epithelio-muscular cells. Contraction of the muscle tails exerts a pressure on the fluid in the sub-epithelial space. This force assists changes in cell shape. (c) Vertical section through body wall of sea anemone: left, circular muscle fibres relaxed; right, fibres contracted. m, direction of force exerted by muscles. p, pressure of coelom on endoderm. X, direction in which wall is extended. (d) Sphincter muscle of a, *Actinia*, b, *Metridium*, c, *Calliactis* in vertical section. (e) Folding of muscle fields in mesogloea: Top, simple condition; middle, folding of mesogloea increases volume of contractile tissue; bottom, muscle may become enclosed in mesogloea. actph. circ. musc., circular muscles of actinopharynx; bas. musc., basal muscle fibres; col. circ. musc., circular muscle fibres of column; dir. mes., directive mesentery; disc circ. musc., circular muscles of oral disc; ect., ectoderm; end., Endoderm; long. ret. musc., longitudinal retractor muscles of mesenteries; mes., mesogloea; m. fi., contractile tails of epithelio-muscular cells; par-bas. musc., parieto-basal muscles; rad. musc., radial muscles of the mesenteries (mesentery and muscle partly cut away at bottom); sph. musc., sphincter muscle.

known about how such inhibition or co-ordination is effected in anthozoans. McFarlane (1974) has shown that impulses in the endodermal slow conducting system (SS2) are responsible for inhibition and relaxation of the longitudinal and circular muscles of the body wall, and the ectodermal slow conducting system (SS1) is responsible for causing the radial muscle to relax during expansion of the oral disc in pre-feeding behaviour (McFarlane, 1970).

It also follows that all the muscle fields must be in a state of tension sufficient to counteract the slight positive pressure in the coelenteron. This fact is clearly demonstrated in an anaesthetized anemone; the body loses its turgidity and becomes completely flaccid.

In addition to the large coelenteron there is another fluid-filled cavity, the sub-epithelial space, within the body wall between the epithelio-muscular cells (Fig. 10.10(b)). Its volume is not very great – Robson (1957) estimated that it lies between 1% and 10% of the epithelial volume - but it undoubtedly plays an important hydrostatic role. Robson suggested that it assisted in the rapid change of shape of the epithelial cells that must occur whenever the anemone undergoes extensive changes in size. When a muscle contracts, the sub-epithelial fluid will push the epithelial layer away from the mesogloea, thus allowing the cells to adjust their shape quickly. Another suggestion is that pressure exerted on the sub-epithelial fluid assists extension of the 'muscle tails' (Fig. 10.10(b) and (c)). Internal coelenteric pressure will be transmitted to the sub-epithelial fluid and, provided antagonistic muscles do not prevent it, extend the muscles.

The physical properties of the mesogloea largely determine the effect muscle contraction will have on other parts of the body. Although it contains collagen, the mesogloea does not form an inelastic connective tissue corresponding to that of other metazoan animals, for it is extremely extensible and can be stretched two or three times its resting length before resisting further extension.

Gosline (1971) has shown that extensibility and elasticity depend upon the properties of an amorphous polymer network in the mucopolysaccharide matrix to which the collagen fibres are bound. This would account for the extensibility of the mesogloea. In vertebrate animals, collagen fibres are cross-bonded and this resists extension. Gosline considered that ionic groups on the collagen fibrillae, which normally form the cross-links, are blocked by inorganic ions in the fluid component of the mesogloeal matrix.

When a force is applied to the mesogloea, it will stretch to a length proportional to the weight applied, but it does so in an unusual way. An ideal elastic substance would stretch immediately to a length proportional to the stress applied. With mesogloea, however, there is an initial rapid lengthening, but this soon slows down and it may take hours before the final proportional length is reached. When the stress is removed, there is an initial rapid shortening followed by an extremely slow return to its original length

(Alexander, 1962). It is clear that the mesogloea is elastic and can return to its original shape after stretching, but the time course for this is profoundly modified by the viscous forces in the matrix. At first most of the work is done in overcoming viscous forces in the matrix, but as the mesogloea elongates, more energy is absorbed in stretching the fibrous component.

Several important conclusions can be drawn from a knowledge of the properties of the mesogloea. Firstly, rapid muscular contraction will be opposed by relatively large viscous forces, and the mesogloea will tend to resist deformation. The same would apply to external forces such as wave action suddenly pushing against the column. On the other hand, a very slow sustained contraction (or external force) encounters no viscous resistance, and the mesogloea acts as an elastic substance. The slower a muscle contracts, the less energy it requires to maintain a given tension; moreover, tension can be maintained without high oxygen consumption (Davson, 1959). Thus, from the point of view of energy expenditure, slow muscular contraction is more efficient than fast contraction.

The elastic properties of the mesogloea have an important bearing on the functioning of the siphonoglyphs. A particular anemone will have a certain volume at which the mesogloea is neither stretched nor compressed – the neutral volume. Above the neutral volume, the siphonoglyphs are, in effect, pumping against the elasticity of the mesogloea (viscous forces can be ignored because inflation is so slow). Theoretically inflation could continue until the force of elastic recoil equalled the pressure generated in the siphonoglyphs; at that point the anemone would have attained its maximum size. It is, of course, possible to inflate an anemone beyond this point by injecting water into the body cavity, but the anemone's own muscular contractions cannot draw water into the coelenteron. Direct measurement of the pressure generated by the flagella of a siphonoglyph has proved difficult because of mucus secretion, but the force required to stretch the mesogloea can be measured relatively easily. Alexander (1979) has shown that it only requires a force of $0.4\ \mathrm{N\ m^{-1}}$ to double the length of an isolated strip of mesogloea 1 cm wide. In a sea anemone 4 cm in diameter, such a tension would require a coelenteric pressure of *c.* $20\ \mathrm{N\ m^{-2}}$. Batham and Pantin (1950a) found an average coelenteric pressure of $26\ \mathrm{N\ m^{-2}}$ in unstimulated *Metridium senile*, and it can be assumed that the siphonoglyph can function beyond this pressure, so the siphonoglyph seems capable of inflating an anemone to two or three times it neutral volume.

The mesenteries fulfil a number of important functions in sea anemones, but it is difficult to decide which are primary and which secondary specializations. The selective advantages that originally led to the evolution of a partitioned coelenteron can only be guessed.

Looked at from a mechanical point of view, it is clear that the actiniarian polyp could never have reached the relatively enormous size that we see today without mesenteries. A single polyp of *Clava* (Fig. 11.1(d)) scaled up to the size of a sea anemone would be little more than a thin-walled bladder scarcely able to maintain its shape against movements of the surrounding water. Either it would have to develop a thick body wall or it would have to maintain a very high coelenteric pressure to give it turgidity. A thick body wall might lead to problems with oxygen supply, and a high turgidity is incompatible with the function of the coelenteron as a digestive cavity, which has to be opened to admit food. The radial partitions arranged symmetrically about the column axis give stability in all radial directions, yet allow bending, extension and contraction without high coelenteric pressures.

A feature common to anthazoans is the longitudinal mesenteric retractor muscles. Their symmetrical arrangement round the vertical axis ensures that the oral disc is pulled down evenly whilst still allowing bending. Conversely, the same muscles are placed in the most effective position to prevent bulging of the disc, or eversion of the actinopharynx, when circular muscles contract and increase the pressure within the coelenteron. It would be impossible for longitudinal muscles in the ectoderm of the column to fulfil either of these functions.

One possible reason for the origin of mesenteries is the evolution of the actinopharynx. There can be little doubt about the value of extending the mouth inwards with a flagellated tube to pass food into the coelenteron. The device acts as a valve which allows the mouth to be opened to admit food without water escaping from the coelenteron (and hence, drop in pressure); however, there is no way in which longitudinal or circular muscles of the actinopharynx could open the tube against the pressure of the coelenteron to allow food in. The solution to this problem has been to provide radial muscles which lie in the mesenteries. It is significant that during development, the first mesenteries appear at the time the actinopharynx starts to form. Primary mesenteries arise as infoldings of the body wall that grow towards the actinopharynx during development of young anemones (Fig. 10.4(b)).

Originally the flagella lining the actinopharynx may have been concerned with passing food to the coelenteron. Only a small modification would be necessary to utilize them to pass water into the body for respiratory purposes, and from there the evolution of the siphonoglyph as a pump for the hydrostatic skeleton is not difficult to envisage. It seems highly probable that mesenteries evolved as dilators to a tubular, valve-like mouth, and only later took on the function of supporting the retractor muscles. The separation of the functions of digestion and hydrostatic skeleton for the coelenteron has been achieved by confining the enzyme-secreting cells to the gastric filaments.

Given that the mesenteries allow sea anemones to grow to a very large size, other advantages accrue from their presence. By adding more mesenteries, the internal surface area : volume ratio is not altered during growth. This may be important in maintaining adequate supplies of nutrients and oxygen to all tissues of the body, for it has allowed anemones to grow to a very large size without increasing the thickness of the body wall or developing a true vascular circulatory system. The flagellated bands on the gastric filaments, as well as the general presence of flagella on the endoderm cells of the body wall and internal wall of the tentacles, are responsible for creating a general circulation of water throughout the coelenteron. The mesenteries also provide a very large area on which the gonads develop. The sexes are usually separate and the ova or sperm tubes develop within the mesogloea of the mesenteries.

By concentrating the enzyme-secreting cells on the free edge of the mesenteries, the efficiency of extracellular digestion is greatly improved. When food is ingested, it is completely surrounded by the gastric filaments, which secrete mucus and enzymes with strong trypsin- and chymotrypsin-like activity over it. This would improve the efficiency of extracellular digestion by concentrating the enzymes round the food and preventing dilution by the general coelenteric fluid.

Finally it might be noted that the mesenteries and their musculature enable complete closing down of the body, to expel stagnant water from the coelenteron as well as to void undigested remains of food. In this connection, Hand (1966) has made the important observation that in scleractinian corals, the mesenteries of a pair lie on either side of a septum, thus when the retractor muscles contract, they pull the disc right down and expel all water from the coelenteron. This might not be possible if the mesenteries were unpaired.

10.5 BEHAVIOUR OF SEA ANEMONES AND ITS CONTROL

The method of co-ordination and control of the very slow movements of sea anemones is poorly understood, for although the structure and location of both sensory and nerve cells has been studied, no very clear picture has emerged of nervous pathways. This is largely because the fibres of nerve cells only run a few millimetres and are rarely seen running to an effector organ, thus it is not possible to trace long axons to their destination in ganglia or muscles, as has been done in other invertebrate animals.

Bipolar, tripolar and multipolar neurones occur in the endoderm of sea anemones, where they form a net-like nervous system with synaptic connections at points where their processes cross one another. Nerve impulses pass through the nerve net, but the branches of nerves cannot be classed as axons or dendrites, for there is evidence that they can transmit impulses in either direction (i.e., they are not polarized). There is no nerve net in the ectoderm, although sensory cells are known to be present in the column epithelium (Lawn, 1980). Histological and electrophysiological studies indicate that nerves pass through the mesogloea from ectoderm to the endodermal nerve net (Robson, 1957; Lawn, 1980). There is no aggregation of neurones to form a ganglion in sea anemones, therefore there can be no centralized control of behaviour.

Other nerve cells producing secretions or hormone-like substances that control growth and regeneration have been found in *Hydra* (Burnett *et al.*, 1964; Lentz, 1968). Similar nerve cells may occur in sea anemones.

Little is known about how complex patterns of behaviour are co-ordinated, and the following facts have to be explained before even simple activities can be understood. Stimulation anywhere on the column can initiate a whole pattern of behaviour: nerve impulses must spread quickly to all the mesenteries if all the retractor muscles are to contract simultaneously and lower the oral disc symmetrically: at least three discrete muscle systems, each capable of independent action, are combined in the protective withdrawal response: because pressure within the coelenteron is distributed equally in all directions, contraction of any one muscle requires that all the other muscles in the body act as antagonists or synergists, this implies inhibition and co-ordination. Finally, because there is considerable variation of plasticity in response, it is unlikely that patterns of sea anemone behaviour can be explained as simple reflexes.

The physiological properties of the nerve net and muscles of sea anemones were investigated in the classic studies of Pantin (1935a,b,c) on *Calliactis parasitica*. He found that the properties of nerve conduction are very similar to those of other animals. That is: conduction is all-or-nothing; impulses travel at a constant velocity and are all of the same magnitude for a particular nervous pathway; there is a refractory period during which a second impulse cannot pass; and there are synaptic junctions between nerve cells which cause delays. Synapses in the nerve net have been confirmed from electron microscopical studies (Westfall *et al.*, 1970).

The nerve net in the colum shows no evidence of polarization – it conducts impulses in any direction – but there is some evidence of polarization in the tentacles, oral disc, and actinopharynx, although the nature of this polarization is not understood (Pantin, 1935c; McFarlane, 1975).

The velocity of conduction in the nerve net varies in different parts of the body (Pantin, 1935a). The fastest conduction was found in the mesenteries (1.2 m s^{-1}) and sphincter (1.0 m s^{-1}). This has been called the through-conducting system, and it is responsible for the rapid withdrawal behaviour of the tentacles and disc. Batham *et al.* (1960) have identified a network of large bipolar cells in the mesenteries which they believe to be responsible for fast conduction. In the column, the velocity of conduction was much slower (*c.* 0.1–0.15 m s^{-1}). These conduction velocities may appear slow, but it is misleading to compare them with the axons of other animals, for they may involve many synapses in the path over which they were measured.

More recent studies have revealed other conducting systems besides the endodermal nerve net. McFarlane (1969a) has demonstrated conduction in the ectoderm of the column with a velocity of 0.04–0.06 m s^{-1} and in the oral disc at 0.12 m s^{-1}. This he has called the ectodermal slow conducting system (SS1). The absence of a nerve net in the ectoderm of anemones suggests that conduction is neuroid (impulses passing through the epithelial cells), although Mackie *et al.* (1984) pointed out that gap junctions are not found in the Anthozoa and could find little evidence for neuroid transmission. It has been shown that the SS1 system is responsible for detachment of the basal disc (McFarlane, 1969b), pre-feeding behaviour (McFarlane, 1970), and chemoreception in the column (Lawn, 1975) A second slow conducting system with conduction velocity of 0.04 m s^{-1} has been found in the endoderm. This has been called the endodermal slow conducting system (SS2). Whether it is nervous or neuroid is not known. It has been shown that it is responsible for controlling mouth opening (McFarlane, 1975) and settling behaviour of *Stomphia* (Lawn and McFarlane, 1976).

The response of sea anemone muscle fields to electrical stimulation is very different from that of other animals. The first peculiarity is that a single stimulus does not cause contraction, but it does sensitize (facilitate) the muscle field to subsequent stimuli, provided they arrive within a certain time after the first stimulus. The number of stimuli and their frequency play an important part in determining the extent of contraction. When the first of several stimuli arrives at a muscle, some fibres will be facilitated so that they will contract on arrival of the second stimulus, which in turn facilitates more fibres, and so on, thus a 'staircase' effect can be obtained by applying a series of shocks to an anemone and recording the contraction they produce. If the time interval between stimuli is very long (>2 s), a slow steady rise in tension is produced. At higher frequencies (intervals of 0.5 s or less), contraction is more rapid and reaches a higher tension.

Different muscles respond to different frequencies of stimulation, thus Pantin (1935b) showed that a train of stimuli at a frequency of one every 3 s caused slow contraction of the parietal muscles, whereas one every 2 s caused slow contraction of both parietals and retractor muscles. At a frequency of one every 1.1 s, the sphincter muscle contracted slowly, as did the other two, and at one every 0.6 s, there was rapid contraction of the sphincter and retractor muscles.

The same muscle may respond to different frequencies of impulse in quite different ways. At very low frequencies (one shock every 3–10 s, the sphincter muscle of *Calliactis* responds with a slow tonic contraction. With shocks only 0.5 s apart, rapid contraction takes place, with immediate relaxation when stimulation stops.

In all these experiments it is assumed that the muscles respond to impulses triggered by shocks in the same way that they would respond to impulses from sense receptors or cells of the nerve net. Little is known about the way the muscle is excited to contract. Some evidence for neuromuscular junctions has been obtained from electron microscopy (Westfall, 1970), but these junctions do not appear to be frequent. It is possible that excitation passes from the contractile process of one cell to another. In the intact sea anemone, impulses passing through the nerve net, SS1 and SS2 systems must be initiated by specific neurones, which generate a train of impulses with a particular temporal pattern (i.e., pacemaker neurones).

Although a knowledge of the properties of the neuromuscular system helps to explain individual movements, it falls far short of explaining the intricate patterns of behaviour seen in anemones. Some of these can be described in terms of fixed action patterns (stereotyped behaviour) and reflexes, but there are difficulties in doing this.

It is a matter of common experience that there is an element of variability in all actiniarian behaviour that makes it unpredictable. When a number of similar anemones are fed, it is not uncommon to find a few that will not accept food and may even close up completely when it touches their tentacles. Sometimes this can be attributed to recent feeding or, in the case of *Actinia tenebrosa*, to incubation of embryos in the coelenteron, suggesting that the internal state of the anemone may influence its behaviour. This is not an unusual phenomenon, for it is well known in most other animals, but it must not be assumed that it is initiated or controlled by similar mechanisms.

Many patterns of anemone behaviour can be identified which are more or less repeatable and follow clearly-defined sequences. They can be regarded as fixed action patterns. The pre-feeding and feeding responses can be taken as two examples of well-defined patterns of behaviour that show both similarities and differences between species.

Food extract (from crushed tissue of, say, mussel), when added to the water, usually causes pre-feeding behaviour which seems to put the anemone in a 'receptive' condition for feeding. The tentacles of *Actinia tenebrosa* are spread to their fullest extent by expansion of the oral disc, and the mouth may open partially so that the ridged actinopharynx protrudes (Fig. 10.2(a)). A similar sequence has been described for other species. In *Metridium senile* (Fig. 10.1(f)) the column slowly elongates. Eventually it bends over and performs a swaying movement so that the tentacles sweep the ground (Batham and Pantin, 1950c).

It has been shown in *Urticina felina* that chemoreceptors, located in the ectoderm of the column, are excited by substances in the food extract and cause very-low-frequency impulses (one every 10–20 s) to pass through the slow conducting system of the ectoderm (SS1) (Lawn, 1975). These impulses cause relaxation of the radial muscles in the oral disc and sphincter (presumably by trans-mesogloeal conduction, since the latter muscles are endodermal), so that the slight internal pressure of the coelenteron causes the actinopharynx to bulge through the mouth. Chemoreceptors activating this system are not found on the tentacles or oral disc. It is interesting to note that

McFarlane (1970) was able to elicit the whole sequence of pre-feeding behaviour in *Urticina felina* by electrically stimulating the SS1 with shocks at intervals of between 5 and 30 seconds.

There are three distinct phases to the feeding response. Initially, spirocysts and nematocysts arrest any prey organism that happens to collide with a tentacle. The tentacle now shortens and bends towards the mouth by contraction of longitudinal muscles on the oral side which is presumably brought about by stimulation of tactile and chemoreceptors on the tentacle. Local contraction of radial disc muscles at the base of the tentacle draws the mouth and tentacle closer together. The mouth opens slowly and asymmetrically so that the actinopharynx bulges towards the contracting tentacle. Once the food contacts the oral disc or actinopharynx, it is drawn smoothly to the middle by the flagella and appears to sink into the actinopharynx, the walls of which roll in to close behind the food. The mouth itself may not close for some time after the food has been swallowed.

The role played by the flagella of the actinopharynx appears to be independent of tentacle activity, but it is influenced by the nutritional status of the anemone. The actinopharynx can be cut out and opened to study the flagella. In the siphonoglyphs the flagella beat towards the coelenteron, but elsewhere they beat in the opposite direction, towards the mouth. When food is placed on them, they reverse their beat and carry it towards the coelenteron. Parker (1917) observed that flagella on an actinopharynx taken from a starved anemone reversed their beat and passed inert material (filter paper) towards the coelenteron. After a number of food particles had been passed in the same direction, filter paper no longer caused reversal, implying that some form of sensory adaptation had occurred. Reversal of the flagellar beat took place even in the presence of narcotizing agents, showing that their reversal was not controlled by a nervous mechanism.

This stereotyped feeding response is common to nearly all anemones, and yet it is subject to considerable variation, even within a single individual. It may be very rapid or it may take up to an hour. Large food items may cause many tentacles to be involved, and the edge of the disc may roll in or the disc may be withdrawn. Sometimes the flagellar beat is reversed and the food rejected. It sometimes takes a long time for food to pass into the coelenteron, and opening of the mouth accompanied by eversion of the actinopharynx followed by withdrawal and closure may occur at regular intervals of 5 or 10 minutes in *Corynactis*. The reason for this repetitive activity is not clear as muscular contractions are not necessary for swallowing – food placed on the mouth of a narcotized anemone will be carried down the actinopharynx by the flagella. *Metridium senile* may feed in the above manner, but it has another mode of feeding. The numerous fine tentacles are flagellated and possess mucous gland cells in the ectoderm. Small planktonic organisms caught by the tentacles are carried in mucus to the tips, from whence they fall onto the oral disc and are carried to the actinopharynx. The feeding response can be modified to suit circumstances.

McFarlane (1975) has shown that chemoreceptors in the vicinity of the mouth are extremely sensitive to certain substances, such as proline and reduced glutathione. They appear to be responsible for increasing impulse activity in the endodermal slow conducting system (SS2), which causes inhibition or relaxation of circular muscles in the disc and actinopharynx, thus allowing the mouth to open and the actinopharynx to be protruded by coelenteric pressure.

Feeding behaviour does not stop at this point. A feeding–digestion cycle may last 2 d or more and pass through a sequence of distinct phases. These are clearly shown in the case of *Metridium senile* described by Batham and Pantin (1950c). After swallowing food, the oral disc and tentacles relax and spread. This is followed by slow peristalsis of the column, during which there is a marked increase in height and volume. Some hours later, bending and swaying commence, similar to that described for pre-feeding behaviour. Further inflation results in a loss of activity. Within 24–48 h, considerable anti-peristaltic activity occurs, and this culminates in the mouth and actinopharynx being opened maximally. The mucus-covered undigested remains of the meal are now egested (in *Actinia tenebrosa* this appears to be done by reversal of the flagellar beat). Immediately after egestion, the anemone contracts completely, expelling all the water from the coelenteron. It then starts to slowly reinflate. How these events are co-ordinated is not known, but they clearly show that the controlling mechanism is complex and could result from inherent rhythmic activity in the nervous system.

The apparent inactivity of anemones is deceptive. In *Metridium senile* there is a constant interplay of slow movements, described as inherent activity, which continues under constant environmental conditions where stimulation from external influences is minimal or absent. Sea anemones are seldom, if ever, completely at rest. Batham and Pantin (1950b) found that some specimens of *Metridium senile* showed unmistakable diurnal rhythmic activity which continued over a period of many days. In some specimens kept in constant darkness, the rhythm persisted with a free-running period of slightly less than 24h. These individuals showed entrainment to a reversed light/dark cycle, which strongly suggests that they possess a true circadian rhythm.

Other cases of spontaneous activity and inherent rhythms of shorter periodicity have been noted. For example, Jones *et al.* (1977) found a regular periodic deflection in their oxygen consumption recordings for *Actinia equina* and *Metridium senile* (mean periods 34 min and 28 min, respectively) which was attributed to regular periodic loss of water from the coelenteron through the mouth. Batham and Pantin (1950a) found that an isolated strip of circular muscle from *Metridium senile* exhibited a rhythm of spontaneous contractions (with a periodicity of *c.* 25 min) when subject to a slight tension, but the rhythm was not seen in the intact animal. McFarlane (1975) has recorded spontaneous low-frequency impulses (one every 60 s) in the endodermal slow conducting system (SS2) of *Calliactis parasitica*. Thus there are grounds for suspecting that one or more spontaneously firing neurones or pacemakers in the nerve nets are responsible for maintaining muscle tone and play a part in the inherent activity of the body.

Actiniarian behaviour might be explained as a sequence of reflexes, one activity providing the stimulus for the next, but there are serious difficulties with this idea, not the least of which is the extremely long delay that may occur between one activity and the onset of the next. To start with, it is necessary to ask what the term 'reflex' might mean in terms of anthozoan neural organization. It is impossible to equate any part of the classical reflex arc (sensory receptor – sensory neurone – interneurones – motor neurone – specific muscle) with what we know of the anemone neuromuscular system. Sensory receptors are abundant in both ectoderm and endoderm, but as far as is known, they are not connected to any specific sensory or motor pathway. Over most of the conducting systems, there is no polarity in direction of impulse propagation, thus the impulses that arise from stimulation of a particular sensory receptor will spread in all directions. Which muscles contract is determined by the frequency of impulses, not be specific motor connections. All the evidence points to the absence of clearly defined pathways from sensory receptors to specific muscles (with the possible exception of the through-conducting system and the sphincter muscle), and so the analogy with reflex arcs in other animals breaks down.

Even in those cases where a sense receptor and its adequate stimulus have been established, the response called forth is seldom a single movement, but rather a sequence or pattern of movements (fixed action patterns), thus there is seldom a clear one-to-one relationship between cause and effect. A further problem concerns the repeatability of the response. There are few responses that can be repeated time and time again in exactly the same manner; for most there is considerable variability.

The intertidal sea anemone *Actinia tenebrosa* can move its position on the rock surface. Detached specimens, freshly collected from the shore and placed in an aquarium, will attach themselves to the bottom. If they come to lie on their side or with the tentacles downwards, the base is slowly dilated until part touches the substratum. This region adheres and very slowly the anemone rights itself and attaches the rest of the base. The whole process may be accomplished in an hour or it may take up to 2 d. Some specimens move up the side of the aquarium and come to rest above the water line in a contracted state. There is also a general tendency for individuals to move away from light. This vertical migration takes place in darkness and progress is slow (usually *c.* 15–20 cm overnight). The side of the column in the direction of progress is periodically inflated so that the base is stretched 1–2 mm over the surface. It is assumed that basal muscles then pull the rest of the base foward. This behaviour enables specimens on the shore to change to a new site, presumably if the present position is in some way unsatisfactory, but how an anemone can detect light and move away from it is not known. The whole pattern of behaviour is, indeed, complex.

10.6 PHYSIOLOGICAL ADAPTATIONS OF SEA ANEMONES TO SHORE LIFE

The problems faced by suspension feeders living on the middle and upper shore have already been discussed (see page 39). However, few sea anemones feed on fine plankton or particulate material. They may be regarded as passive raptorial feeders, for, apart from pre-feeding behaviour involving maximal extension of the tentacles and possible lowering of the threshold for discharge of nematocysts, cnidarians do not seek out or pursue their prey. Spirocysts and nematocysts do not discharge unless stimulated to do so by potential food actually touching the tentacles.

Although sea anemones take planktonic crustaceans (copepods, nauplii), many capture relatively large items of food such as crustaceans, particularly amphipods, and polychaetes. There may be a distinct advantage to living on the middle and upper shore, for it is well known that many species of intertidal amphipod show a peak of swimming activity as they are covered by the rising tide. Many other animals show a similar surge of activity at this time, therefore the advancing tide carries an unusually rich supply of large potential food items. The biomass of one amphipod or small syllid worm will be equivalent to that of many copepods or planktonic larvae and will represent a saving in feeding time which would compensate for the longer periods of emersion.

An interesting comparison between the food intake of individuals of *Anthopleura elegantissima* living high on the shore and those at low tide level has been made by Zamer (1986). High-level anemones averaged only 3 h immersion per tidal cycle, yet they contained about three times as much food (by weight) in their coelenteron as low-level individuals able to feed almost continuously. This appeared to be the result of a higher rate of capture resulting from increased receptivity among high-level anemones. They not only absorbed food more efficiently, but they had a higher daily growth rate. Even on a low natural food intake (1% of the anemone's dry body weight per day), they reached reproductive size earlier than low-level anemones.

During normal digestion of larger food items, extracellular digestion takes place between the gastric filaments as already described (see pages 198, 211). The enzyme secretion of *Metridium senile* has a strong chitinase activity

(Elyakova, 1972), which would break down the cuticle of crustacean prey. Proteolytic enzymes will reduce the tissues to small particles, macromolecules and amino acids. The last are absorbed by cells of the cnido-glandula tract (Fig. 10.4(a)). The ectoderm of the column and tentacles is also capable of absorbing these smaller molecules. Macromolecules and particles up to *c.* 10 μm are taken up by phagocytic (digestive) cells in the general endoderm of the column and mesenteries as well as the intermediate tracts on the gastric filaments (Van-Praët, 1985).

Many Anthozoa possess symbiotic dinoflagellates belonging to the genus *Symbiodinium* (= *Gymnodinium*) in the endodermal cells of the body and tentacles. The symbionts are present as zooxanthellae that can carry out photosynthesis in the host's tissues when exposed to sunlight (Yonge, 1963; Muscatine, 1974). Various lines of evidence show that both host and symbiont derive some benefit from this mutualistic relationship, although both can live independently. Among the benefits to the host are complex organic substances (carbohydrates, amino acids), translocated from the zooxanthellae, which can be used as an energy source. Growth of freshwater green hydras (*Hydra viridis*) was little affected by a shortage of normal food, whereas aposymbiotic animals (i.e., those deprived of their zoochlorellae) showed a marked decrease in growth rate (Muscatine and Lenhoff, 1965). Goreau (1959) found that the rate of deposition of calcium carbonate in corals was higher in sunlight than in total darkness. This was attributed to the photosynthetic activity of the zooxanthellae, although the mechanism involved is not clear, for the symbionts in the endoderm are not near the site of mineral deposition.

It is usually assumed that one of the principal roles of the zooxanthellae is to remove and assimilate ammonium ions that arise from the host's metabolism, but Rees (1987) has pointed out that all the evidence for this is indirect. Host glutamate synthetase levels may be high, and this suggests that the host anemone, not the symbiotic algae, assimilates the ammonium ions.

During photosynthesis, zooxanthellae synthesize more nutrients than they require for their own metabolic needs, so the excess can be passed to the host. Trench (1971) made the surprising discover that freshly isolated zooxanthellae, when incubated in a ^{14}C-labelled carbon source, released massive amounts of nutrients to the medium on addition of tissue extracts from the host anemone. Up to 42% of the total ^{14}C-labelled material passed out of the symbiont, but without the host's tissue extract, little passed into the medium. This introduces a new facet on symbiosis, for it indicates that the host can induce the release of nutrients, such as glycolate and alanine, into its own tissues. The advantages to the host are clear. Not only does it receive a significant part (if not all) of its nutritional needs from the zooxanthellae, but it does not have to expend energy on food collection or digestion, as the nutrients pass directly into its own metabolic pathways.

It might be anticipated that some species will come to rely entirely on their symbionts for their energy source, and lose the ability to capture food in the normal way. This has occurred in certain xeniid alcyonaceans, which have numerous zooxanthellae in their tentacles but few nematocysts.

Investigating the mode of nutrition of *Heteroxenia fuscescens* from Madagascar, Schlichter (1982) found that nematocysts were scarce and the gastric filaments were much reduced, suggesting that *H. fuscescens* did not capture and digest food in the normal way of cnidarians. Polyps were not observed to feed under experimental conditions, but zooxanthellae in the endoderm carried out photosynthesis, and there was evidence that nutrients passed to the host. *H. fuscescens* appeared to be able to digest its zooxanthellae intracellularly, and extracellularly if they were extruded into the coelenteron. However, in the latter case, nutrients might pass out through the actinopharynx into the surrounding water. Some were absorbed by the actinopharynx, but glucose and amino acids were also absorbed by the ectoderm of the tentacles and column. Schlichter suggested that these nutrients might first be absorbed onto mucopolysaccharides secreted by the ectodermal cells. Under experimental conditions, up to 80% of the soft coral's energy needs could be met by the absorption of nutrients dissolved in the seawater alone. This ability, together with the nutrients derived by translocation from the zooxanthellae in the endoderm, appears to be more than sufficient to satisfy the metabolic demands of this animal.

The ectodermal cells of the tentacles in *Anemonia viridis* possess microvilli and are capable of absorbing L-amino acids and D-glucose from natural concentrations of these substances in seawater (Schlichter, 1975). Endoderm cells, however, absorbed very little. It was considered that up to 50% of the anemone's basal metabolic requirements could be met by absorption from the seawater. *Anthopleura elegantissima* possesses zooxanthellae in the endoderm of its tentacles and column. Some high-level anemones may not be able to feed for several consecutive days, and it is probable that their much-reduced metabolic demands can be met by nutrients translocated from the symbionts. Shick and Dykens (1984) found that anemones taken from the lower shore received *c.* 40% of their total assimilated energy from the zooxanthellae, the remainder coming from captured zooplankton, whereas high-level individuals that had been exposed for 15 h received only 17% of their energy from the symbionts. This was due to a reduction in photosynthesis, caused by withdrawal of the tentacles and partial closure which shaded the zooxanthellae. *A. elegantissima* appears to be able to regulate the rate of photosynthesis of its zooxanthellae by withdrawal behaviour and attaching debris to the column. This is necessary to protect the animal from the harmful effects of high oxygen levels, free oxygen radicals and photosensitization caused by high levels of photosynthesis during periods of peak insolation.

Zamer and Shick (1987) have calculated energy budgets for high- and low-shore sea anemones. Most of the energy was expended on respiration and excretion, but mucus secretion was estimated to account for 30% of the total energy budget. The remainder was designated 'scope for growth'. This was higher (29%) for high-level anemones than low ones (20%), showing that *A. elegantissima* is well adapted physiologically to life in the upper intertidal zone.

During emersion, high-level anemones showed lowered metabolic demands, but unlike *Mytilus edulis* (see page 106), *A. elegantissima* does not switch to anaerobic metabolism, and no oxygen debt is incurred. Even when the anemone is partly contracted, water is constantly circulated in the coelenteron and may be pumped in and out through the mouth onto the oral disc, where it will be reoxygenated (Shick and Dykens, 1984; Zamer and Shick, 1987). In addition, oxygen will be produced in the tissues by photosynthesis of the zooxanthellae, thus there is likely to be sufficient oxygen available during emersion for normal aerobic respiration.

Anthopleura elegantissima may not be typical of sea anemones in general as far as respiration is concerned, for a number of species have been shown to incur an oxygen debt under anoxic or hypoxic conditions. *Bunodosoma cavernata*, for example, can survive 6 weeks without oxygen. Ellington (1982) has shown that alanine, glutamate and propionate accumulate as end products during anaerobiosis, and well-defined oxygen and ammonia debts occur after hypoxia in this sea anemone. Alanine is broken down to ammonia during the period of recovery on return to normal aerobic conditions.

Although the problem of hypoxia during emersion is unlikely to be as acute for sea anemones as it is for *Mytilus edulis*, which closes its shell, it is likely that many anemones have the ability to switch to anaerobic pathways similar to those of molluscs. Support for this idea comes from the fact that the enzyme octopine dehydrogenase, which is a catalyst for the formation of octopine from arginine and pyruvate, occurs in the tissues of *Bunodosoma cavernata* and *Metridium senile* (Ellington, 1980).

SUMMARY

Several species of sea anemone live on the lower part of the middle shore. They have no hard protective shell, but the body is covered by a thin film of mucus which may prevent abrasion or microbial attack. The base is broad and attached to the substratum by a mucopolysaccharide cement. Most species can move slowly to a new location. During emersion or turbulent wave action the tentacles are withdrawn and the body contracted to a smooth spherical or conical shape. A large volume of water may be retained in the body cavity. Direct sunlight is avoided by living in the shade of rocks, and nematocysts in the ectoderm provide protection from predators.

Three forms of discharging capsule are found: spirocysts, nematocysts, and ptychocysts. Spirocysts and nematocysts are used in food capture; nematocysts may be used in defence, territorial behaviour and adhesion. Ptychocysts form a tube in which the anemone lives. Different types of nematocyst may be located in different parts of the body. In hydrozoans they are confined to the ectoderm, but in sea anemones they also occur in the gastric filaments. The discharge of some nematocysts is under the control of the nematocyte, but the discharge of others may be modified by nervous intervention. During formation of the nematocyst, the tubular nematoneme is inverted into the capsule; during discharge it is everted. Eversion of the nematoneme is partly by hydration which alters the molecular configuration, and partly by osmotic influx of water into the capsule. Many nematocysts contain substances in their internal fluid which are toxic to prey and predators. Food items are relatively large (consisting of crustaceans, fish, etc.) and so interruption of feeding by the tide is not of great importance. Some species derive nutrients from symbiotic zooxanthellae in their tissues, while other species are known to be able to absorb dissolved organic matter (amino acids) from the seawater.

Eggs are incubated in the coelenteron by a few species and this may be an adaptation to shore life.

Metabolism is reduced during emersion. Some species are able to use atmospheric oxygen when out of the water, by circulating fluid from the coelenteron over the oral disk and back. Anaerobiosis occurs in certain species with the formation of propionate. The oxygen debt is repaid on return to normal conditions.

11

Hydroids
(*Cnidaria: Hydrozoa*)

Their small size, delicate structure and absence of a hard protective shell make hydroid polyps (order Hydroida*) vulnerable to desiccation, and few species are encountered on the middle shore except in tide pools or sheltered crevices. Many are well adapted to turbulent conditions and reach their greatest abundance in the sublittoral region.

Unlike that of sea anemones, the typical life history of hydroids involves an alternation of generations between a sessile, asexually reproducing polyp and a sexually reproducing gonophore that may be a free-swimming medusa or a sessile structure on the polyp. By analogy with the compound ascidians, the free-swimming medusa is a blastozooid which produces either spermatozoa or ova (the sexes are separate in hydroids). There is a brief larval phase which settles and grows into the primary polyp or oozooid. This reproduces asexually to give rise to many new polyps or the next generation of blastozooids (medusae) to complete the cycle. This simple alternation of generations is subject to many modifications, particularly in the suppression of free-swimming medusae, which take the form of sessile gonophores on the oozooid or polyp phase.

Australomedusa baylii from SE Australia illustrates the typical life cycle of hydroids. At certain times of the year small jellyfish (2–5 mm in diameter) are common in the more saline region of the lagoons formed at the mouth of seasonal rivers by a sand bar (Fig. 11.1(b)). The medusae develop as buds on small polyps (3–5 mm high) found attached to submerged logs or stones (Fig. 11.1(a)). Each polyp has a ring of tentacles round the body and a terminal mouth. The mouth region is remarkable for its mobility, for it can be extended like an elephant's trunk to swallow copepods captured by the tentacles. Buds just below the tentacles develop into medusae, which are nourished by the voracious parent polyp. They grow rapidly and soon start to

* See Appendix for classification of hydroids.

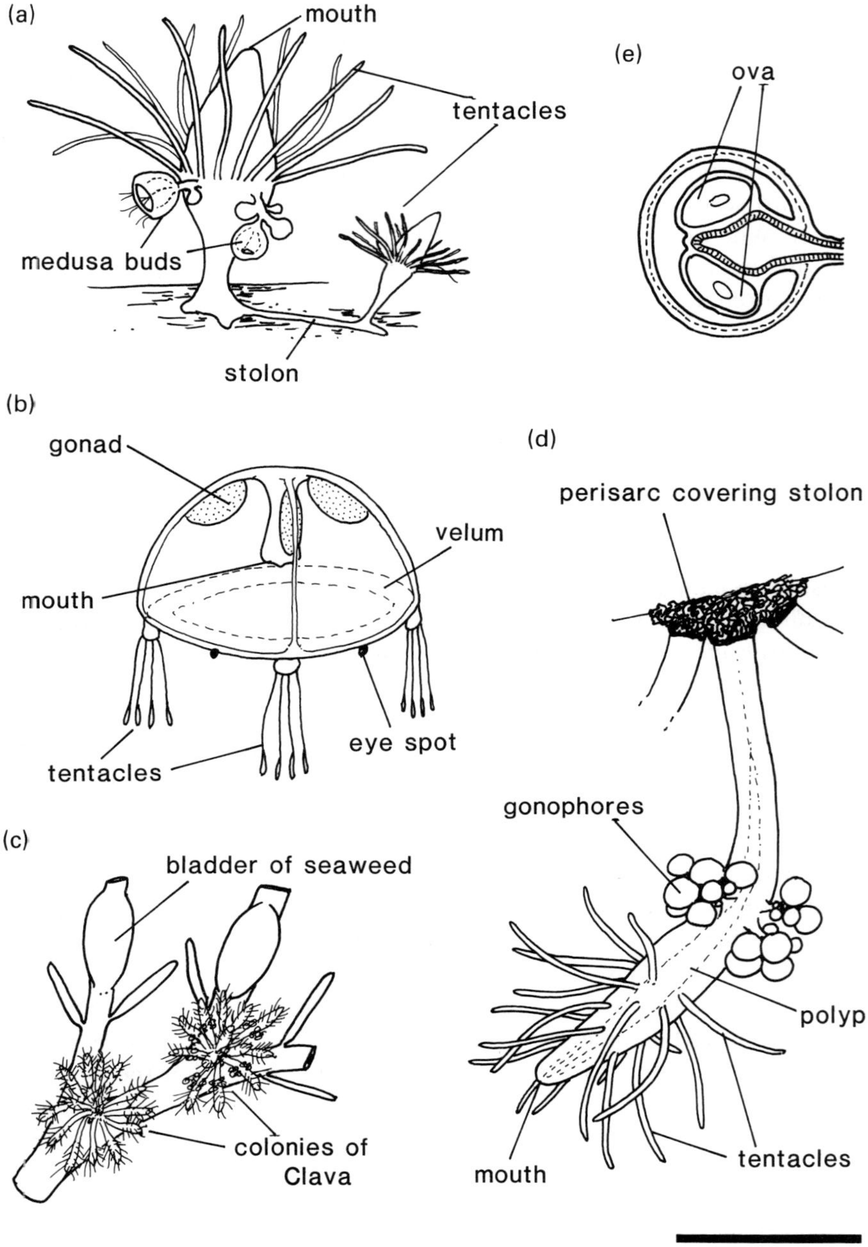

Fig. 11.1 (a) and (b) *Australomedusa baylii*: (a) polyp. (b) Medusa. (c) to (e) *Clava multicornis*. Natural appearance on *Ascophyllum*. (d) Single female polyp. (e) Gonophore of (d). Scale bar: (a) = 2.5 mm, (b) and (d) = 3.5 mm.

pulsate. Eventually they break free, swim away, feed, grow, and develop gonads. Up to two medusae per day may be released from a polyp.

All the medusae derived from the same polyp are of the same sex. This has led to a curious distribution of the species. In some lagoons only female medusae are found, while in others only males occur. Under these conditions sexual reproduction cannot take place, but it must be presumed that where the two sexes do occur in the same lagoon, sexual reproduction will take place. Where only one sex occurs in a lagoon, the species must be maintained by asexual reproductive methods. From the base of the polyp one or more tubular outgrowths, the stolons, grow over the surface and at intervals give rise to new polyps.

This species has been named from the medusae only (Russell, 1970, 1971). The polyp phase has not been described previously.

11.1 STRUCTURAL ORGANIZATION AND EXAMPLES OF INTERTIDAL HYDROIDS

A species of *Eleutheria* (= *Clavatella*) is extremely abundant about mean tide level on the seaweeds *Hormosira banksii* and *Lobophora* sp., growing in shallow tide pools on the East Coast of Australia. The medusae cannot swim; they can walk about or hold on to the surface of seaweed by means of six highly modified, tubular tentacles, each of which ends in an adhesive organ. *Eleutheria* can cling tenaciously to the seaweed in turbulent water (Fig. 11.2(a)–(d)). Half-way along each tentacle, there is a branch bearing a ball of large nematocysts, the acrosphere, used to capture copepods. Their appetite is gargantuan; they can swallow three copepods as large as their own body in quick succession (Fig. 11.15(b)).

The sessile polyp phase of *Eleutheria* consists of a horizontal stolon firmly attached to the substratum by a hard protective perisarc. Each polyp that arises from the stolon has a tall cylindrical body with four horizontal tentacles, each bearing an acrosphere charged with nematocysts (Fig. 11.2(c)). Buds form near the base of the polyp and grow into medusae, which eventually break free to lead an independent life. The medusae reproduce asexually and produce buds on their dorsal (aboral) surface (Fig. 11.2(b)). Budding is prolific; a new bud arises each day. It takes 2 d to grow into a medusa, and 2 d later this medusa starts budding, thus when food is abundant, *Eleutheria* can undergo a population explosion. Medusae also reproduce sexually. Female medusae retain their fertilized eggs in a dorsal brood chamber (Fig. 11.2(a)), where they develop into barrel-shaped planulae, which swim about for up to 2 d after release. On settling, the planula undergoes a reorganization of its cells and secretes a protective covering, the perisarc. Later a new polyp with four tentacles develops and

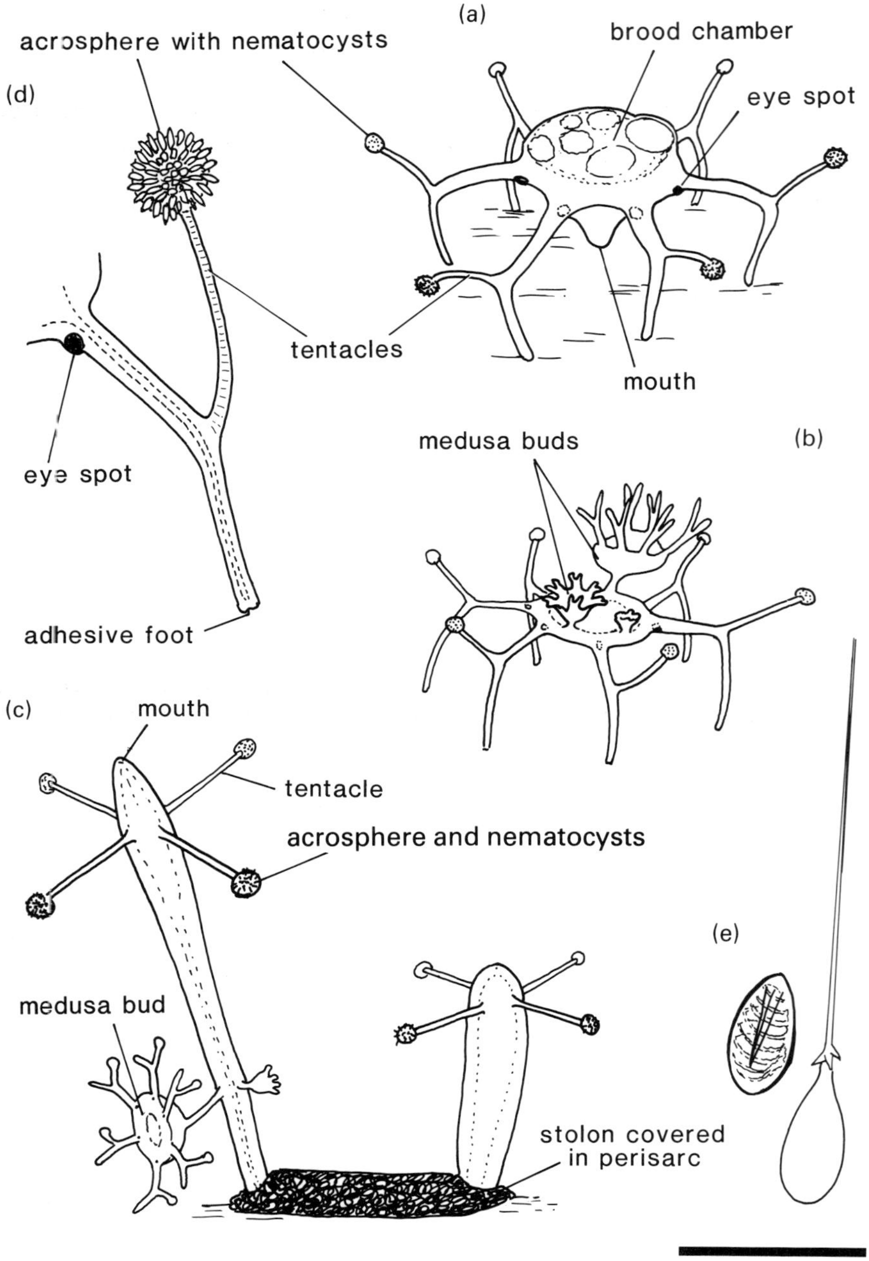

Fig. 11.2 Eleutheria sp. (a) Female medusa with developing eggs in brood pouch. (b) Budding medusa. (c) Polyps. (d) Tentacle of medusa. (e) Nematocysts from acrosphere (undischarged and discharged). Scale bar: (a), (b), (c) = 0.5 mm, (e) = 7 μm.

starts feeding. The stolon grows and produces more polyps before medusa buds appear.

Both *Australomedusa baylii* and *Eleutheria* belong to the suborder Athecata or Gymnoblastea, in which there is no protective cuticle or perisarc round the polyp or gonophores.

Silicularia bilabiata (Fig. 11.3(a)) is a very different hydroid, well adapted to surf action during low tide on rocky reefs in Australia and New Zealand. It is common on the basal parts of the seaweeds *Phyllospora comosa* and *Cystophora* spp. at low water of spring tides. The hydroid phase is firmly attached to the seaweed by an extensive stolonic network, from which short vertical stalks arise at regular intervals, each bearing one polyp or hydranth. Hydranths are partly protected by a very thick, two-lipped, asymmetrical cup or hydrotheca of perisarc, which is attached to the stalk by a very flexible joint (Fig. 11.3(b) and (c)). About 30 solid tentacles with rings of small nematocysts form a ring round the mouth region or hypostome. These hydranths are incapable of any form of reproduction, and are only concerned with feeding.

Flat purse-shaped structures, the gonothecae, also arise from the stolon on very short stalks. They contain a highly modified polyp, the blastostyle, which produces just one medusa at a time by asexual budding (Fig. 11.3(d)). The bell-shaped medusa is poorly developed and, although it can swim by pulsation, it has neither mouth nor tentacles with which to feed, so the gametes must receive their nourishment indirectly from the hydranth polyps. The gonads develop on the medusa and produce eggs or sperm (according to the sex of the hydroid) whilst it is still in the gonotheca. The medusa is short-lived and probably sheds its gametes soon after leaving the gonotheca; indeed, many if not most medusae are not released, and shed their gametes whilst still in the gonotheca. Imperfect medusae like this are said to be eumedusoid.

Several features of *Silicularia bilabiata* are well suited to life in turbulent waters. The extensive stolon, with its low profile and lateral rhizoids, provides a strong attachment to the substratum, while the short pedicels and prostrate gonothecae reduce drag in high-velocity currents. The flexibility of the hydranth joint allows the hydrotheca to bend with the current regardless of its direction, and the massively thick perisarc of the cup, pedicels and stolon probably prevents crushing by the thallus of the seaweed as it is tossed about in the surf. Its rigidity holds the pedicels upright. The partial supression of the free-swimming stage may be related to the unsuitability of turbulent inshore waters for delicate medusae. When a specimen of *Silicularia bilabiata* is disturbed (for example, by being transported to the laboratory), the feeding hydranths regress. Their substance is resorbed into the stolon, the empty hydrotheca breaks away, and perisarc seals the distal end of the pedicel. New hydranths may develop from the old pedicels, but when they do, a new

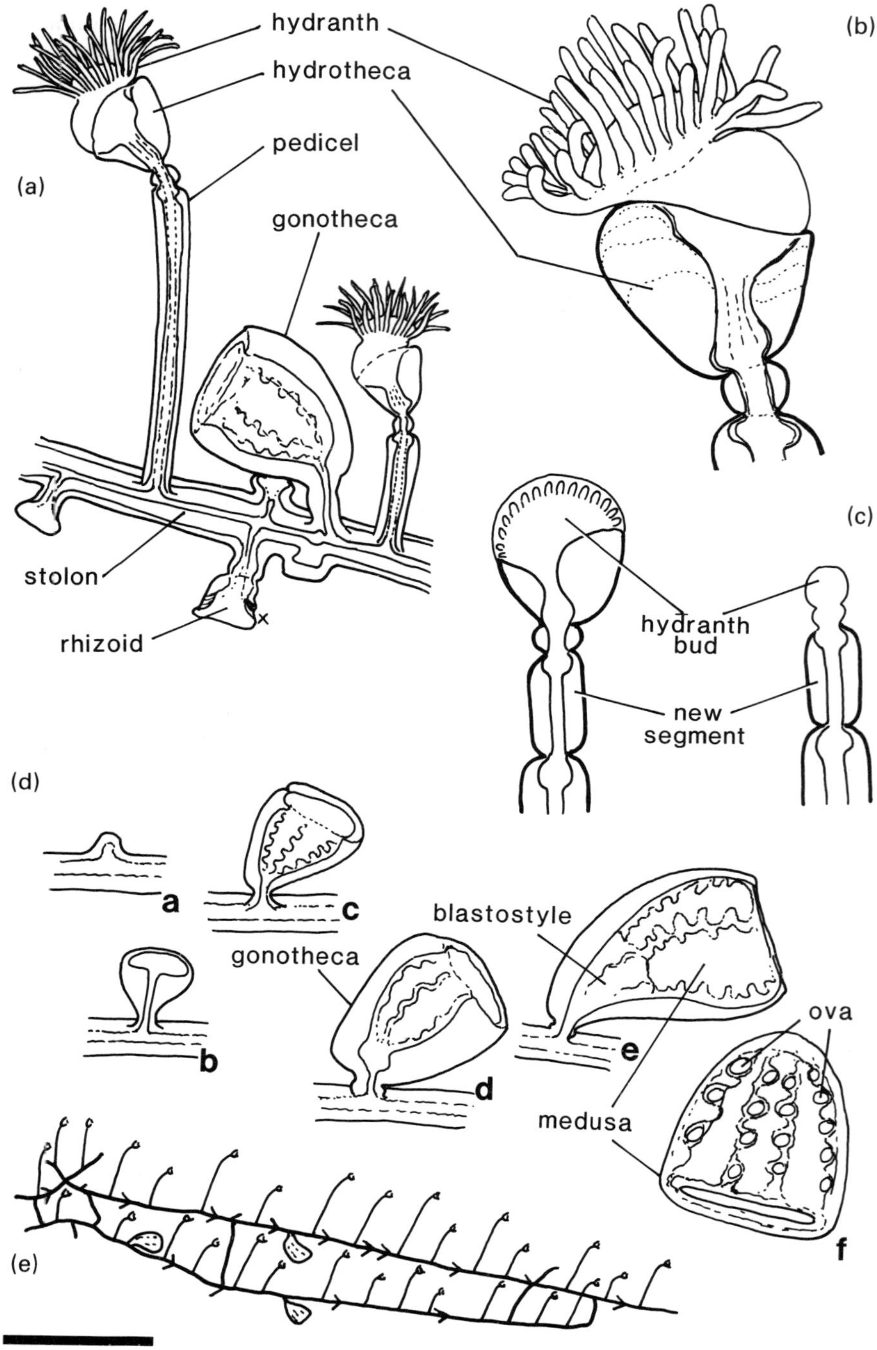
hydranth
hydrotheca
pedicel
(a)
(b)
gonotheca
stolon
rhizoid
(c)
hydranth bud
new segment
(d)
a
b
c
gonotheca
d
blastostyle
e
ova
medusa
f
(e)

segment is added (Fig. 11.3(c)). When an extensive growth is examined closely, a small proportion of the hydranths will be found in various stages of regression or regeneration, showing that this is a normal process and that hydranths are replaced from time to time.

In contrast to athecate hydroids, the medusa of *Silicularia bilabiata* is produced on a specialized polyp (called the blastostyle) within a cup-like gonotheca of perisarc. Accordingly it is placed in the suborder Thecata or Calyptoblastea.

A common hydroid encountered up to mean tide level on the shores of western England is *Dynamena pumila*. It is often found growing near the base of the thallus on *Fucus vesiculosus* and *F. serratus*. A closely related species from Australia and New Zealand, *Amphisbetia minima* (Fig. 11.4(b) to (d)), occupies a similar position, where it is equally abundant on the underside of rocks, hanging from the roof of caverns or growing on seaweeds such as *Ecklonia radiata* at low tide level. The stolon forms a rectangular network over the surface and gives off vertical hydrocauli, 10–15 mm high, at regular intervals. Hydranths are arranged in opposite pairs in sessile hydrothecae, which are fused along one side of the hydrocaulus. Both hydranth and hydrotheca are asymmetrical. A pear-shaped gonotheca is attached just below the first pair of hydranths, but free-swimming medusae are not produced. Near the base of the hydrocaulus, there is a conspicuous hinge joint which allows the whole stem to fold over in strong currents (Fig. 11.4(d)). The small hydranth can be completely withdrawn to the bottom of the strong tubular hydrotheca for protection, and a thin flap of perisarc passively closes the aperture to retain a drop of seawater. This may explain why *Amphisbetia minima* can survive regular emersion at low tide in sheltered places.

Many families of both thecate and athecate hydroids have members living at low tide level and in the sublittoral, but few can be considered specially adapted for life in the intertidal zone, apart from modifications for living in turbulent waters. Nevertheless some show points of general interest to hydroid biology and will be included in the following discussions. Central to this account of hydroids are personal studies of an unusual species, *Thyroscyphus marginatus*, from Australia, which help to clarify some aspects of hydroid organization and behaviour.

Fig. 11.3 Silicularia bilabiata: (a) Portion of hydroid, X marks point at which thickening of the chitinous perisarc occurs (page 239). (b) Hydranth (preserved specimen). (c) Regeneration of hydranth. (d) a–f: Formation of medusa in gonotheca; f, free medusa. (e) Pattern of growth. Scale bar: (a) = 1 mm, (b) = 0.4 mm, (d) = 0.7 mm.

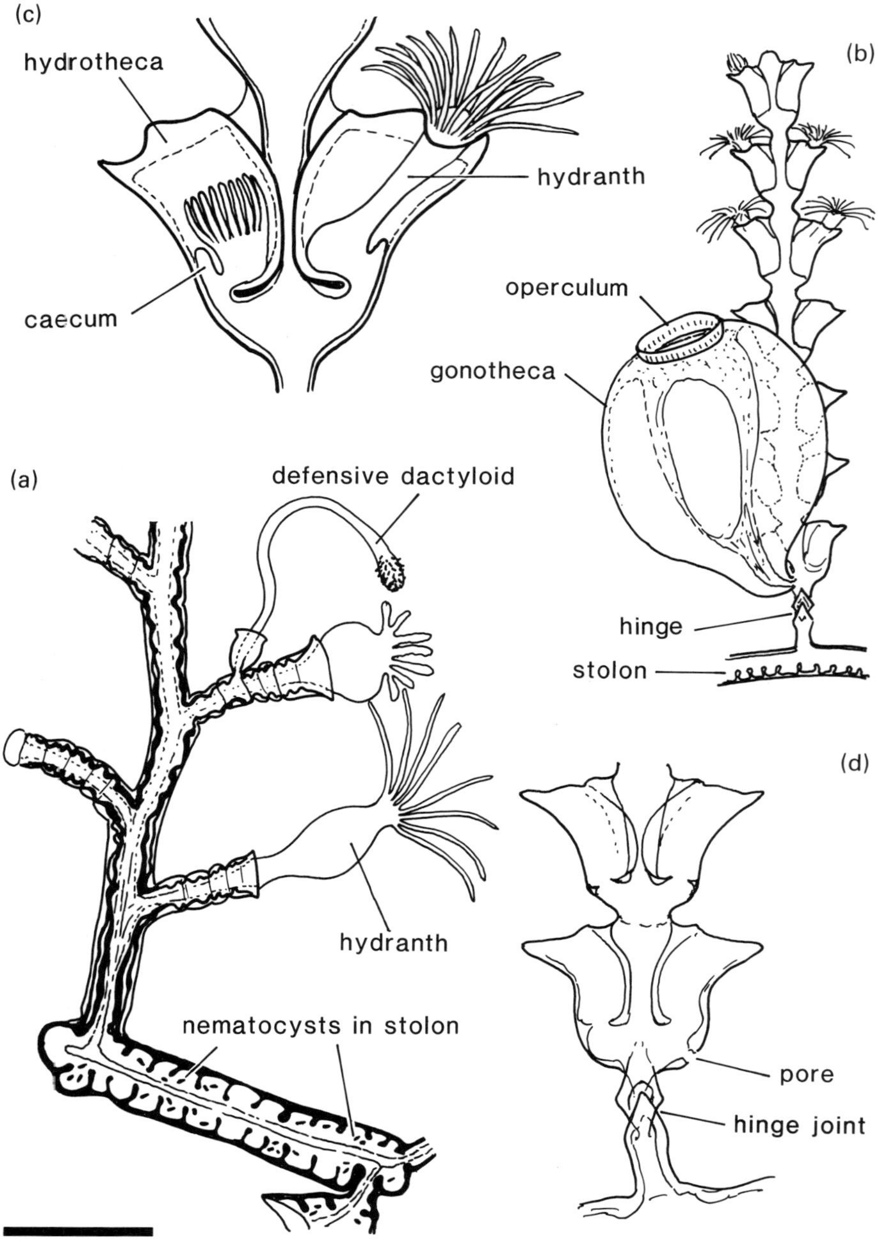

Fig. 11.4 (a) *Phylactotheca armata*. Portion of hydroid, approximately × 12. (b) to (d) *Amphisbetia minima*: (b) single hydrocaulus with gonotheca. (c) hydranths; (d) articulation at base of hydrocaulus. Scale bar: (a) = 0.5 mm, (b) = 0.6 mm.

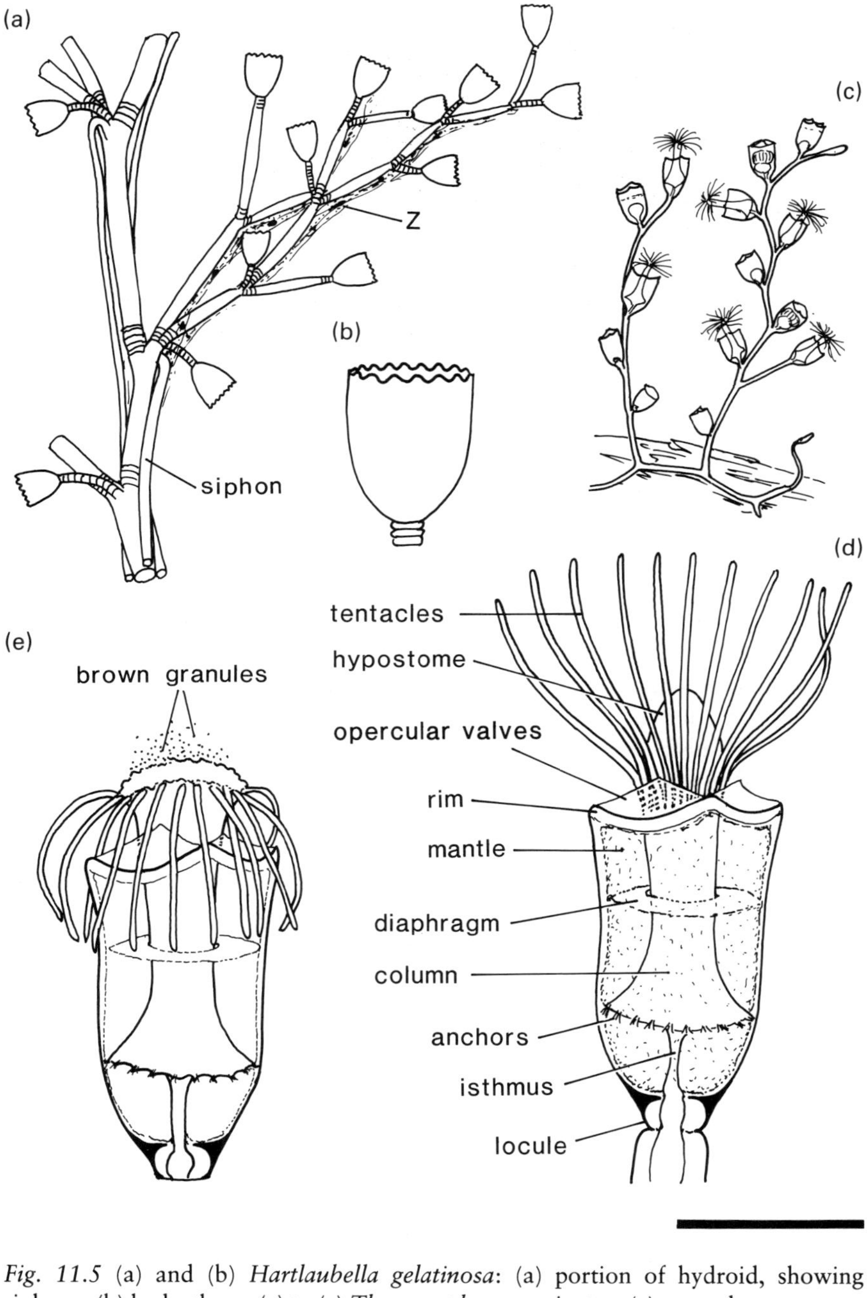

Fig. 11.5 (a) and (b) *Hartlaubella gelatinosa*: (a) portion of hydroid, showing siphons; (b) hydrotheca. (c) to (e) *Thyroscyphus marginatus*: (c) natural appearance. (d) expanded hydranth; (e) process of egestion. Scale bar: (a) = 8 mm, (c) = 5 mm, (d) and (e) = 0.75 mm.

Thyroscyphus marginatus (Fig. 11.5(c)) has relatively large pink hydranths (*c.* 4 mm when expanded) which make it an ideal animal for study. The irregular stolon is usually embedded in other encrusting organisms and gives off vertical hydrocauli, 30–40 mm high, which may be branched, but have no joints or annulations. The hydrocauli bear alternate hydranths. Each hydrotheca is attached by a short, inflexible pedicel, and its aperture is passively closed by four delicate opercular valves (Fig. 11.5(d)) when the hydranth is withdrawn.

The fine structure of hydroids seen in histological sections is difficult to interpret for several reasons. Expansion and contraction of the polyp causes great distortion of cell shapes, and cyclic processes, such as secretion, digestion and regression, constantly change the appearance and staining properties of the cells. There are also marked differences in the appearance of cells between their developmental and functional stages. The following account is based on *Thyroscyphus marginatus* and Fig. 11.6 shows the appearance and position of the various cell types.

A non-living perisarc covering all parts of the stolon and hydrocaulus provides both protection and support. The living tissue inside (called coenosarc) is a tube composed of two layers; the ectoderm and endoderm, separated by an extremely thin non-cellular mesogloea. In the stolon the cytoplasm of both ectodermal and endodermal cells is granular. Among the ectodermal cells there are large orange amoeboid cells and scattered clusters of developing nematocysts. Flagellate cells occur in the endoderm and many of the other cells contain large granules, suggesting that they store nutrients. The central cavity communicates with all parts of the hydroid, including the digestive cavity of the feeding hydranths; it is called the gastro-vascular system, and fluid periodically moves backwards and forwards in it.

The histological appearance of the tissues in the hydranth is radically different from that of the stolon owing to functional and regional differentiation of the cell types.

The principal cell type is the epithelio-muscular cell, already noted in sea anemones, which is present in both ectoderm and endoderm. The main part of the cell appears empty because of a large vacuole, the cytoplasm is clear, and a thin layer of mucopolysaccharide covers the external surface. Adjacent cells are joined by septate junctions to form an epithelium (Haynes, 1973). Muscle tails containing contractile fibres are attached to the cell in the plane of the mesogloea. At either extremity they form junctions with adjacent muscle tails, forming a contractile network over the inner and outer surfaces of the mesogloea. Muscle tails of ectodermal cells run longitudinally (parallel to the axis of the polyp), and those of endodermal cells run circumferentially. This network of contractile fibres is really part of the ectodermal and endodermal epithelium, and is not to be confused with the

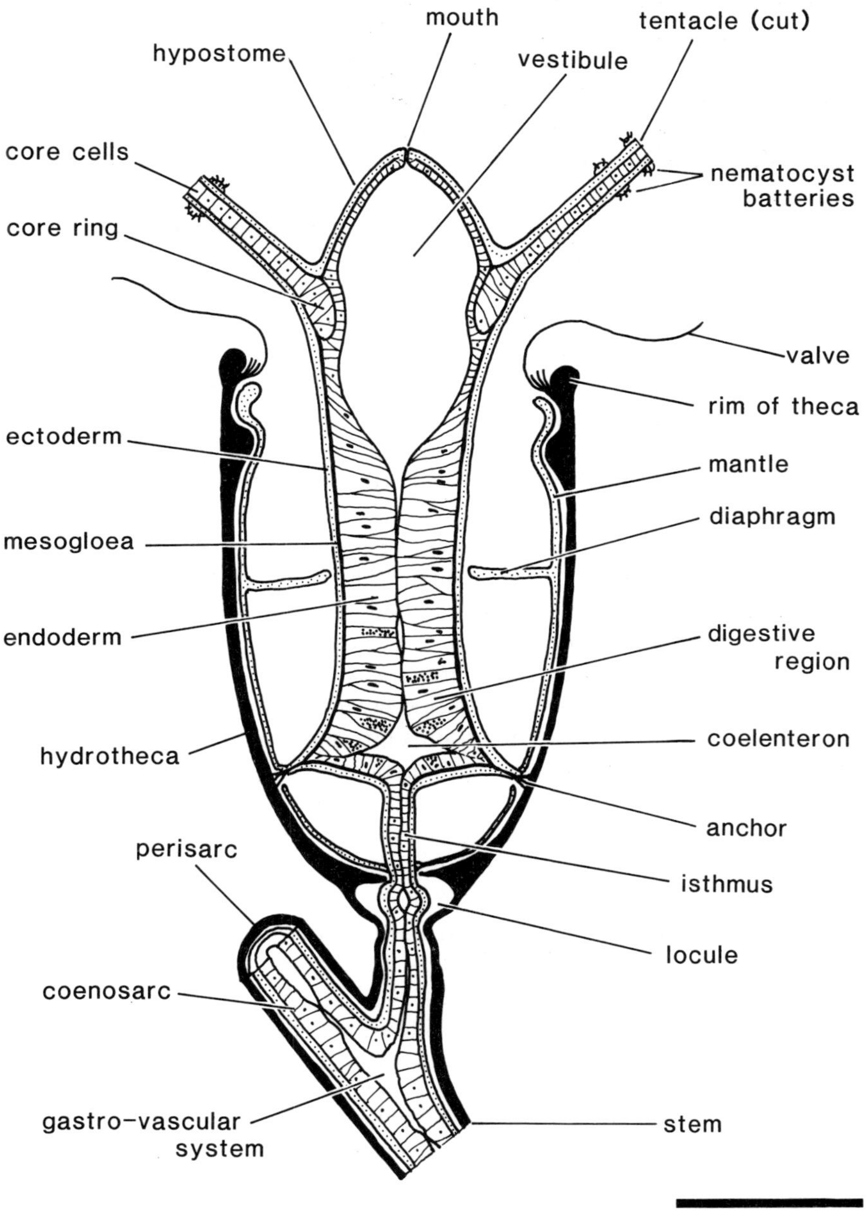

Fig. 11.6 Structure of *Thyroscyphus* (semi-diagrammatic). Scale bar = 0.4 mm.

myocytes or muscle cells found in other Metazoa; nevertheless, for convenience in the following account they will be called longitudinal and circular muscles.

Because the epithelio-muscular cells narrow towards the base, a fluid-filled space is formed in the ectoderm between the cells (Fig. 11.7(a)). There are two possible functional reasons for this arrangement. Elongation and contraction cause tremendous changes in the shape of the epithelial cells. Their narrow base and the fluid space will enable these changes to take place rapidly and smoothly. Movements of the body will also cause displacement of this fluid, which could exert hydrostatic pressures elsewhere in the ectoderm, as discussed for sea anemones (see page 208 and Fig. 10.10(b) and (c)).

On the hypostome the ectodermal muscle fibres run radially towards the mouth. The epithelio-muscular cells of the tentacles are flat (squamous), and the contractile fibres run both lengthwise and transverse to the tentacle.

During development, ectoderm cells migrate up the inside of the hydrotheca in *T. marginatus* to form a mantle – a structure unique to this species. The function of the mantle is not known, but it usually forms a diaphragm round the hydranth (Fig. 11.5(d)) and may completely enclose the hydranth when the latter is withdrawn. There is no mesogloea or endoderm in the mantle, and muscle tails are absent. The various changes in shape and position of the mantle appear to be the result of migratory amoeboid movements of the component cells.

Other cells are found in the ectoderm. Conspicuous among these are orange amoeboid cells that can be seen streaming from the hydrocaulus into the developing hydranth bud. Some pass into the mantle, while others move up the column and accumulate in the hypostome, but they do not migrate onto the tentacles. These amoeboid cells are elongate whilst moving, but under strong illumination (from the microscope lamp) they stop and shorten to an oval outline (Fig. 11.7(c)). It is the amoeboid cells that give *T. marginatus* its pink colour. Each contains about 20 orange-red spherical inclusions whose function will be discussed later (see page 238).

Fig. 11.7 Fine structure of hydroids. (a) Body wall, showing epithelio–muscular cells (diagrammatic). (b)–(e) Histological structure of *Thyroscyphus*, vertical sections through hydranth; (b) L.S. tentacle; (c) Section through hypostome; (d) Section through rim of hydrotheca; (e) Section through digestive region. Scale bar: (b)–(e) = 0.04 mm. br. gr., brown granules; circ. musc., circular muscle fibres; dig. c., digestive cells; ect. c., ectodermal epithelio-muscular cells; end. c., endodermal cells; f. vac., food vacuole; flag., flagellum; hth., hydrotheca; l. nem., large nematocyst; long. musc., longitudinal muscle fibres; mant. c., mantle cells; mes., mesogloea; mu. c., mucous cells; nem., nematocysts; or. c., orange cell; rad. musc., radial muscle fibres; vac. core c., vacuolated core cells; y. c., yellow cells; zy. c., enzyme-secreting cell with secretory granules.

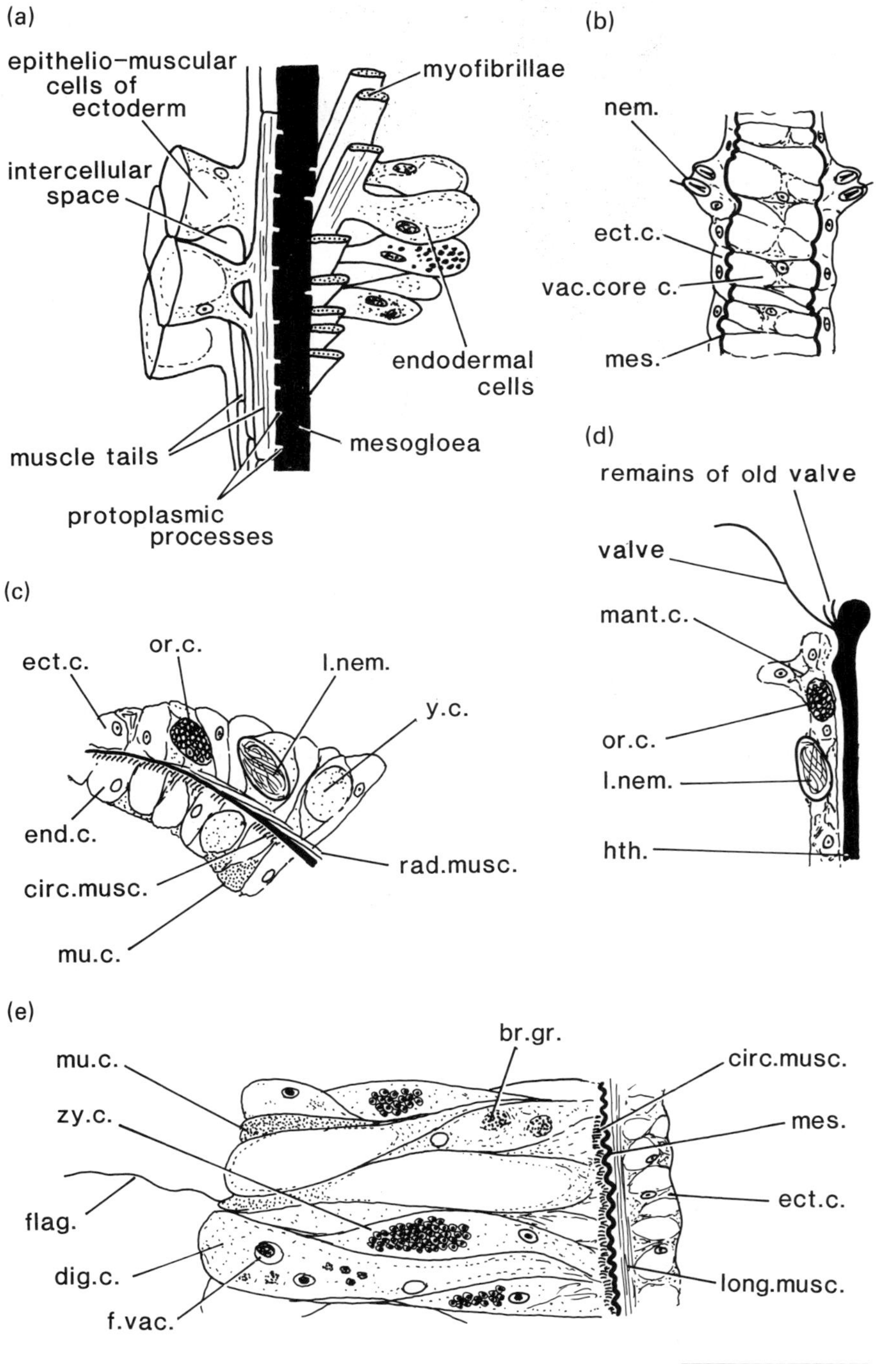
(a)
epithelio–muscular cells of ectoderm
intercellular space
myofibrillae
endodermal cells
muscle tails
mesogloea
protoplasmic processes
(b)
nem.
ect.c.
vac.core c.
mes.
(c)
ect.c.
or.c.
l.nem.
y.c.
end.c.
circ.musc.
rad.musc.
mu.c.
(d)
remains of old valve
valve
mant.c.
or.c.
l.nem.
hth.
(e)
br.gr.
mu.c.
zy.c.
circ.musc.
mes.
ect.c.
flag.
dig.c.
f.vac.
long.musc.

During development, large numbers of nematocytes, each carrying a fully formed nematocyst, can be seen migrating from the stolon and hydrocaulus into the hydranth bud on living specimens. Two types can be distinguished: those with nematocysts only 6 μm long, which migrate exclusively to the tentacles, where they take up position in rings of 30 or 40, and those with larger nematocysts 26 μm long (Fig. 11.7(b) and (c)). The latter never pass onto the tentacles but accumulate on the hypostome and in the mantle. The mechanism responsible for this segregation of nematocysts is not known, but it is not alone, for similar phenomena have been observed in other thecate hydroids. Large nematocysts are conspicuous in the stolon of *Phylactotheca armata* (Fig. 11.4(a)). This strange hydroid, found on *Ecklonia radiata* and other seaweeds at low tide in Australia, has snake-like protective dactyloids as well as hydranths, and during development of a new hydranth bud, the large nematocysts pass exclusively to the dactyloid. Only small nematocysts pass to the tentacles of the hydranth. Similarly, in *Aglaophenia plumosa* (another Australian species), large elongate nematocysts pass into the nematothecae (Fig. 11.8(b)), not the hydranth, during development or regeneration.

In histological sections, large oval cells with granular cytoplasm and a pale nucleus accumulate at the base of the tentacles and on the hypostome. They stain yellow with orange G stain and appear to be involved in excretion, for they are lost along with the nematocysts during regression of the hydranth (see page 260). A ring of bipolar nerve cells is located in the ectoderm at the base of the tentacles. At the base of the hydranth, a ring of anchor-like structures attaches the polyp to the hydrotheca that protects it (Fig. 11.6).

The mesogloea of the hydranth differs from that of sea anemones in being much thinner and devoid of cells. Suitable staining reveals both circular and longitudinal fibres in the mesogloea, but whether it has the same visco-elastic properties as in sea anemones is uncertain. When the longitudinal muscle contracts, the mesogloea is thrown into fine folds (Fig. 11.7(e)).

The endoderm consists of highly vacuolated epithelio-muscular cells as already noted, but nearly all are further specialized as mucous, enzyme-secreting or digestive cells (Fig. 11.7(e)). The mucous cells are more abundant in the region of the hypostome and vestibule, where they accompany normal epithelio-muscular cells. Enzyme-secreting cells release their secretory spherules into the coelenteron, where the enzymes break down food organisms by extracellular digestion to small particles, which can be phagocytosed by the digestive cells. The final stages of digestion are carried out intracellularly within the food vacuoles of the digestive cells. Brown waste particles may accumulate in the cytoplasm. Histological evidence suggests that the digestive cells break down and their contents pass into the coelenteron or through the gastro-vascular system to other parts of the organism.

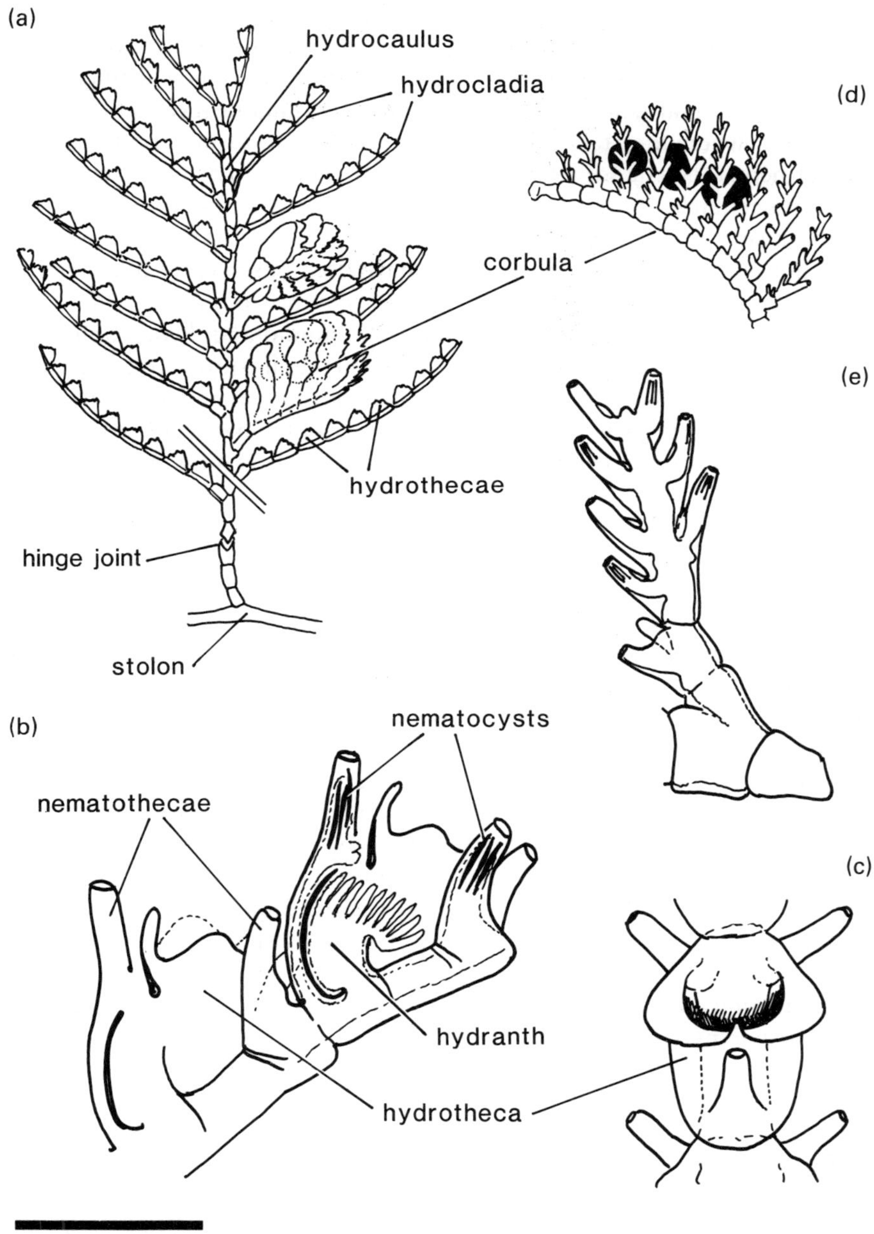

Fig. 11.8 Aglaophenia plumosa. (a) Part of plume-like hydroid. (b) Sessile hydrothecae with nematothecae (side view). (c) Same as (b), viewed from above. (d) Corbula of female *A. acanthocarpus*. (e) Single gonohydrocladium from corbula, to show nematothecae. Scale bar: (a) = 1.5 mm, (b) = 0.15 mm.

The tentacles have a solid core of large, highly vacuolated cells with a central nucleus that are derived from the endoderm during development (Fig. 11.7(b)). They contain oblong crystals which are strongly birefringent in polarized light and give the tentacles a silver-white appearance. At their base, the core cells of the tentacles join a ring of similar vacuolated cells, which encircle the hydranth just below the hypostome (Fig. 11.6).

11.2 IMPORTANCE OF THE PERISARC TO HYDROIDS

The tough cuticular perisarc covering the stolon and hydrocauli is a characteristic feature of hydroids, and in the Thecata it covers the blastostyle. A tubular or bell-shaped hydrotheca may protect the hydranth when contracted, but this is not always the case, as will be seen in *Phylactotheca armata* (Fig. 11.4(a)). The perisarc performs a number of functions besides protection.

Perisarc structure and function

The external surface of athecate polyps like *Clava* (Fig. 11.1(d)) and *Sarsia* (= *Syncoryne*) *tenella* is covered by a thin flexible film of mucoprotein which acts as a form of cuticle. The true perisarc is more complex. It is composed of chitin and protein. Histochemical tests show that the outer layer of the bilaminate perisarc of *Sarsia* is a mucopolysaccharide and the inner layer is a complex of chitin and protein (Wineera, 1972).

Dopamine, the precursor of quinone substances, is present as spherical inclusions in the amoeboid tanning cells found in the ectoderm of *Laomedea flexuosa* (Fig. 11.12(a)), and the enzyme *ortho*-phenol oxidase has been detected in both the tanning cells and the perisarc (Knight, 1968, 1970). Thus there is reason to believe that sclerotization of the chitin–protein complex by tanning can occur in hydroids. The older part of the perisarc in some species is brown, indicating tanning, but in most specimens the perisarc is colourless and transparent, suggesting that some other form of sclerotization, perhaps β-sclerotization, takes place. The orange amoeboid cells in the ectoderm of *Thyroscyphus marginatus* show the same staining properties as the tanning cells of *Laomedea flexuosa* and are presumed to be identical, but this leads to a difficulty, for the orange cells are present all the time, not just when perisarc is being produced, and they collect on the hypostome, where there is no perisarc. It is possible that the quinone substances fulfil some other function, such as acting as an antibiotic or preventing other organisms from settling on the hydroid.

The perisarc is usually considered to be produced by the ectoderm, but this needs qualification, for the ectoderm of a feeding hydranth is unable to repair any damage done to its hydrotheca. The hydrotheca is secreted at an early stage in the development of a new hydranth bud, long before the ectodermal cells have undergone differentiation to epithelio-muscular cells. In a bud of *T. marginatus*, it is possible by means of their staining properties to distinguish a ring of cells round the periphery of the cap that are responsible for the secretion of the rim and opercular valves. It is likely that the same cells secrete the entire hydrotheca as the bud increases in size. Once the ectodermal cells of the bud differentiate into epithelio-muscular cells, they lose their capacity to secrete cuticle. In species like *Phylactotheca armata*, the laminated appearance of the hydrocaulus suggests that the ectodermal cells of the hydrocaulus retain their capacity to secrete new perisarc, and this takes place periodically on the inside of the old perisarc (Fig. 11.4(a)).

Attachment to substrate

It is the perisarc of the stolon that attaches the hydroid to the substratum. The nature of the attachment is seen best in species that live on the smooth surface of seaweeds. In *Silicularia bilabiata*, for example, the perisarc of the stolon is extended laterally to give a broad area of contact with the surface. The growing tips of the stolon and rhizoids terminate in a thin pellicle, closely applied to the surface. As no mucous gland cells are seen in histological sections of the tip, it is likely that the pellicle itself is adhesive, and possibly a mucopolysaccharide, similar to that found on the outer surface of *Sarsia* perisarc, binds the stolon to the subsratum. The chitinous perisarc is laid down later, a little behind the growing tip (point X in Fig. 11.3(a)).

An unusual specialization for attachment is seen in *Hartlaubella gelatinosa* and a number of other hydroids. The hydrocaulus branches dichotomously, with four annulations at the base of each segment (Fig. 11.5(a)). At the distal end of each segment on the main hydrocaulus, an unsegmented tubular outgrowth resembling a stolon, called a siphon, runs back towards the base, closely pressed to the main stem. Consequently the basal part of the organism is strengthened by numerous tubes. The polysiphonic condition appears to be an adaptation to soft substrata, and shows a striking parallel with the kenozooids of *Zoobotryon* (Fig. 9.8(a)). Specimens of *Hartlaubella gelatinosa* have been collected on sand flats at extreme low tide at Dungeness, England. The siphons branch out basally into the substratum and end blindly.

In fast-flowing water, there is a danger that the soft polyp of an athecate hydroid like *Clava multicornis* (Fig. 11.1(c) and (d)) could be torn away from the stolon. This is prevented by a ring of anchor-like rivets that bind the mesogloea to the rim of the stolonic perisarc (Fig. 11.9(g)). The anchors are dead structures formed at the bud stage by cells called desmocytes. They are composed of chitin and protein, and are believed to be sclerotized by tanning (Chapman, 1969).

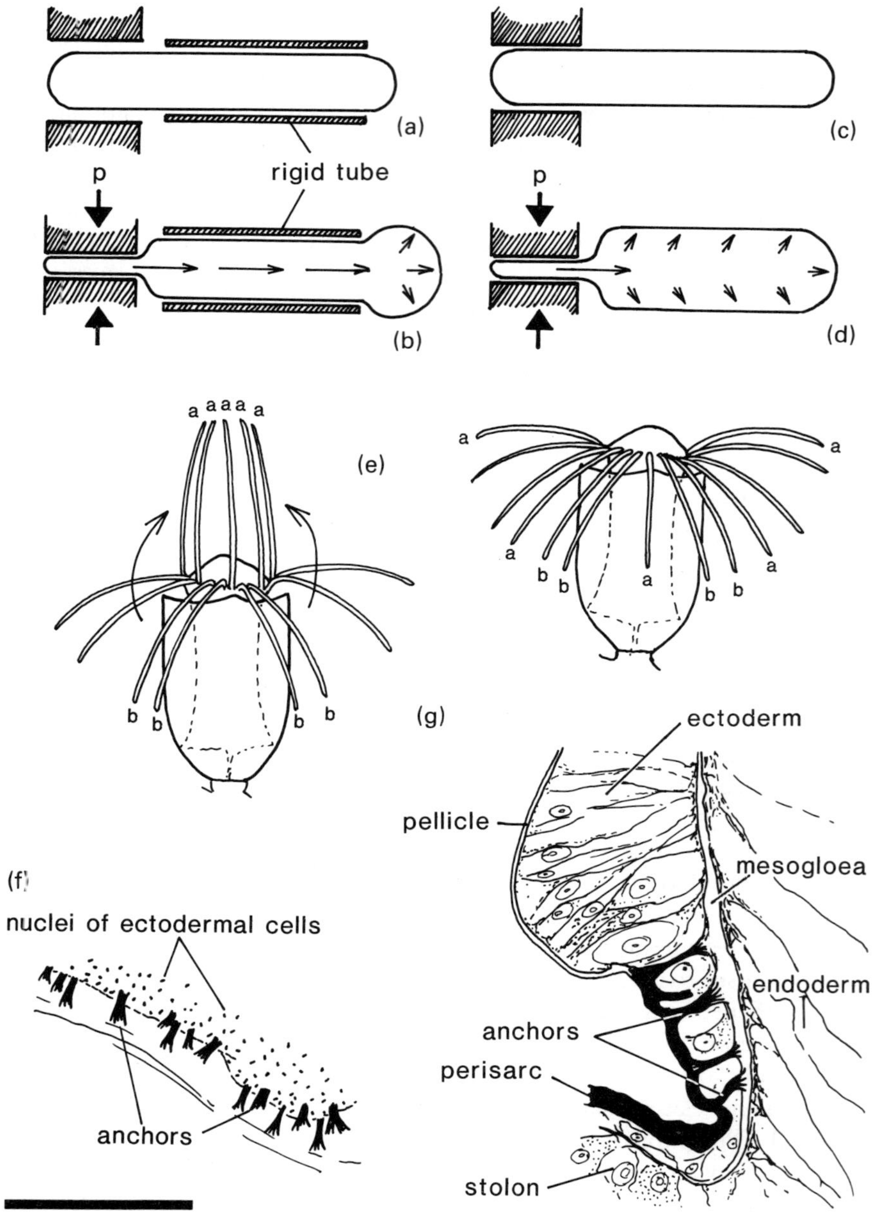

Fig. 11.9 (a) to (d) Function of tubular perisarc (see text for explanation). (e) Tentacle twitching of *Thyroscyphus*: right, relaxed position; left, certain tentacles (marked a in this particular hydranth) are suddenly raised simultaneously and then returned while the remaining tentacles (marked b) do not move. (f) Anchors at base of hydranth (*Obelia geniculata* stained with chlorazol black). (g) Anchors at base of polyp (*Clava multicornis*, azan stain). Scale bar: (g) = 0.01 mm.

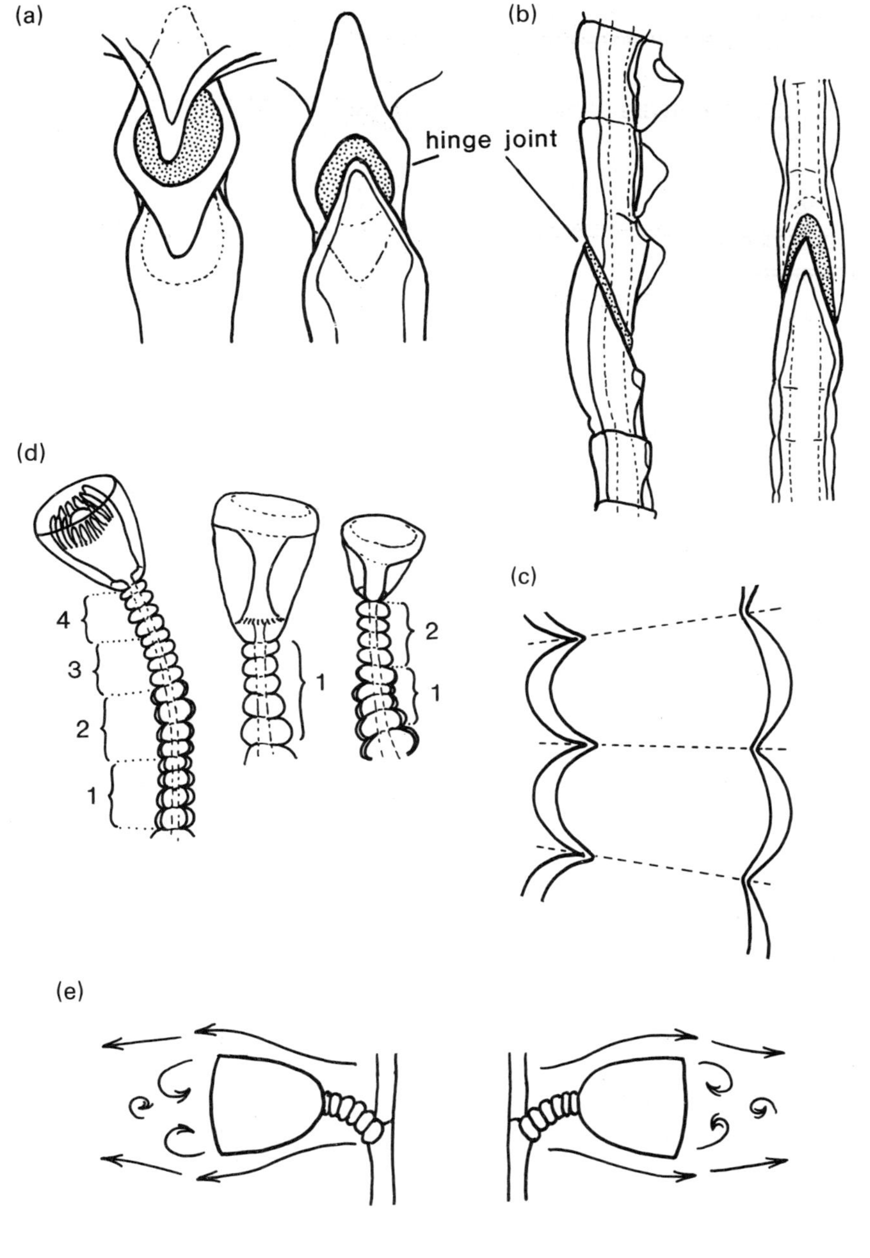

Fig. 11.10 Bending of hydroids. (a) Hinge joint of *Amphisbetia* (back and front view). (b) Hinge joint of *Aglaophenia acanthocarpus* (side and front view). (c) Annulation of perisarc, showing distortion during bending. (d) Addition of new annuli to hydranth pedicel in *Laomedea flexuosa* at each hydranth cycle. (e) Effect of flow on bell-shaped hydrotheca. Note eddies (small recurved arrows).

Similar anchors are found at the base of thecate hydroids (Figs 11.5(d), 11.9(f)). There are *c.* 80 in *Obelia geniculata* (Fig. 11.11(a) and (b)), and half that number in *Thyroscyphus marginatus*, where they are fan-shaped. They ensure that the hydranth is pulled down into the hydrotheca when the longitudinal muscle contracts. Apart from these anchors, the ectoderm of hydroids is not attached in any way to the perisarc, consequently muscles can never cause flexion or movement of the tubular perisarc at any joints or annulations. In this respect, the perisarc of hydroids differs from the cuticle of arthropods.

Flexibility

A branched, upright growth form has the advantage that the feeding polyps are extended into the water column and away from the substratum, but this shape subjects them to the forces of flowing water. At low flow rates, shear forces will cause bending but little strain on the stem. As velocity increases and the branches are bent over, their resistance to flow produces a pulling force or strain on the basal part of the stem.

Tubular construction of the perisarc gives maximum strength and rigidity for a given quantity of skeletal material, but there is a danger that the tube will kink or collapse if the force causing bending exceeds the mechanical strength of the tube. This is unlikely to happen in hydroids, because the perisarc is progressively thickened towards the base and various devices have evolved to allow some degree of bending.

A common way of providing flexibility in the perisarc is by annulations. These occur in the hydrocaulus and in the pedicels of the hydranths (Fig. 11.10(d)). The perisarc reaches its maximum thickness in the central, convex region of each annulus and is thinner between the annuli. There is a possibility that this thinner perisarc may not be tanned, for Knight (1970) found that tanning cells were absent under this region. Bending will take place in the zone between annuli. A shearing force applied just above an annulus will cause most bending in the zone between annuli (Fig. 11.10(c)), but slight changes of curvature in the rings (decreasing on the side of the applied force and increasing on the opposite side) may also contribute to bending. Owing to the elasticity of the perisarc, a righting force of equal magnitude, but opposite direction to the applied force, is induced in the perisarc, thus when the applied force is removed, the restoring force (elasticity) bends the stem back to its original position. The evolution of such a system shows functional economy for no muscles or metabolic energy are required to maintain the upright posture of the hydrocauli.

The taller and more branched a hydroid grows, the greater will be the resistance to water flow. As the velocity of flow increases, the stem will bend over and shear forces will give way to a pull or stress on the base of the stem. Elasticity is now of little importance, for it is the tensile strength of the perisarc that has to oppose the strain caused by the applied stress. Since the pulling force increases towards the base, one might predict that the perisarc ought to be thicker near the base. This is so in tall, branched hydroids and species like *Phylactotheca* that live in turbulent water.

A bell-shaped hydrotheca is characteristic of the family Campanulariidae, to which *Silicularia bilabiata*, *Obelia geniculata* (Fig. 11.11(a)) and *Laomedea flexuosa* (Fig. 11.12(a)) belong. The annulated, flexible pedicels allow the hydrotheca to move in any direction. Because of its shape, the bell will always point with its mouth downstream, i.e., in the direction that offers minimal resistance to water flow (Fig. 11.10(e)). Calculation predicts that flow past the bell must be smooth, possibly with eddies where the tentacles lie. Currents of low velocity, say $0.01\ \mathrm{m\,s^{-1}}$ or less, could assist feeding by delaying or retaining a food organism in the eddy whilst nematocysts are discharged. Some nematocysts take more than 1 s to evert completely, whereas others may require only a few milliseconds, but in a current, the organism that triggered a nematocyst to discharge would be carried out of range in that time, whereas organisms caught in an eddy would be more likely to be captured. Hydroids are probably more efficient at capturing food in calm water.

An important adaptation to life in an oscillating current is found in some sertularians and plumularians. In these families the hydrothecae do not have pedicels and may be attached to horizontal branches called hydrocladia. Annulation is not common and so these hydroids have a more rigid structure than say, *Laomedea flexuosa*. The flexible pedicel of a hydranth may be a point of weakness in strong currents, and this might be a reason why sessile hydranths have evolved in certain groups (Figs 11.8(a), 11.16(a)).

In *Amphisbetia minima* there is a diamond-shaped double hinge joint below the first pair of hydrothecae that allows the upper part of the stem to twist nearly 90° to the left or right of the plane of bilateral symmetry (Fig. 11.10(a)). The joint will also allow the hydrocaulus to bend forwards *c.* 45°, but will not allow it to bend backwards. The hinge joint consists of three diamond-shaped thickenings of the perisarc, set obliquely in the sagittal plane and connected by a relatively thin perisarc. This allows bending in certain directions but not others. A similar hinge joint (sometimes more than one) is found in *Aglaophenia plumosa* (Fig. 11.10(b)), where it allows the whole feather-like structure to turn passively with a gentle current or fold down in a strong current. When water movement stops, it returns to its resting position.

It can be shown by the simple experiment of cutting the stem above or below the joint that restoration of the normal posture is not due to pressure or turgidity within the cavity of the perisarc. As muscles are absent, it must be

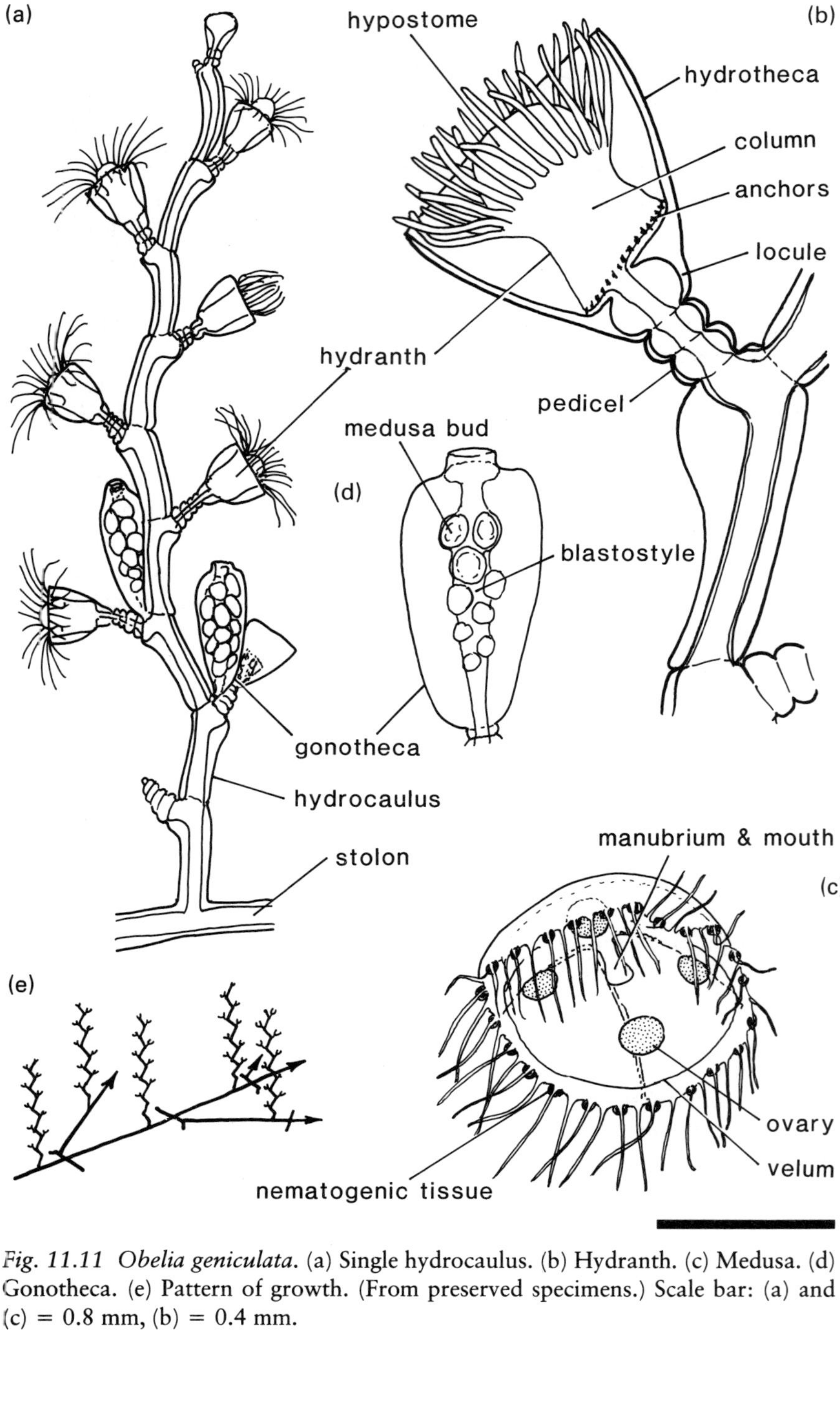

Fig. 11.11 *Obelia geniculata.* (a) Single hydrocaulus. (b) Hydranth. (c) Medusa. (d) Gonotheca. (e) Pattern of growth. (From preserved specimens.) Scale bar: (a) and (c) = 0.8 mm, (b) = 0.4 mm.

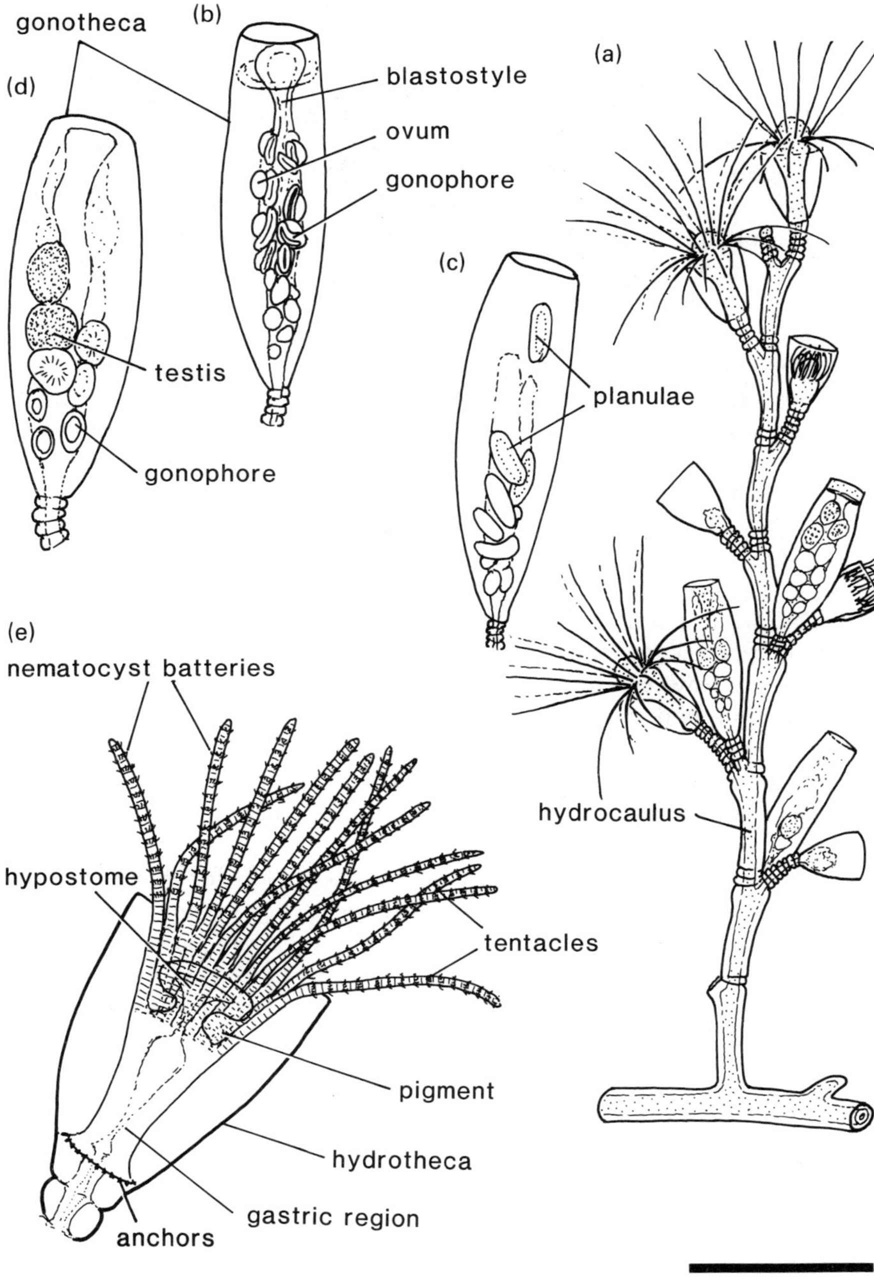

Fig. 11.12 Laomedea flexuosa. (a) Single hydrocaulus. (b) Blastostyle of female hydroid, with developing ova. (c) Planulae being released from gonotheca. (d) Blastostyle of male hydroid. (e) Hydranth. Scale bar: (a) = 1.5 mm, (e) = 0.4 mm.

concluded that it is the elasticity of the thinner perisarc in the joint that provides the restoring force.

The plume or feather-like shape of *Aglaophenia plumosa* (Fig. 11.8(a)) has evolved as an adaptation to efficient food gathering in low-velocity oscillating currents, where upright rigidity, large food-collecting area and ability to turn with the current are desirable characteristics, but with suitable safeguards against high-velocity currents. It provides a fine example of the way coevolution of rigidity, bilateral symmetry, sessile hydranths evenly spaced over a wide area, and of a basal hinge have resulted in an organization of high efficiency and harmonious function.

The combination of strength, rigidity and flexibility confers a unique advantage on the hydroid perisarc in its role as a skeletal supporting structure. Its plasticity has enabled a wide array of shapes and growth patterns to evolve, each suited to some particular need or environmental constraint. This is reflected in the extremely wide range of adaptive radiation shown by the class Hydrozoa.

Protective function

The perisarc also fulfils a number of functions that can be grouped under the heading of protection. Abrasion and crushing have already been mentioned. The thalli of seaweeds are constantly rubbing against each other because of wave motion, and species like *Silicularia bilabiata* and *Obelia geniculata* which grow on seaweeds must be able to withstand such crushing. It is significant that *Silicularia bilabiata* is only found on the basal 20–30 cm of *Phyllospora*, and not on the long distal fronds, where rubbing would be severe.

The perisarc would provide protection from water loss for species (such as *Amphisbetia minima*) living under boulder rocks or in caverns, where they are commonly exposed to air for brief periods at extreme low tides, but it would be inadequate to withstand longer periods of exposure and drying. This clearly places a limit on the ability of hydrozoans to become true littoral species.

Far more important is the protection the perisarc gives against attack by bacteria and marine fungi. Old stolons and stems of *Thyroscyphus marginatus* and the dead remains of *Amphisbetia minima* often have other organisms growing on them (e.g. suctorians, thecate and colonial ciliates, foraminiferans, diatoms, algae, fungal hyphae and juvenile stages of sponges, bryozoans and even other hydroids), but seldom are such organisms found on healthy active hydroids. Ralph and Thompson (1968) recorded that at certain seasons, heavy growths of diatoms formed on *Obelia geniculata* in Wellington harbour. The general impression is that hydroids have some way of preventing other organisms from settling on their perisarc. How this is achieved is not known, but the mucoprotein outer surface, or possibly quinone substances, may be involved.

The degree to which the perisarc gives protection against predators is uncertain. It is ineffective against browsing nudibranchs, which are undeterred by the nematocysts, and specialized predators such as pycnogonids can thrust their proboscis into open hydrothecae to eat the hydranth. The delicate opercular flaps of *T. marginatus* and sertularians would be no protection against such predators. This prompts the question, what is the function of these opercula? It is probable that a predator relies on chemical stimuli to detect its prey. If the operculum seals off the hydranth so that no chemical emissions escape, predators would be unable to detect the hydranth inside, thus an animal might achieve protection by isolating itself from a predator's sensory system.

Little is known about the permeability of the perisarc to water and ions, but some thought must be given to the vital process of obtaining oxygen and disposing of carbon dioxide. It is known that the cuticle covering the gill surface of crustaceans is freely permeable to both gases, and circumstantial evidence seems to indicate that the same is true of the hydroid perisarc. Under adverse laboratory conditions, all the hydranths of *Thyroscyphus marginatus* may regress completely into the hydrocaulus. The pedicels are sealed off with perisarc so that all the tissues of the coenosarc are completely isolated from the surrounding water by perisarc. However, this is not a time of dormancy or cessation of metabolic activity, for fluid is still moved back and forth through the gastro-vascular cavity, and oxygen must still be required. This can only be obtained by diffusion through the perisarc. In *Phylactotheca armata*, which has an unusually thick perisarc over most parts of the hydrocaulus, the outer (exposed) surface of the stolon is relatively thin. From these lines of evidence it might be concluded that the perisarc is permeable to gases. The expanded hydranths have a large surface area in contact with the water; diffusion will provide sufficient oxygen for the ectoderm and the digestive cells, where metabolic demand is likely to be high.

When food is swallowed, some water will be taken into the gastric cavity. Since this is connected to the whole gastro-vascular system, it is likely that the fluid that is passed back and forth has the same osmotic and ionic composition as seawater, thus osmotic and ionic regulation will not be a problem for the hydroids.

An inevitable consequence of having a dead perisarc over all parts of the body is that the organism is isolated from most forms of sensory information from the environment. If any part of the hydranth of *Thyroscyphus marginatus* is touched with a glass rod or jet of seawater, an immediate response is elicited. This may range from movement of a tentacle to complete withdrawal of the hydranth. However, no response is called forth when the perisarc is touched, provided it does not vibrate the water.

11.3 MOVEMENTS AND BEHAVIOUR OF HYDROIDS

Despite their apparent simplicity, hydroids are capable of a number of complicated movements concerned with feeding or, in the case of medusae, locomotion. These movements are produced by the epithelio-muscular cells already described.

Sensory receptor cells have been identified in the ectoderm of hydroids (Jah and Mackie, 1967) and many activities probably arise as direct responses to environmental influences detected by these receptors. Withdrawal of a hydranth in response to touch, or the adduction of the tentacles and opening of the mouth in response to food would fall into this category, but some activities appear to be endogenous, such as the rhythmic twitching of the tentacles in *Laomedea flexuosa* and *Thyroscyphus marginatus*, or the regular reversal of hydroplasmic flow. Spontaneous and regular bursts of action potentials, presumably from neurones acting as pacemakers, have been recorded from both the polyp and stem of the hydroid *Tubularia* (Kruijf, 1977).

Cnidarians are renowned for their exceedingly slow changes in shape or posture. The mechanism for such slow movement is not understood and may not be caused by muscular contraction. Campbell (1980) has suggested that amoeboid activity of the muscle tails of the epithelio-muscular cells may cause slow morphological changes by slowly creeping along the mesogloea. When the muscle processes are fully stretched, amoeboid movement might exert a small force on the mesogloea and deform it. The additive effect of vast numbers of cells doing this might cause slow changes in shape.

A muscle fibre can only contract. Its original length has to be restored passively by some other agency, such as the hydrostatic pressure of fluid in the coelenteron (sea anemones: see page 207), antagonistic muscles like the adductor and abductor muscles of the tentacles in *Thyroscyphus marginatus*, or an elastic element as in the thick mesogloea of medusae and, to some extent, the mesogloea of sea anemones. The mesogloea of the hydranth is very thin and muscle contraction throws it into fine folds, suggesting that it has little elasticity and may play only a minor role in restoring body shape. However, turgidity of the core cells appears to impart elasticity to the solid tentacles of hydroids and restore them to a neutral or resting position when the hydranth is expanded. The arrangement of muscle tails of the ectodermal and endodermal cells at right angles to one another has already been described (see page 232). Acting on the fluid coelenteron as hydrostatic skeleton, they antagonize each other, the contraction of one extending the other.

The hydranth of *Thyroscyphus marginatus* is not a simple fluid-filled cylinder (Fig. 11.6). When fully extended, the vestibule is a large cavity filled with seawater and surrounded by the thin body wall, but the coelenteron in the lower gastric region is nearly obliterated by the highly vacuolated gastric endoderm. When food is not present in the coelenteron, these cells occupy 85–90% of the volume. However, the vacuolated cells can be deformed easily, thus pressure exerted on them by the muscles will have the same effect as if the coelenteron were a fluid-filled cavity, except that there will be no displacement or loss of fluid from the region. It would be possible for elongation and contraction of the gastric region to take place in response to muscle contraction, even when the mouth was wide open. This explains why the hydranth can change its shape when fully contracted and without water in the vestibule.

Because the hydranth is firmly attached to the hydrotheca at the base by the anchors previously described (see page 242), contraction of the longitudinal muscles draws the hydranth down into the hydrotheca. During the digestive phase following ingestion of food, peristaltic waves pass along the column of the fully-extended hydranth (Fig. 11.13(a)). These changes in shape are caused by sequential contraction of circular muscles of the endoderm and do not alter the height of the column.

Movements of the tentacles and mouth

Volume or pressure changes in the coelenteron have no effect on the length of movements of the tentacles in thecate hydroids. Unlike sea anemones, their tentacles are solid and contain a core of turgid cells which give rigidity, but prevent contraction to less than about two thirds their extended length. The epithelio-muscular cells in the tentacle ectoderm are responsible for elongation, shortening, and S-bending, but rapid adduction and abduction (movement of the tentacle towards or away from the mouth) are brought about by muscles at their base.

A feature shown by many hydroids is spontaneous 'twitching' of the tentacles, in which alternate tentacles suddenly move to a vertical position and back again (Fig. 11.9(e)). In *Thyroscyphus marginatus*, tentacles appear to be grouped. Sequences such as ABBABB and ABBBABBB are common (where A denotes the tentacles that move and B the stationary tentacles). The significance of this behaviour is not known, but it indicates a complex pattern of nervous co-ordination and may indicate a higher state of receptivity to stimulation by prey.

The mouth is opened and closed as follows. The endodermal muscles of the hypostome are arranged concentrically round the mouth, thus the mouth is closed when they contract. Ectodermal muscles are arranged radially, but it does not follow that their contraction will open the mouth: they may merely reduce the diameter of the hypostome. However, vacuolated core cells identical to those of the tentacles form a skeletal ring round the column at the base of the tentacles. The radial muscles of the hypostome are attached to this ring, consequently they pull the mouth open when they contract. Extreme abduction of the tentacles alters the direction in which the adductor muscles of the tentacles exert their pull, and this increases the extent to which the mouth can open.

Expansion, egestion and feeding behaviour

The initial expansion of the newly formed hydranth is a very slow process, taking 1 h or more. As hydroids do not have a siphonoglyph to pump water into the coelenteron, expansion cannot be due to increased hydrostatic pressure, and osmotic swelling of the tissues may be responsible. During the final stages of differentiation, in which cells of the bud are transformed into definitive epithelio-muscular cells, large intracellular vacuoles appear which increase the cell volume. Moreover, at this time the sub-epithelial space appears between the cells and mesogloea. Both of these processes would increase tissue volume and hence the size of the hydranth.

It is impossible to predict when a particular hydranth of *Thyroscyphus marginatus* will emerge, for each acts independently of others on the hydrocaulus. Expansion from the fully contracted position is slow and 'hesitant', with many apparently-spontaneous retractions which delay full emergence. The bunched tentacles and hypostome push past the mantle and valves as the column elongates, but the tentacles are not fully spread. During this period the hydranths appear very sensitive to outside disturbances, such as slight vibrations or water movements, and respond with immediate withdrawal. It may take many minutes before they re-emerge. At this time they are insensitive to food, for no attempt is made to capture or swallow brine shrimp nauplii that collide with their tentacles.

Suddenly a new pattern of behaviour appears. All the tentacles are synchronously thrown back to touch the hydrotheca and the mouth opens maximally (Fig. 11.5(e)). Glistening brown and amber particles can be seen in the mouth and vestibule, but they are not forcefully ejected, and after 10–20 s the mouth slowly closes, leaving some particles outside. This activity (called egestion) may be repeated many times, at intervals of two or three minutes, until all the particles have been lost to the exterior.

These observations are highly significant, for they show that there is no difference of pressure between the vestibule and the outside seawater. Moreover, no detectable change in volume occurs when the mouth is opened.

When egestion stops, the hydranth slowly extends to its maximum height and the tentacles are fully spread. It is now less sensitive to disturbance and does not respond to a jet of water from a pipette.

The hydranth is now in what might be termed the feeding posture. A brine shrimp nauplius that touches a tentacle is immediately arrested by the small nematocysts, and the tentacle responds by slight shortening, S-bending and adduction (movement towards the mouth). All the other tentacles remain motionless unless the prey struggles, in which case a few adjacent tentacles may bend over and assist. As soon as the nauplius touches the hypostome, the mouth opens maximally, with simultaneous abduction of all the other tentacles, just as in egestion, and the nauplius sinks slowly into the mouth cavity (Fig. 11.13(a) a,b)). The mouth now closes. If the nauplius has not been completely swallowed in the first attempt, the mouth will open again 20–30 s later in exactly the same manner. With larger prey, such as metanauplii, the mouth may have to be opened several times, and a number of tentacles used to restrain or bear down upon the food, before it is swallowed. Prey larger than the hydranth (> 1.5 mm) cannot be swallowed and is rejected. *Thyroscyphus marginatus* has been observed to capture and swallow prey without any assistance from the tentacles when it collided with the hypostome. The mouth opened instantly, the food sank into the vestibule, and the mouth slowly closed behind it.

Food in the vestibule is slowly moved down to the gastric region by peristaltic contraction of the circular muscle. During this period, which may last 1–10 min, the swallowing response is inhibited. When the vestibule is again empty, a normal feeding response can be elicited and a second nauplius may be swallowed.

The hydranth again changes its behaviour as it passes into the extracellular digestive phase. The hydranths are noticeably unresponsive and the tentacles can be touched with no reaction or only a slight S-bending. Very slow peristaltic waves pass food back and forth between the gastric region and the vestibule at intervals of 3–10 min (Fig. 11.3(a)). It is believed that proteolytic enzymes are released from the enzyme-secreting cells into the coelenteron to carry out partial digestion of the food.

After a variable period (usually *c.* 2–3 h, the indigestible remains of the prey is egested in exactly the same manner as described previously.

Other methods of ingestion

The method of food capture and ingestion just described is fundamentally different from that described for *Hydra littoralis* by Rushforth and Hofman (1972). It also differs from the way the hypostome is used in most other members of the Hydroida. Feeding in *Eleutheria* will be described as an example of the more typical method.

Careful observation has failed to reveal any evidence that the medusae of *Eleutheria* can detect the presence of suitable prey. It would seem that

(a)

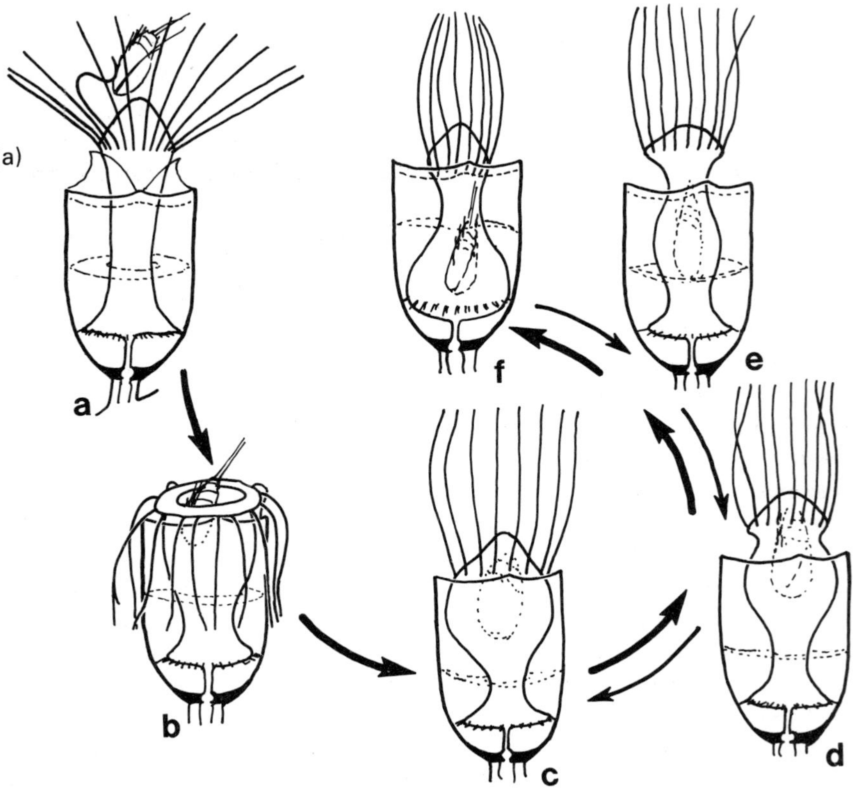

(b)

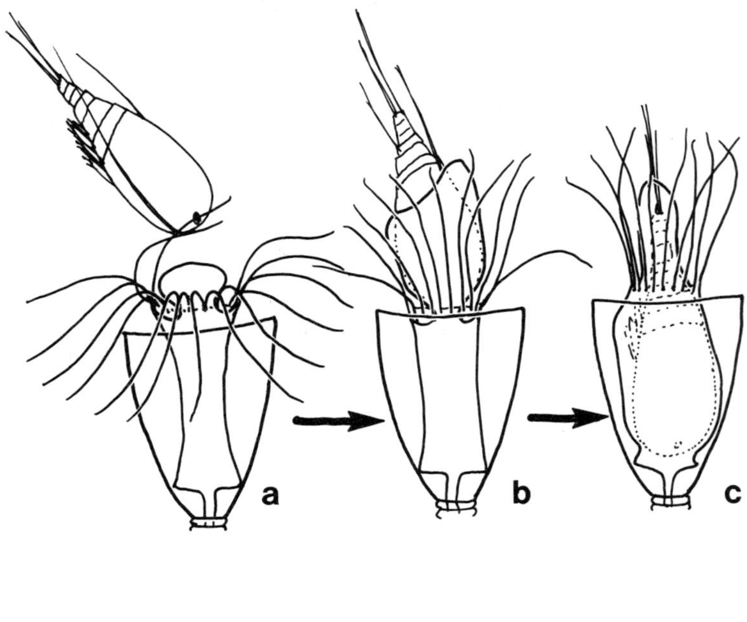

although it can move about, *Eleutheria* still takes a passive role and waits for prey to collide with its capitate tentacles in much the same way as all other sessile hydrozoans.

The position of the tentacles is such that they can seldom transfer food directly to the mouth, and so the manubrium (or the hypostome in polyps) is extended towards the prey (Fig. 11.15(b)). As soon as it touches the food, the mouth opens and the lips appear to creep over the food, enlarging the mouth as they do so. The manubrium (or hypostome) is capable of incredible enlargement and can swallow a copepod as large as the medusa. In this method of feeding, the tentacles play little part other than arresting the prey. A similar pattern of feeding is seen in *Laomedea flexuosa* (Fig. 11.13(b)).

It has been noticed that during the digestive phase, *Eleutheria* continues to kill copepods that touch the tentacles, but makes no attempt to ingest them, suggesting that the swallowing response is suppressed, as in *Thyroscyphus marginatus*.

The manner in which the lips of the manubrium (or hypostome) expand and glide over the prey has not been satisfactorily explained. Both circular and radial muscles are present, but their contractions will not cause expansion. However, epithelio-muscular cells are known in which myofibrillae run from the outer surface of the cell to the basal muscular process (Chapman, 1974). Contraction of these fibres would reduce the height of the cell, with corresponding lateral expansion. They would also compress the fluid that lies between the bases of the epithelial cells, forcing it elsewhere. If such vertical fibres are present in cells of the hypostome, displacement of this fluid could cause further extension of the lips and explain the remarkable way in which the manubrium or hypostome can slowly enlarge and slide over the prey.

11.4 HYDROPLASMIC FLOW

A vital function performed by the perisarc is to enable slight pressure differences to be transmitted from one part of the organism to another. To understand this important role, an elastic balloon can be imagined, partly encased in a rigid tube (Fig. 11.9(a)). If the balloon is squeezed at one end (p in Fig. 11.9(b)), the opposite end will increase in size because fluid can only pass along the tube. If the rigid tube is removed, squeezing one end causes an

---

Fig. 11.13 Feeding behaviour of hydroids. (a) Sequence of ingestion (a, b) and peristaltic movements during digestion (c to f) in *Thyroscyphus*. (b) Capture and ingestion of food in *Laomedea*.

increase in diameter with little flow from one end of the balloon to the other (Fig. 11.9(c) and (d)). A beautiful illustration of this principle was provided by a polyp of *Eleutheria* growing in the laboratory (Fig. 11.2(c)).

Originally the specimen consisted of a single hydranth attached to the substratum by a short stolon encased in perisarc. Eventually a second hydranth grew rapidly at the other end of the stolon. It was now noticed that each polyp underwent regular contraction and expansion, the contraction of one coinciding with the expansion of the other. Under the microscope, particles could be seen passing from the contracting polyp through the stolon to the other one, indicating fluid flow. This occurred with great regularity for several weeks. Contractions lasted 2–3 min, followed by an irregular period of rest, commonly 3–5 min, although after food ingestion, flow stopped for 2–3 h. In this case muscular contraction must have been responsible for the pressure difference that caused fluid to flow from one polyp to the other. During digestion it was noticed that food was passed up and down the digestive cavity of the feeding polyp by peristalsis, but no fluid passed along the stolon, indicating that the hydranth can isolate its coelenteron from the stolon. All this implies a degree of behavioural co-ordination far more complex than is generally admitted for hydrozoans.

The regular pumping is clearly concerned with distributing food, particularly to the cells of the stolon, but it could also distribute oxygen or materials that have been synthesized by the polyp or the stolon. The rhythmic flow of fluid backwards and forwards through the gastro-vascular tube is a characteristic feature of all polypoid hydrozoans and has been termed hydroplasmic flow (Berrill, 1961; Wyttenbach, 1968).

This is a particularly simple example which is easy to interpret. As the complexity of the hydroid increases by branching, the course of fluid movements becomes more difficult to follow because of the increased number of possible routes it can take. In athecate hydroids like *Eleutheria* or *Australomedusa baylii*, the coelenteron of the developing medusa bud is connected directly to the gastric cavity of the parent polyp, but in thecate hydroids, medusae or gonophores develop on a new structure, the blastostyle, which, because it has no mouth or tentacles, cannot catch its own food and must rely on nutrients being passed to it from other hydranths via the gastro-vascular system. Periodical increase and decrease in the volume of the blastostyle can be followed on living specimens of species like *Silicularia bilabiata* or *Obelia geniculata*, showing that nutrients pass from the feeding hydranths to the blastostyle in the fluid flowing in the gastro-vascular system.

Hydroplasmic flow is a universal feature of the polyp phase of all hydrozoans (except the aberrant hydras) and occurs throughout their life. Flow is not constant. In *Thyroscyphus marginatus*, fluid will flow in one direction through parts of the stolonic network for a short period, and then stop for a slightly longer period before starting again in the opposite direction. The periods of flow and counter-flow usually last *c.* 30–60 s and show considerable regularity. The interval between them usually lasts 1–3 min. A particular rhythm may continue for hours with great regularity. Occasionally there may be a change of route through the stolonic network and, in extensive networks, there may be more than one regional flow, each being isolated temporarily from the other.

Hydroplasmic flow is not dependent upon the presence of the hydranths, for it continues with equal regularity during adverse conditions when all hydranths have regressed and the stolon is completely sealed off with perisarc. Sudden change in temperature, damage to the stolon, or gross disturbance, as when a specimen is removed from the substratum, usually causes a temporary halt to the flow. In *Eleutheria* it stops during the extracellular phase of digestion; this may occur generally in hydroids.

The exact mechansim causing fluid to pass back and forth is not understood. Muscle tails have not been detected by histological methods in cells of the stolon or hydrocaulus of *Thyroscyphus marginatus*, but Bouillon (1968) illustrates muscle tails to the endodermal cells in the stolon of *Sarsia*. If this is of general occurrence among hydroids, it could help to explain hydroplasmic flow in stolons that have no hydranths (i.e., in regressed organisms). Muscle fibres in the endoderm would run circularly and be antagonized by pressure in the gastro-vascular cavity.

Flagellate cells occur in the endoderm of the gastro-vascular cavity, but it can be shown that these are not responsible for fluid flow, for the flagella continue to beat when flow stops prior to reversal. Their activity only stirs the fluid locally and probably keeps particles suspended so that they do not block the gastro-vascular tube.

Two lines of evidence point to muscular contraction of the hydranth being involved in hydroplasmic flow.

By taking a series of time-lapse photographs at 20 min intervals, it was possible to follow the very slow changes in volume of *Thyroscyphus marginatus* hydranths. Changes in the volume of the gastric cavity could be calculated and compared with the volume of fluid moved along the gastro-vascular channels in the stolon. The results were unexpected. Fully expanded feeding hydranths show very little volume change over long periods. When the mouth is opened, fluid does not pass out, indicating that there is no positive pressure in the gastric cavity of expanded polyps; moreover, fluid does not ebb and flow from the gastric cavity of the hydranth into the gastro-vascular system. This is puzzling, for the expanded polyp would seem to be an ideal mechanism for pumping by virtue of its column muscles. On the other hand, volume changes were considerable in developing buds. A stage is reached when a rapid expansion in size of the bud occurs, and this appears to result entirely from fluid flowing in from the gastro-vascular system. It is tempting to think that this expansion is caused by a slight positive pressure in the fluid; however, this has not been demonstrated experimentally. Throughout the development of the hydranth bud there is the same regular ebb and flow, but the net movement of fluid must be into the bud.

After a period of active feeding, hydranths may regress and pass into the 'bulb stage' (Fig. 11.14), during which there is a regular movement of fluid into and out of the gastric cavity. Calculation of changes in volume during the bulb stage shows that considerable volumes of fluid are passed into the gastro-vascular system. The term 'regression' is misleading, for the histological appearance of the gastric region is normal, with active digestive cells present. Clearly in *Thyroscyphus marginatus* the bulb stage is not a period of degeneration; it is an active phase concerned with nutrition of other parts of the hydroid.

The functional significance of the tubular perisarc is now apparent: its strong wall permits slight differences of pressure to develop between various parts of the organism so that fluid will flow down the pressure gradient. Muscular contractions in 'bulb' hydranths, blastostyles and the coenosarc itself cause the regular and persistent hydroplasmic pulsation that circulates nutrients to all parts of the hydroid.

11.5 GROWTH OF THECATE HYDROIDS

Stolonic growth

Medusae show a continuous increase in size that is very similar to growth in other animals, but growth of the hydroid phase shows a number of peculiarities.

The stolon of many species may grow continuously, spreading the hydroid over the available space, or it may remain quiescent. Growth of the stolon takes place at the tip, which is covered in a very thin pellicle. Contrary to what might be expected, mitosis and the production of new cells does not take place at the tip in *Laomedea flexuosa*. The growth zone is located a short distance back, and the tip appears to be pushed forward (Campbell, 1974). Forward growth is not smooth and continuous. Wyttenbach (1968, 1969) has shown that the tip makes an extremely regular cycle of advance and retraction at an interval of *c.* 6–7 min (the exact periodicity is characteristic of a particular genetic stock), and this results in a maximum growth rate of 1.9–6.2 mm d^{-1} (again depending upon the stock). By cutting off the tip, Wyttenbach was able to show that isolated tips continue for a long time to advance and retreat at the same periodicity as the parent stolon, thus demonstrating that these movements and the growth cycle are an intrinsic property of the tip and are not caused by pressure of the hydroplasmic flow. However, hydroplasmic flow does augment forward growth, for the stolon advances further during periods when flow is towards the tip than when it is away from the tip.

Increase in height of the hydrocaulus is fundamentally different from stolonic growth, and proceeds periodically by the addition of a new segment or internode. A terminal bud develops rapidly into a new internode, a new hydranth and another terminal bud. It then passes into a resting period, the duration of which may be very regular, suggesting an endogenous timing mechanism, but external factors such as temperature also affect duration. Thus growth of the hydrocaulus is quantal (i.e. proceeds by jumps). Crowell (1957) found that new internodes were added to the hydrocaulus of *Laomedea flexuosa* at intervals of 1.5 d at 17 °C. The production of the first internode of a new hydrocaulus in species such as *Obelia geniculata*, *Plumularia setacea* (Fig. 11.16(a)) or *Amphisbetia minima* is strictly related to stolon growth, but the addition of subsequent internodes to the hydrocaulus follows its own time course which is not related to stolon growth.

Hydranth development

Hydranths and their polymorphs (blastostyles, nematophores, dactyloids) are produced by budding or blastogenesis. All the cells from which the initial bud develops must come from the polyp, stolon or hydrocaulus. First, undifferentiated ectodermal cells form a bulge which, as it grows outwards, carries with it undifferentiated endoderm cells (Berrill, 1949). The bud is enclosed in a thin perisarc and has a cavity which connects with the gastrovascular system of the hydrocaulus. The bud increases in size until the hydrotheca is fully formed, and then a process of differentiation starts in which cells undergo a change in cytological structure and definitive organs take shape. Finally certain cells increase in volume and the hydranth becomes a functional trophic unit. Eventually the hydranth will regress. It may disappear completely (its substance being resorbed into the stolon or passed to blastostyles) and a new bud form at the old site; this bud will develop into a new hydranth (Crowell, 1953). In other cases the hydranth does not regress completely, but regenerates into a new hydranth.

The development of a hydranth in *Thyroscyphus marginatus* (Fig. 11.14) is similar to that described for *Obelia* by Berrill (1949), except that there are no annulations on the very short pedicel. The bud is almost spherical to start with, but a second bulge appears which gradually elongates and flattens distally to form a disc or cap. It is this second growth that forms the main hydranth; the first bulge gives rise to the basal locule (Fig. 11.14(a) and (b)). Ectodermal cells forming a ring round the periphery of the disc are histologically distinct and appear to be responsible for secreting the perisarc of the hydrotheca, which now starts to take on a bell shape. At about this time

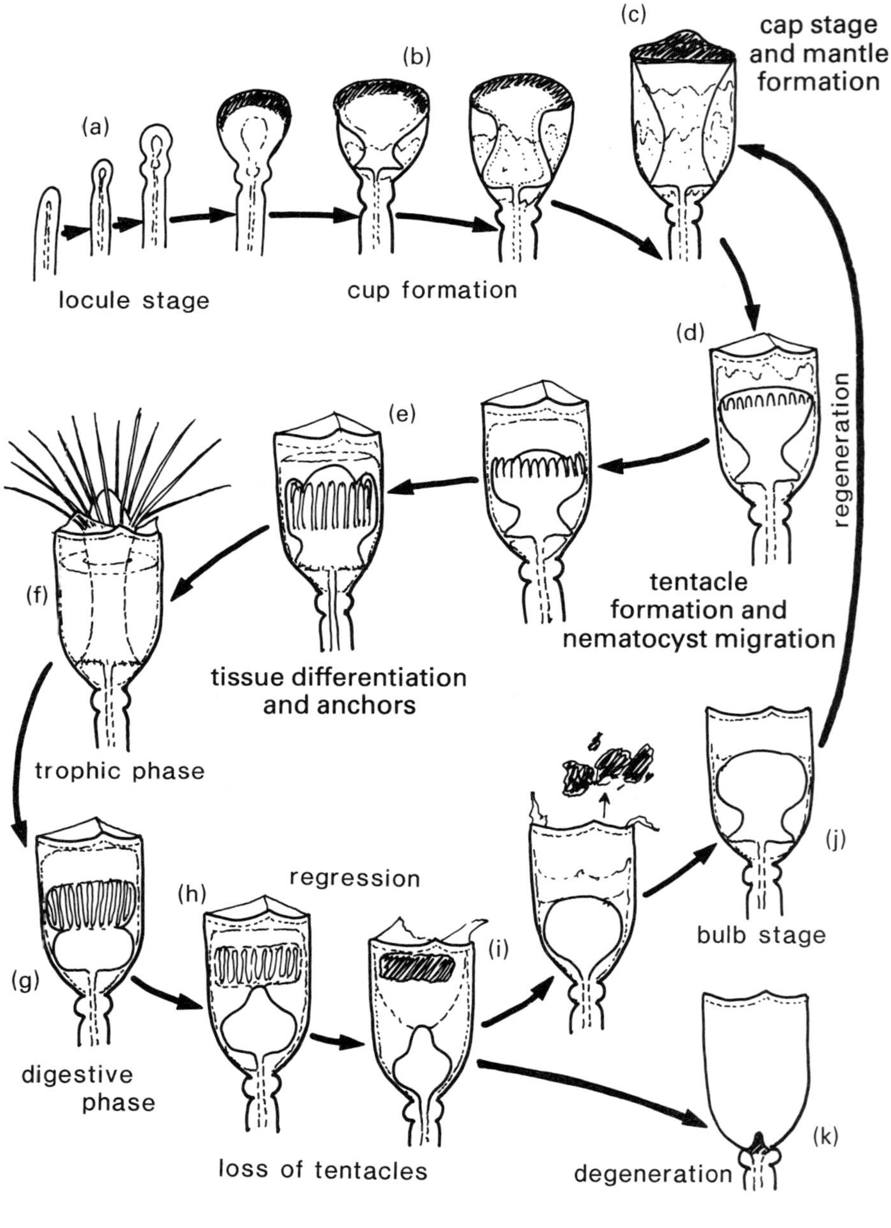

Fig. 11.14 Growth of hydranth and hydranth cycle in *Thyroscyphus marginatus*. See text for explanation.

the side walls shrink away from the perisarc, except for one region where the desmocytes of the ectoderm bind the mesogloea to the perisarc (the anchors will form here). Below this point the tube remains narrow (the isthmus). Slow peristaltic contractions of the developing hydranth elongate the body and gradually extend the hydrotheca. Fluid enters and leaves the bud during normal hydroplasmic flow, but there must be a net flow into the bud, for it increases in volume. It is doubtful whether the pressure of hydroplasmic flow is sufficient to inflate the bud. Berrill (1949) found that peristaltic pulsations continued in *Obelia* after the bud had been severed from the stalk, and concluded that they resulted from muscular contraction of the bud.

Hydroplasmic fluid is not the only material entering the hydranth bud. By watching the narrow isthmus it can be seen that vast numbers of cells migrate into the bud through the ectoderm. The most obvious are the orange amoeboid tanning cells and the nematocytes carrying either small or large nematocysts, but cells of the ectoderm and endoderm also pass into the bud, and these must be partly responsible for the increase in volume.

The disc now takes on the shape of a low pyramid and secretes the four triangular opercular valves, and round its perimeter the thickened rim with four cusps. Because the disc is red (owing to the amoeboid cells migrating into it) and is shaped somewhat like a cardinal's cap, this stage will be called the cap stage. This disc is now withdrawn so that a space exists between it and the valves, and it can be seen that the hydranth is completely enclosed in perisarc (Fig. 11.14(c) and (d)).

Differentiation of the cells of the bud now takes place and rudiments of the 26 tentacles appear round the perimeter of the cap. Histological sections show that a ring of endodermal cells with dense cytoplasm start to differentiate as the core cells of each tentacle and a ring at the base of the tentacles. The central part of the cap becomes the hypostome. Both ectoderm and endoderm cells now begin to take on their familiar vacuolated appearance, and staining shows that the anchors have been formed. Migration of ectodermal cells up the inside of the hydrotheca forms the mantle, which completely surrounds the developing polyp. A similar layer of ectoderm lines the chamber below the ring of anchors where the isthmus lies.

Cells in the future digestive region above the isthmus enlarge, and the original cavity of the bud becomes the vestibule (Fig. 11.6). The tentacles are bent over or 'n'-shaped and are only about one quarter of their final length (Fig. 11.14(e)).

The hydranth now passes into its final phase of development. The initial opening and expansion of the hydranth is a slow process which takes 30 min or more. Elongation causes the hydranth to push through the mantle and rupture the opercular valves, which because of their shape always close behind the hydranth. The tentacles now straighten, slowly elongate to three or four times their original length and stiffen. The mechanism for this, and the trigger that initiates the process, are not known, but the core cells undergo a marked increase in volume – their cytoplasm becomes extremely vacuolated and they resemble a pile of coins. Presumably their turgidity is the result of some osmotic phenomenon. It has been noted that birefringent crystals appear in the vacuoles of these cells at this time. Increase in size of the hydranth during the initial phase of expansion must be caused in part by the large vacuoles in the epithelio-muscular cells of ectoderm and endoderm. This also suggests an osmotic mechanism of expansion.

The hydranth cycle

A cycle of feeding, regression and regeneration is a common feature of thecate hydroids. When the undisturbed hydrocauli of species such as *Obelia geniculata*, *Plumularia setacea* (Fig. 11.16) or *Laomedea flexuosa* are examined, the majority of hydranths are expanded and feeding, but a significant percentage of positions will be occupied by developing buds or hydranths that either appear to be losing their tentacles or have been resorbed and the hydrotheca lost. Crowell (1957) showed that this was caused by a cycle of events which were repeated on average every 3.8 d at 21–24 °C and 6.8 d at 10–17 °C for *L. flexuosa*.

The cycle has been studied in *Thyroscyphus marginatus* where the first signs of regression are the tight withdrawal of the hydranth, which becomes enclosed in the mantle (Fig. 11.14(g)). The gastric region is spherical and the tentacles fully withdrawn and folded over, which suggests that the central vacuolated cells are no longer turgid. The tentacles and hypostome detach from the gastric region, which rounds off as a 'bulb', and undergo degenerative changes in which individual cells are set free (Fig. 11.14(h)). These accumulate as a brown mass under the opercular valves, and in the debris it is possible to identify thousands of undischarged nematocysts (large and small), large cells laden with amber or brown particles, and the core cells of the tentacles. The core cells are spherical at first with their crystals, but eventually they collapse and appear as cell 'ghosts' staining in the same way as the mesogloea, this observation suggests that they have an outer envelope of collagen fibres. Eventually all this debris is discarded to the outside world (Fig. 11.14(i)). It is surprising to find that such vast numbers of nematocysts are lost without ever being used.

Histological examination of the bulb shows that the ectoderm and endoderm are normal and functioning at this stage. Fluid can be seen passing into and out of the bulb during hydroplasmic flow, and over a period of time the bulb undergoes considerable changes in volume. This phase may persist for a few hours or days before regeneration commences.

The bulb now behaves like a bud. It elongates and expands distally to form a new disc which secretes four new opercular valves. The old valves may persist for a while, but are eventually lost. The disc then withdraws to develop into a new hypostome and tentacles in exactly the way described for the development of the first hydranth. Because a new set of valves can only be formed during the cap stage, it is possible to determine the number of hydranth cycles that have been passed through by counting the number of rims in transverse sections of the hydrothecae (Fig. 11.7(d)).

Older hydranths near the base of the hydrocaulus may regress completely (Fig. 11.14(k)) and eventually the hydrotheca is lost. Regeneration at these points has never been observed. A summary of events in the hydranth cycle is given in Fig. 11.15(a)).

The hydranth cycle in *Obelia geniculata* differs in three important respects. There is no persistent bulb stage, the whole of the gastric region is resorbed into the hydrocaulus, and the hydrotheca is lost. A new bud is formed which lays down two new annuli and a new hydrotheca for the hydranth. It is possible to tell the number of cycles that have been passed through by dividing the number of annuli in the pedicel by two. In *Laomedea flexuosa*, four annuli are added at each cycle (Fig. 11.10(d)). The gastric region of *Plumularia setacea* is also completely resorbed after the tentacles and hypostome have disintegrated, but the sessile hydrotheca is not lost. The nematocysts of the three adjacent nematophores are also discarded and new nematophores are formed in the old nematothecae. *Silicularia bilabiata* loses both hydrotheca and the terminal annulus, and a new segment plus annulus is added at the next cycle (Fig. 11.3(c)).

In Table 11.1 a comparison is made between samples of *Obelia geniculata* and *Plumularia setacea* to show the proportion of hydranths undergoing

Table 11.1 Proportion of regressing and regenerating hydranths in *Obelia geniculata* and *Plumularia setacea**

Species	*Number of hydranths*	*Number of hydrocauli*	*Average per hydrocaulus*	*Regenerating (%)*	*Regressing (%)*	*Feeding (%)*
Obelia geniculata	646	38	17 (24)‡	8	5†	87
Plumularia setacea	425	23	18 (32)‡	17	19	64

* Source of specimens: *O. geniculata*, Loch Leven, Scotland; *P. setacea*, Broulee, New South Wales, Australia.

† Probably low because of difficulty in identifying early stages of regression without staining.

‡ Maximum number of hydranths on any one hydrocaulus in the samples.

(a)

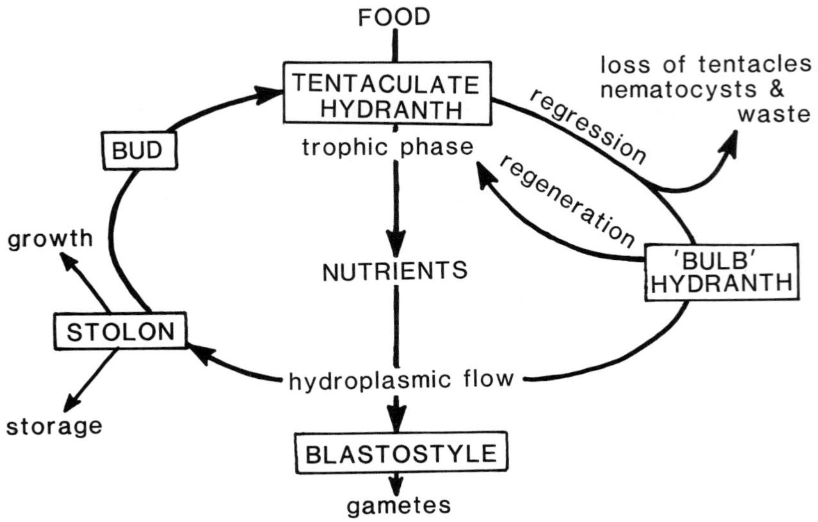

(b)

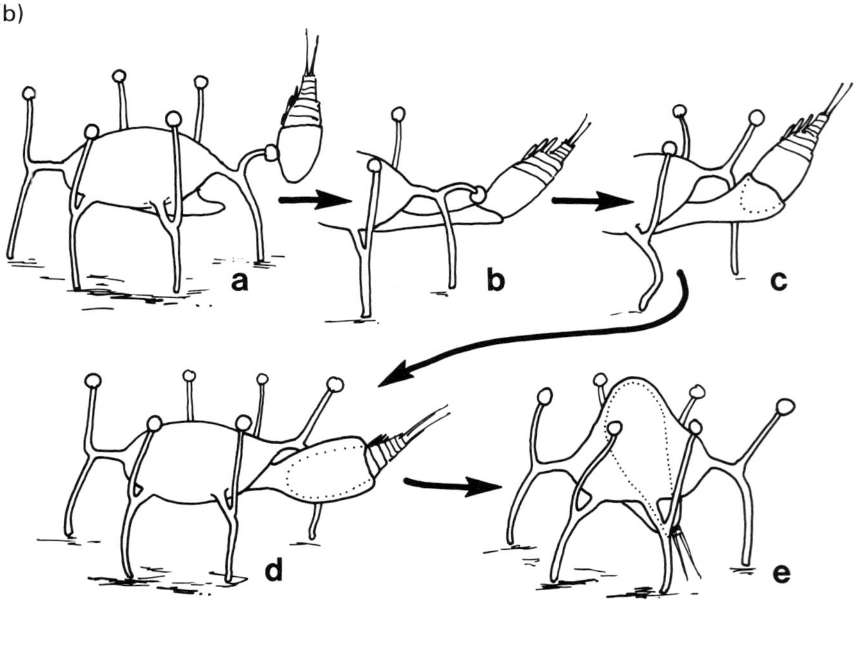

Fig. 11.15 (a) Hydranth cycle of *Thyroscyphus* (see text). (b) Feeding behaviour of *Eleutheria*; capture and ingestion of a copepod by the medusa.

regeneration. In total 13% of the hydranths of *O. geniculata* and 36% of *P. setacea* were either regressing or regenerating (i.e., not in the trophic phase). This raises the question, what is the functional significance of the hydranth cycle? It does not appear to be correlated with the formation of medusae or gonophores, for it occurs in young growths that have not started to produce blastostyles. It has been noted in a number of hydroid species at various times of the year and does not appear to be connected with environmental or seasonal changes. A plausible hypothesis is that the hydranth cycle is a necessary outcome of digestive processes. Each hydranth could be looked upon as a temporary digestive organ formed as part of a physiological cycle. With intracellular digestion the undigested part of the food may accumulate in the digestive cells. Eventually these cells would have to disintegrate to discharge their waste to the exterior. It is possible that this is the source of the brown granules seen during egestion in *Thyroscyphus marginatus*.

Modular growth

Hydroids differ from the animals considered in previous chapters by their capacity to repeat parts of their structure.

When the planula larva of *Laomedea flexuosa* (Fig. 11.12(c)) settles, it secretes perisarc to form a small stolon for attachment, and from this a single vertical segment of hydrocaulus arises which bears a feeding hydranth. One by one new segments, each bearing a single hydranth, are added to the hydrocaulus. This strategy of replicating feeding structures increases the size of the net or collecting area, and hence the feeding capacity.

Individual hydranths cannot be equated with zooids for they are incapable of reproduction. An ascidian zooid might be compared with the minimum hydroid structure capable of carrying out all functions, i.e., a portion of stolon, a hydrocaulus bearing replicate feeding hydranths and replicate blastostyles capable of producing gonophores (blastozooids) asexually. The identity of the individual or zooid becomes obscured when the stolon grows laterally and produces many new hydrocauli with their own hydranths and blastostyles. Each of these new units or modules is capable of carrying out all vital functions, thus the whole complex organism that develops from a single planula can be considered as a modular organism. The minimum repeat unit or module might be defined as the sum of all subsets necessary to carry out all functions characteristic of the species. Thus a hydroid can be regarded as composed of x modules comprising y different subsets (a, b, c, etc.) that are in physiological continuity, are genetically identical and produce gonophores of the same sex. If the modules are not in physiological continuity (as in most compound ascidians), they become members of a clone. The module of *Amphisbetia minima*, for example, comprises a portion of stolon bearing one hydrocaulus with a joint, a pair of basal hydranths bearing a gonotheca with blastostyle and one or two nematophore pores, and a variable number of oppositely arranged sessile feeding structures, the hydranths.

Each module passes through a period of quantal growth until it reaches a size characteristic of the species for a specified set of environmental conditions. When subsets of the module have fulfilled their functional role (that is, feeding and producing gonophores or medusae), their remaining substance is withdrawn into the stolon and the perisarc is discarded. The stolon may continue in existence until environmental conditions induce it to enter a new cycle of modular growth.

Patterns of stolonic branching

Branching of the hydrocauli and stolon of thecate hydroids is governed by well-defined rules which are species specific. Each module follows a pattern of growth in space and time that is little affected by environmental conditions, although temperature may influence the height of hydrocauli, the degree of branching and the number of blastostyles produced (Ralph, 1956; Ralph and Thompson, 1968). The pattern of stolon branching is more easily seen in species that live on seaweeds.

Specimens of *Plumularia setacea* growing on a flat surface like *Eklonia radiata* follow a relatively simple pattern of growth. The flattened stolon travels in a straight line over the substratum, and at intervals of *c.* 1.4 mm hydrocauli grow up vertically. At the same point, typically two (but sometimes only one) lateral branches grow out horizontally at right angles to the main stolon axis. These laterals follow the same pattern, producing a new hydrocaulus and two laterals at the same interval. When one stolon intercepts another derived from the same original hydroid, fusion takes place and a regular quadratic stolonic network develops (Fig. 11.17(a)). The hydrocauli comprise a series of segments, each with an internode, a lateral bulge connected to the hydrocladium (side branch), and a node with an axillary nematophore (Fig. 11.16(b)). Hydrocladia alternate right and left on successive segments. Each hydrocladium is made up of a short basal annulus plus one, three or five other segments that bear either one nematophore and no hydranth or a hydranth and three nematophores on the adcauline or upper side. Gonothecae containing gonophores are attached singly to the internodes of the hydrocaulus. Up to ten gonothecae have been found on one stem. The module comprises a section of stolon (with secondary stolons at right angles) and a vertical hydrocaulus bearing a variable number of hydrocladia and blastostyles (Fig. 11.17(a)).

The pattern of growth in *Silicularia bilabiata* is more easily seen on stolons running parallel to one another (Fig. 11.3(e)). These have a tendency to run catadromically or with the current. At intervals of about 1.2 mm very-thick-walled, vertical pedicels arise, each with a single hydranth that faces in the

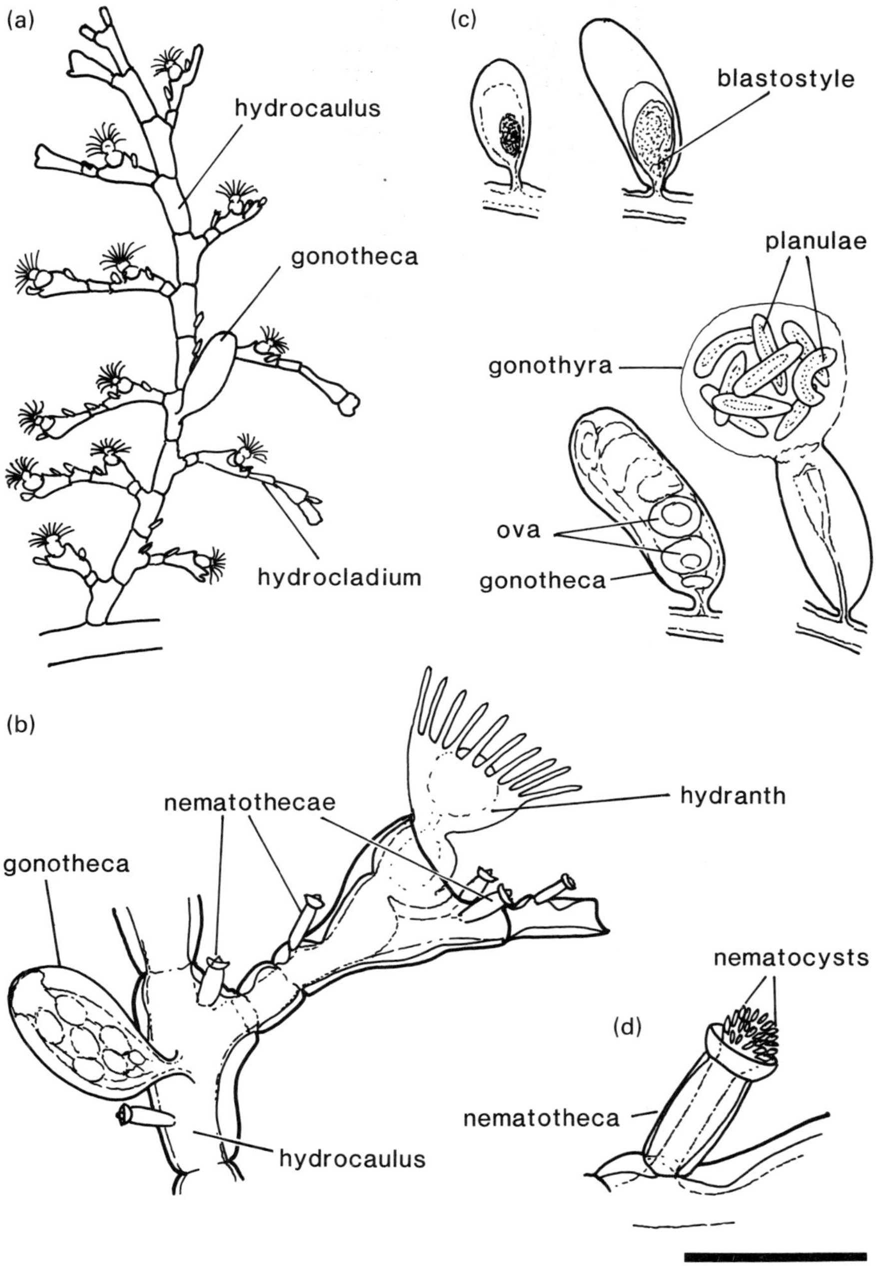

Fig. 11.16 Plumularia setacea (specimens from Australia). (a) Single hydrocaulus. (b) Hydranth and nematophores. (c) Development of female blastostyle and release of planulae. (d) Nematophore. Scale bar: (a) = 1 mm, (b) and (c) = 0.3 mm.

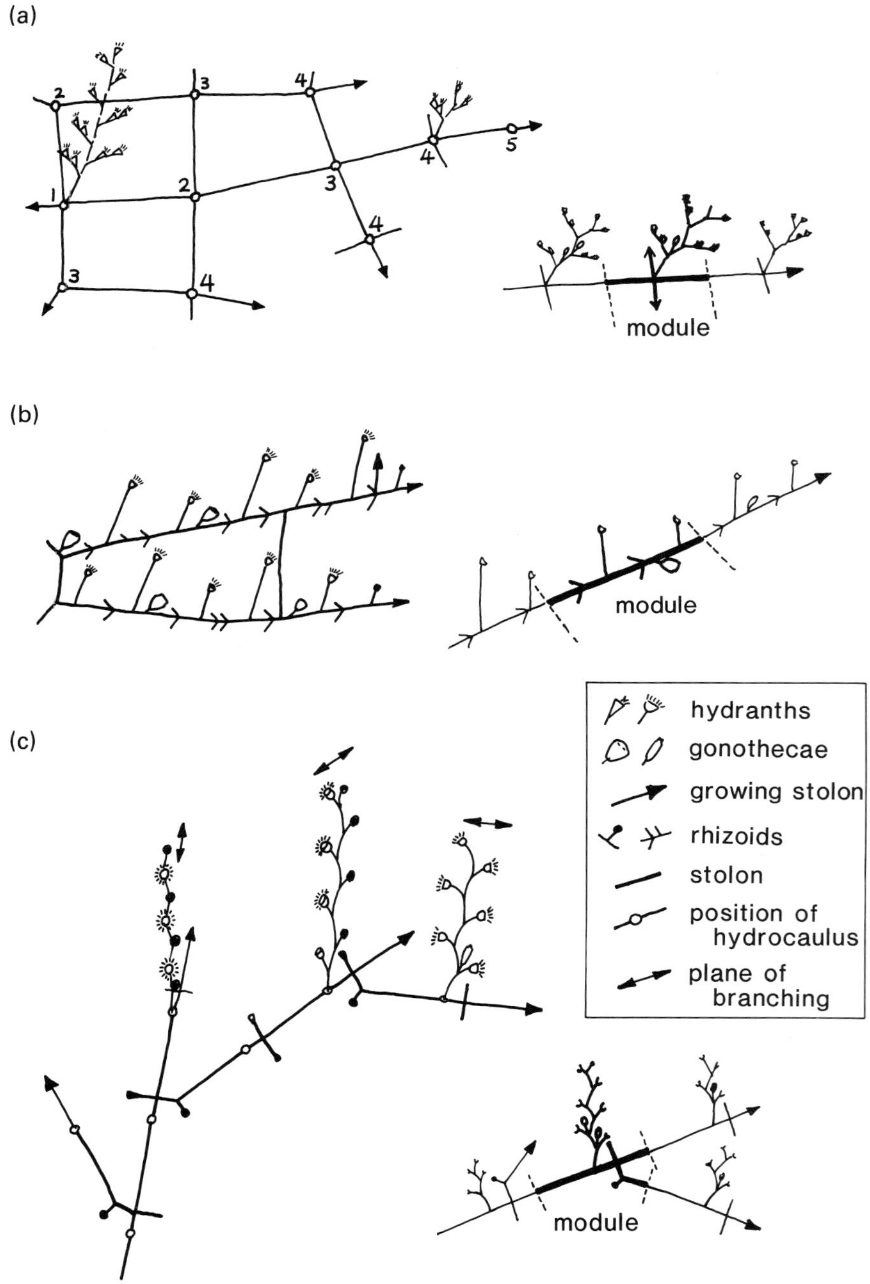

Fig. 11.17 Modular growth and patterns of branching in thecate hydroids (see text). (a) *Plumularia setacea*. (b) *Silicularia bilabiata*. (c) *Obelia geniculata*.

direction of stolon growth. On many specimens there is a regular alternation of tall (1.5 mm) and short (0.8 mm) pedicels. Between the hydranths the stolon gives off three or four lateral rhizoids which are firmly attached to the substratum. They terminate bluntly, rarely grow more than 0.25 mm, and appear to function as additional anchorage for the hydroid. Less frequently, lateral stolons are produced which grow out at right angles to the main axis. They bear hydranths and rhizoids, and when they come into contact with other parts of the same organism they fuse to form a reticulate stolonic network. *Silicularia bilabiata* growing on the deeply ridged surface of *Macrocystis* tends to follow the ridges and form a ladder-like stolon (Ralph, 1956). Sessile gonothecae typically arise from the stolon between pairs of hydranths, but not closer than the 6th hydranth from the tip of an advancing stolon. *S. bilabiata* is well adapted to life in rapidly flowing water. The low profile, thick perisarc, very flexible annulus and extensive rhizoids are specializations for growth in turbulent and fast-flowing water. Each module comprises a portion of stolon with two hydranths, two sets of rhizoids and one gonotheca (Fig. 11.17(b)).

Judging from the frequency with which it is treated in zoological textbooks, *Obelia geniculata* should be one of the best-known hydroids, yet it is surprising how little information exists about growth of the species apart from the account by Ralph (1956). The stolon is straight, and from specimens growing on *Laminaria digitata* appears to run anadromically or against the current. At intervals of 3.9 mm it gives rise to vertical hydrocauli and a pair of short lateral rhizoids. From one or other of the rhizoids a new secondary stolon is produced at an angle of 60° to the primary axis, and this will have its own hydrocauli and rhizoids (Fig. 11.17(c)). Secondary (and tertiary) stolons arise in strict alternation to the right and left of the main stolon, and hydrocauli are orientated so that their branching lies in a plane that runs parallel with the stolon axis. When stolons encounter one another they do not fuse, but stop growing. If, however, they encounter stolons from another hydroid, they grow over the top and continue in a straight line. Hydrocauli consist of characteristically shaped internodes, each bearing a hydranth, and sometimes secondary branching occurs. Urn-shaped gonothecae are axillary and the blastostyles produce medusae (Fig. 11.11(a) and (d)).

In the three species described here, the hydrocauli are strictly orientated to the axis of the stolon from which they arise, but not necessarily to the direction of water flow. This is seen most clearly in *Obelia geniculata* (Fig. 11.17(c)).

11.6 FUNCTIONS OF THE STOLON

The stolon is often ignored in accounts of the hydroids and yet it shows good taxonomic characteristics. The angle of lateral branching and the interval ratios between branches or hydrocauli show very little variation within a species or with changing environmental conditions. The following summary shows that at least eight functional roles can be ascribed to the stolon, making it the most important part of the whole organism.

When the hydroid planula settles, it metamorphoses into a cylindrical structure which secretes perisarc round itself. This is clearly a section of stolon which, in some species, starts to grow over the substratum before any polyps or hydranths are produced.

The stolon is the only living part that remains when the hydroid regresses completely during adverse conditions. It is from this stolonic tissue that new polyps or hydranths will be formed when regrowth starts. In *Silicularia bilabiata*, blastostyles develop directly from the stolon, and so it is clear that the stolon can give rise to all parts of the hydroid.

It is a characteristic feature of both thecate and athecate hydroids that nematogenesis (the formation of nematocysts) takes place in the stolon and possibly the hydrocaulus. In histological sections, clusters of developing nematocysts occur in the ectoderm. In living hydroids, nematocytes containing fully formed nematocysts can be seen migrating from the stolon into the polyp or up the hydrocaulus to the hydranths and nematophores wherever these are developing or being replaced. This migration is most easily followed on living athecates where there is no perisarc to obscure the view.

It is unfortunate that most of our knowledge of nematogenesis comes from studies on the Hydrida, where it occurs in the ectoderm of the polyp's column. This, like so many other features of *Hydra*, must be regarded as aberrant.

An obvious function of the stolon is attachment to the substratum. In some species the stolon may be flattened to give a large area of surface contact or may have short lateral branches or rhizoids. The stolonic perisarc in species like *Amphisbetia minima* or *Phylactotheca armata* has thickened indentations along either side which probably strengthen the wall and prevent the tissue of the stolon from being crushed (Fig. 11.4(a)). Stolons growing over heavily encrusted rocks are often irregular and thread their way between other plants and animals, which provide anchorage for the hydroid.

Growth of a stolon with side branching enables a hydroid to spread rapidly over the surrounding substratum. This exploitation of living space is beneficial, for in most hydroids the amount of food collected will be roughly proportional to the area colonized.

Under adverse conditions the hydranths regress and most of their substance (together with the coenosarc of the hydrocauli) is passed to the endoderm of the stolon; thus the stolon fulfils the role of surviving bad environmental conditions. It can also facilitate a rapid recovery from accidental damage caused by wave action, desiccation or destruction of the polyps by predators.

Endodermal cells of the stolon have a dense granular cytoplasm, indicating that they act as a storage organ for nutrient substances. Many of the cells contain inclusions in vacuoles, suggesting that they carry out intracellular digestion. The fluid constantly flowing back and forth in the gastro-vascular system contains numerous particles, of which some are food particles and others are derived from disintegrating digestive cells in the hydranth. The fact that hydranths periodically regress shows that they cannot store the large amounts of nutrients derived from the food that is digested. This must be passed to the stolon for storage.

The gastro-vascular cavity, which runs through the middle of the stolonic coenosarc and communicates with all parts of the organism, provides a means of transporting nutrients and possibly oxygen. It is also possible that growth-regulating substances could be circulated to co-ordinate patterns of growth or initiate blastostyle development. Thus the stolon provides a vascular system to the hydroid.

Finally it has been shown that the stolon plays an important role in transmitting small pressure differences between various parts of the gastro-vascular system. These pressure differences are the basis for hydroplasmic flow and may be responsible for expansion of hydranth buds.

SUMMARY

Hydroids are small modular organisms attached to the substratum by the perisarc of a branched stolon. The tough, flexible, chitinous perisarc covering all parts of thecate hydroids (except the feeding organs and medusae) makes them well suited to life on the lower shore, but only species with valves closing the hydrotheca can withstand more than an occasional brief exposure to the atmosphere. Athecate species are confined to places where they are not exposed by the tides. Annulations in the perisarc allow tall forms to bend in water currents. A hinge joint, serving the same purpose, is present at the base of larger branched species.

Branching and multiplication of trophic organs (hydranths) increase the area for food collection. Harpacticoid copepods and planktonic larval stages of other crustaceans form the major components of their food, and are captured by discharging nematocysts. Nutrients are passed from the trophic hydranths to other parts of the organism along the tubular gastro-vascular cavity. Trophic hydranths regress periodically, but may be replaced by new ones in a regular cycle.

A wide variety of reproductive strategies are found, but none can be regarded as a specific adaptation to shore life. Hydroids can regenerate from the stolon; this allows parts of the organism, that have been eaten by predators or killed by emersion, to be replaced quickly. Vegetative growth and branching follows a strict pattern characteristic of the species.

12

Sponges
(*Porifera*)

Sponges would appear to be unsuited to shore life as they lack a protective outer cuticle that would prevent water loss during emersion. Nevertheless, several species are found in the littoral zone and some may be found as high as low water neap tide level in suitable places. On the lower shore and into the sublittoral zone, the number of species increases and specimens reach a much larger size. In some areas, for example under kelp, sponges may be the dominant animal covering the rock surface, where green, yellow, orange, red, purple and brown sponges form colourful patches among the encrusting pink coralline algae.

Three attributes of sponges – a cryptic habit, a strong flexible skeleton, and retention of water – help them to withstand strong wave action and the occasional short period of emersion. Few species live where they receive direct sunlight, and the majority of littoral sponges live under rocks, in crevices and caverns, or carpet the floor under the kelp canopy. However, this distribution is a reflection more of larval behaviour at the time of settlement than of adult preferences. Sponges have an internal supporting framework of fibres or spicules which give them a degree of firmness that prevents delicate internal structures from being crushed by wave impact, or internal cavities collapsing when out of the water. These cavities hold a large volume of water which would prevent the delicate internal tissues drying out during brief periods of emersion. Some species from the littoral zone have a firm protective covering provided by vast numbers of spicules just below the surface.

With very few exceptions, sponges living on the lower shore are low encrusting forms that spread out over the rock surface as a smooth layer a few millimetres in thickness. This offers a very large area-of-attachment : volume ratio, which could be regarded as an adaptation to reduce drag forces exerted by rapidly moving water. The capacity for lateral growth makes encrusting species active competitors for space on the rock surface and also enables them to recover quickly from damage caused by predators or physical forces.

Identification of a sponge is not always an easy matter in the field, for both colour and shape are greatly influenced by environmental factors, and it is usually necessary to examine their spicules microscopically to establish their identity. Burton (1963) in his revision of the calcareous sponges considers that analysis of the complement of spicule types is the best way to establish identity, but Bergquist (1978) has drawn attention to the value of biochemical characteristics in their classification and identification.

12.1 STRUCTURE OF SOME LITTORAL SPONGES

Sponges are not easy animals to study microscopically, for hard spicules make it difficult to cut serial sections in order to follow the intricate system of channels and chambers through which water flows, but studies using electron microscopy have radically changed many concepts of sponge structure (review: Simpson, 1984). A detailed account of sponge biology has been given by Bergquist (1978). The following examples illustrate the range of sponge types found in the intertidal zone.

Clathrina cortica (Fig. 12.1(a)) is a small silver-white sponge of relatively simple construction that is of cosmopolitan distribution and may be found on the underside of rocks at extreme low tide. Its appearance is variable; it may be a low incrustation or vase-shaped, but more commonly it is a small mound up to 1 cm high with a volcano-like vent at the summit.

Its basic form is a tube that repeatedly divides and joins up again to give a three-dimensional trellis. Its microscopical structure is simple. The wall of the tube is supported by a mesh of Y-shaped calcareous spicules orientated as shown in Fig. 12.1(b). Lining the interior of the tube, there is a single layer of flagellate 'collar' cells called choanocytes (Fig. 12.1(c)), and on the outside there is a very thin layer of cells (only visible by staining) called pinacocytes. Some of these outer cells have a small hole running through them which leads from the exterior to the interior of the tube: they are called porocytes. In a living sponge, a current of seawater, created by the flagella of the choanocytes, flows through the porocytes into the tube and out through the vent. The choanocytes are responsible for capturing and digesting food particles carried in the water current. In some specimens, some of the tubes may be obstructed by large ova, and water is presumably diverted through other parts of the tubular trellis. The ova develop into a hollow flagellated larva called a coeloblastula (Fig. 12.1(d)).

Scypha (= *Sycon*) *ciliata* (Figs 12.1(e), 12.2(a)) is another very common species of more complex structure. This beautiful little sponge, measuring *c.* 1 cm high, may be found on the underside of rocks among bryozoans, hydroids and other organisms on the lower shore. Its distribution is worldwide.

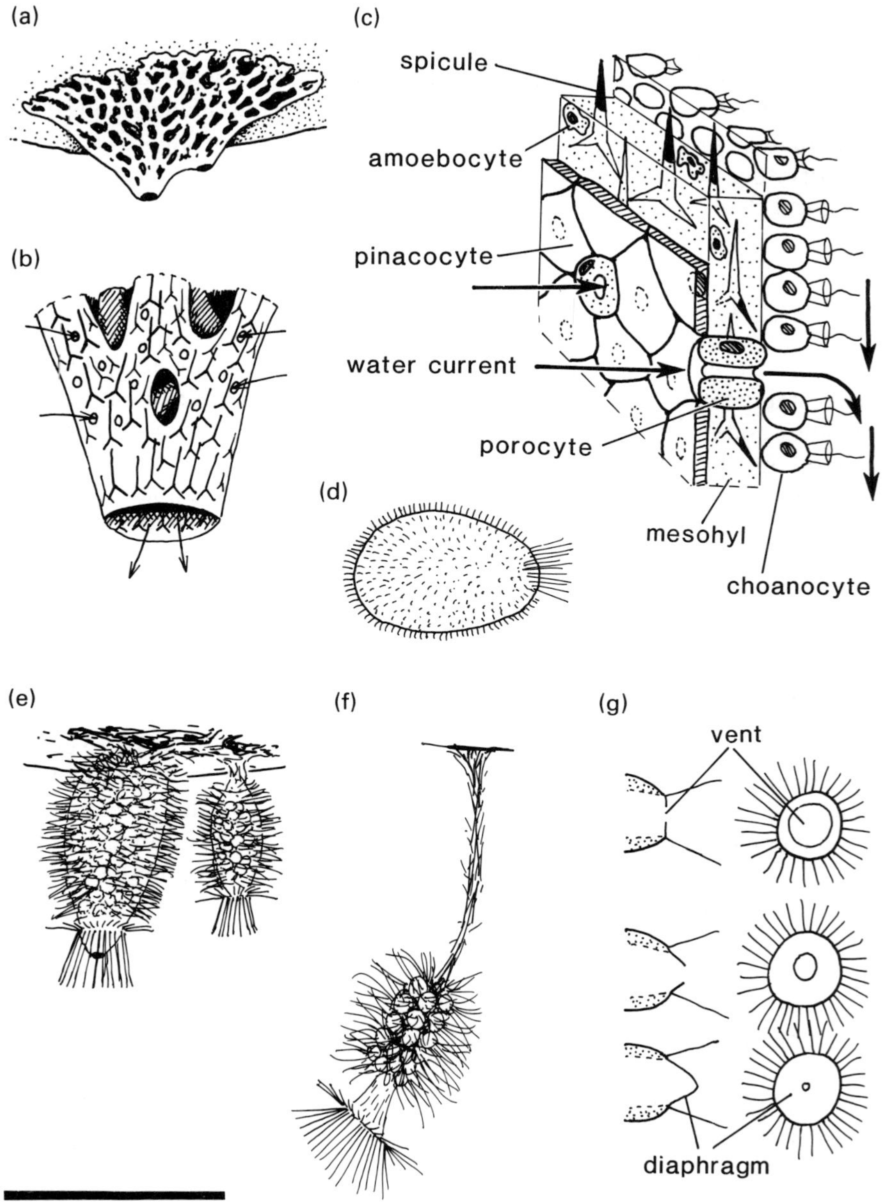

Fig. 12.1 (a) to (d) *Clathrina cortica*, a calcareous sponge: (a) natural appearance. (b) vent, showing orientation of spicules (arrows, water current); (c) body wall, showing water flow (arrows); (d) coeloblastula. (e) to (g) *Scypha ciliata*: (e) natural appearance. (f) very young specimen. (g) Osculum of *Scypha* showing diaphragm (open at top; closed at bottom). Scale bar: (a) = 4 mm, (e) = 8 mm, (f) = 1 mm.

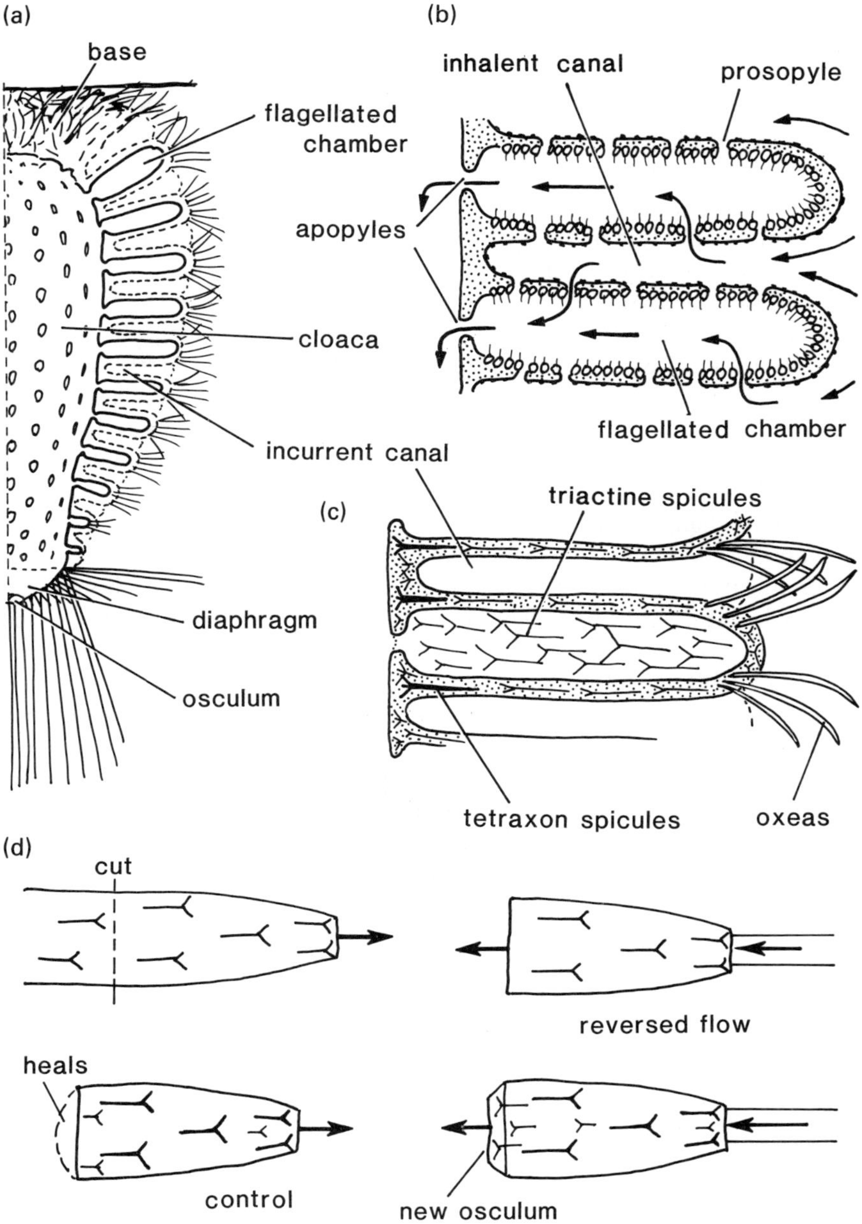

Fig. 12.2 Scypha ciliata (Calcarea). (a) Internal structure, diagrammatic half section. (b) Water flow through flagellated chambers. (c) Arrangement of spicules. (d) Experimental control of spicule growth (see text; arrows, water flow).

A very young specimen looks like the inflorescence of a thistle (Fig. 12.1(f)), with an oval, spine-covered body attached to the substratum by a slender stalk.

A section through the body of an older specimen shows that the flagellated tubes responsible for the water current are arranged radially and discharge through openings called apopyles into a new central cavity, the spongocoel or cloaca. The spongocoel opens to the outside by the osculum which has an adjustable diaphragm and is surrounded by a palisade of needle-like spines (Fig. 12.2(a), 12.1(g)). A basket-like framework of Y-shaped calcareous spicules arranged in a characteristic pattern (Fig. 12.2(c)) supports the flagellated chambers and wall of the spongocoel. The spicules lie in what appears to be a structureless jelly, the mesohyl, between the external pinacocytes and the choanocytes. Electron micrographs, however, reveal a meshwork of collagen fibrillae formed by collencytes in the ground substance of the mesohyl. Suitable staining reveals many other types of cells in the mesohyl, of which archaeocytes are the most common. Between adjacent flagellated chambers, a radial space which opens to the surrounding seawater, called the incurrent canal, is lined with pinacocytes and communicates with the flagellated chambers through small holes, the prosopyles (Fig. 12.2(b)). The low-velocity current generated by the choanocytes is hardly detectable, but it forces the diaphragm up into a cone when the diaphragm closes the osculum (Fig. 12.1(g)).

As the sponge grows in size the spongocoel increases in both height and diameter. The stalk is gradually obliterated and new flagellated chambers are added just below the diaphragm. The structure of *Scypha ciliata* is not static; there is a constant rearrangement of parts to accommodate the new dimensions. Spicules are lost and new ones are secreted to take their place (Burton, 1963; Jones, 1970). Amoeboid cells wander in the mesohyl, probably transporting nutrients from choanocytes to other cells.

Despite its small size and delicate appearance, *Scypha ciliata* can withstand the vigorous water movements that occur between boulders during low tide. Moreover, when the sponge is lifted out of the water, water is held by capillarity in the hair-like covering of spicules, thus the sponge will not dry out, nor will air enter the chambers, on the rare occasions when a specimen is temporarily emersed between waves during extremely low spring tides.

Spicules of *Clathrina cortica* and *Scypha ciliata* are composed of calcite, a characteristic feature of sponges belonging to the class Calcarea*, which are known as calcareous sponges. All the large and highly coloured sponges found on the lower shore belong to the class Demospongiae. Most have a skeletal framework of organic spongin fibres and, where spicules are present, they are composed of silica. The spicules may be attached to, or embedded in,

* See Appendix for classification of sponges.

the fibrous network. The Demospongiae is divided into two major subclasses, the Tetractinomorpha, which may have four-rayed (tetraxonid) spicules, and the Ceractinomorpha, in which spongin fibres are usually present.

The bright orange golfball sponge, *Tethya corticata* (Fig. 12.3(a)), is a common, and in places extremely abundant, tetractinomorph sponge on the Australian coast at about mean low water neap tide level and down into the sublittoral. Rounded individual sponges, 2–3 cm in diameter, have root-like stolons that spread out over the surrounding rock and give rise to more sponges: indeed, in caverns and under rock ledges, the whole surface may be covered by hundreds of sponges, all connected by their stolons to one another and probably derived from one or two founding members.

The sponge feels solid and smooth to the touch, but is able to contract the surface slightly so that it becomes distinctly hispid. When cut in half it displays a characteristic radial arrangement resulting from bundles of long radiating spicules (Fig. 12.3(b)). There is a brightly coloured cortical zone or rind and a paler grey or yellow central medulla. Numerous small pores called ostia communicate with cavities immediately under the thin outer surface, the sub-cortical cavities. From these, numerous inhalent channels pass into the medulla. It is difficult to make out structure in the medullary pulp, for the numerous inhalent and exhalent channels that supply the many millions of small round flagellated chambers are impossible to follow except in serial microscopical sections. Their arrangement is shown in a highly diagrammatical form in Fig. 12.3(c), together with the spicules. The latter are of two types, long straight spicules and vast numbers of very small star-shaped spherasters (microscleres) concentrated in the cortex (Fig. 12.3(d)). All spicules are composed of silica.

Tethya corticata is one of the few marine sponges that are not difficult to keep alive in a marine aquarium, provided they are not disturbed or subject to direct sunlight. Occasionally this sponge undergoes a remarkable transformation, in which large numbers of the radial spines project 1–2 mm above the surface, giving it a prickly appearance. A few hours later the sponge appears normal. The significance of this behaviour is not known, but it appears to be caused by contraction of the sponge and could be related to another curious phenomenon. It is often found that *T. corticata* left in an aquarium for a few weeks produces dozens of radiating strands, 2–3 cm long, that stretch out and make contact with surrounding objects or the side of the aquarium. Each strand has a central axis of one or two extremely long spicules, covered in sponge cells and terminating in a clump of sponge tissue which is the same colour as the parent sponge and contains more spicules and spherasters; it is in fact a bud (Fig. 12.3(e)). This form of budding may be a response to abnormal conditions in the laboratory, such as temperature changes or absence of tides; however, Brien (1973) describes the phenomenon in *T. aurantium* as occurring naturally during the winter

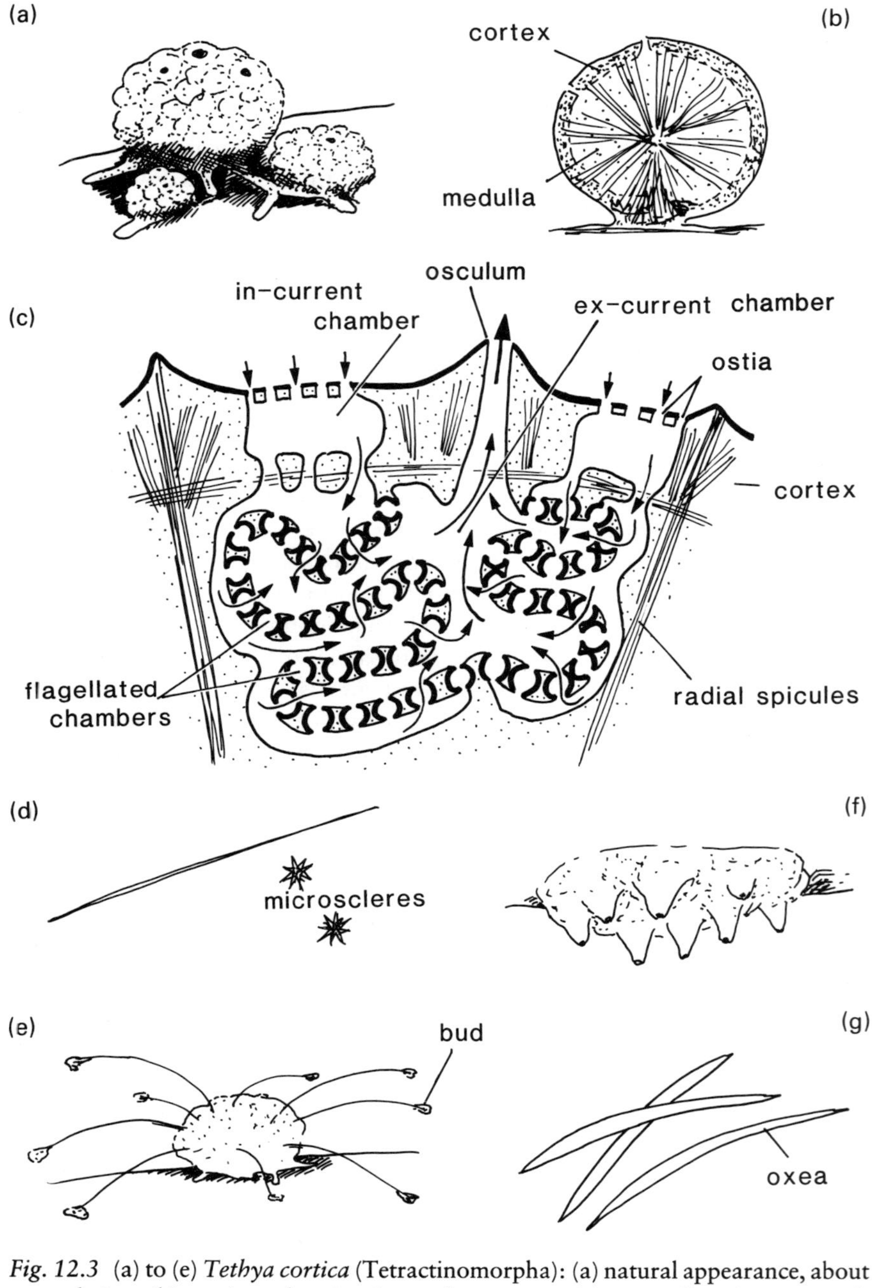

Fig. 12.3 (a) to (e) *Tethya cortica* (Tetractinomorpha): (a) natural appearance, about natural size; (b) section, showing internal structure; (c) water flow (arrows); (d) spicules; (e) budding. (f) and (g) *Halichondria panicea*: (f) natural appearance, slightly reduced; (g) spicules.

months in Europe. In the sub-Arctic White Sea (Beloye More) it is intense, and this suggests that it is a response to low temperatures. The buds are formed in the cortex and are then carried by elongation of the spicules several centimetres from the sponge (Connes, 1967). This is a form of asexual reproduction quite distinct from stolon formation.

A characteristic feature of *Tethya* species and other tetractinomorph sponges is the release of their eggs directly into the sea, where development proceeds to a flagellated larva called a parenchymella. Ceractinomorph sponges, on the other hand, are viviparous; their eggs are retained within the parent and develop to a parenchymella larva.

Halichondria panicea (Fig. 12.3(f)), the crumb-of-bread sponge, is a dirty yellow or greenish encrusting sponge belonging to the Ceractinomorpha that is common at low tide levels in many parts of the world. It is sometimes the dominant species covering the rocks under kelps like *Laminaria digitata* in the sublittoral, and may occur under *Ascophyllum nodosum* on the lower shore. Unlike typical ceractinomorph sponges it lacks spongin fibres and crumbles easily when squeezed. Only one type of siliceous spicule is present (Fig. 12.3(g)) and there are no microscleres. The surface has a very thin crust, formed by a layer of spicules, and below this there is a sub-cortical cavity. The main part of the sponge consists of vast number of small flagellated chambers (each *c.* 30 μm in diameter and containing 50–100 choanocytes), and the small inhalent and exhalent channels. Larger exhalent channels collect the water and convey it to the oscula that open on volcano-like projections. *Halichondria panicea* lacks the orderly radial arrangement of *Tethya corticata*.

Several other members of the Ceractinomorpha possessing both siliceous spicules and spongin fibres are found at low tide levels and in rock pools as brightly coloured incrustations. *Myxilla incrustans*, forming pale yellow or orange growths 2–3 cm thick, has a folded appearance with irregularly spaced oscula level with the surface. It possesses characteristic spicules and microscleres (Fig. 12.4(c)). *Ophlitaspongia seriata* and a species of *Microciona* found on British and Australasian shores form very thin (1–2 mm) orange or bright red incrustations with characteristic microscleres and spicules (Fig. 12.4(b)). In these species the spongin forms a basal network to which the spicules are attached; some may protrude through the surface to give the sponge a slightly rough or prickly feel.

An encrusting form of the bath sponge, *Spongia officialis*, can be found below low tide on Australian shores, where it forms growths 1–3 cm thick, usually grey or pale brown, with an unpleasant odour, resembling carbon disulphide, caused by the presence of terpenols. Siliceous spicules are absent from this sponge, but a complex three-dimensional framework of primary and secondary spongin fibres forms a skeletal support for the millions of delicate flagellated chambers and the passages that lead to and from them

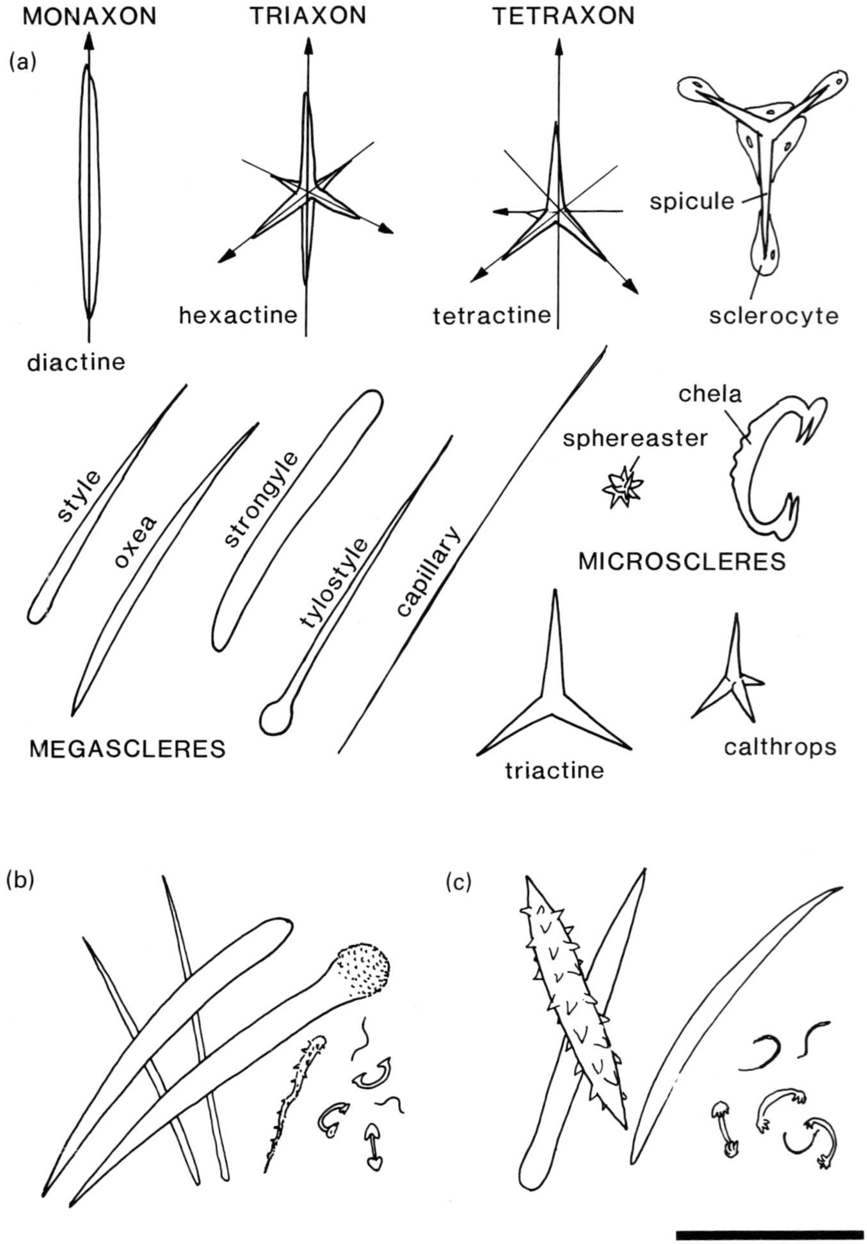

Fig. 12.4 Common types of sponge spicules. (a) Methods of nomenclature. (b) Spicules of *Microciona* and *Ophlitaspongia*. (c) Spicules of *Myxilla*. Scale bar: (b) and (c) megascleres = 0.2 mm, microscleres = 0.1 mm.

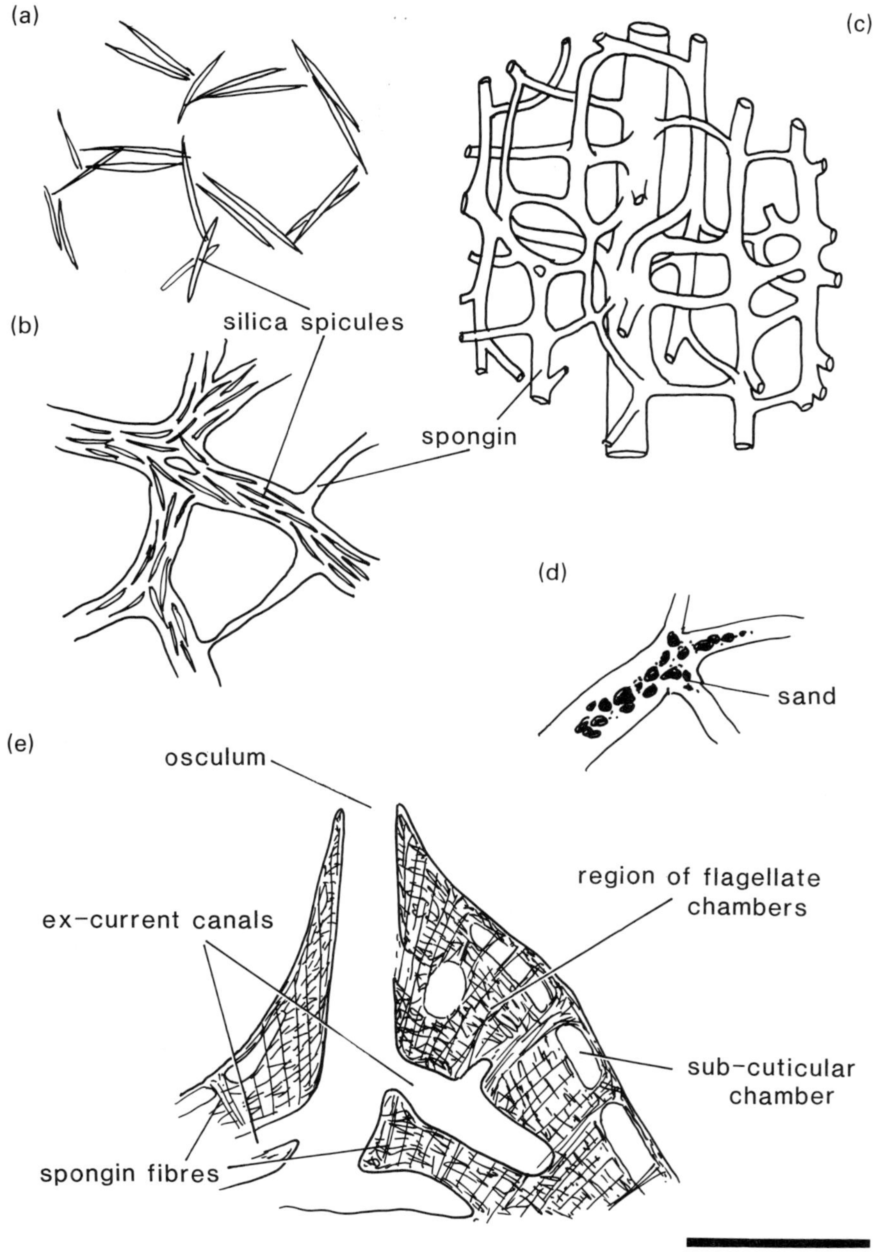

Fig. 12.5 (a) Spicules of *Halichondria*. (b) Siliceous spicules in spongin network. (c) Small portion of spongin network. (d) Sand grains in spongin. (e) Section through piece of *Spongia*. Scale bar: (c) = 0.8 mm, (e) = 5 mm.

(Fig. 12.5(c) and (e)). The oscula open at the top of a few conspicuous volcano-like mounds.

12.2 THE SKELETAL SYSTEM OF SPONGES

Spicules are a characteristic feature of sponges. They may be calcareous, siliceous or organic, and their chemical composition has been used as the basis of sponge classification. Their primary function is to form a skeletal support that prevents the collapse of the millions of delicate flagellated chambers and water channels within the sponge. The mesohyl lacks the thicker collagenous network of fibres found in the mesogloea of sea anemones, and there is no basement membrane binding choanocytes into a firm epithelium. Consequently, sponges lack the cohesion of normal metazoan tissues and rely upon the skeletal system to fulfil this function. In the Demospongiae, siliceous spicules are usually attached to, or embedded in, the spongin meshwork (Fig. 12.5(b)), rendering it more rigid, and in some species grains of sand (i.e., foreign material) are incorporated (Fig. 12.5(d)). Secretion of new spicules or spongin enables relatively rapid changes to be made in the architecture of the aquiferous system in response to changes in pressures and water flow. In addition to their supporting role, calcareous and siliceous spicules may offer some protection from small predators and browsers.

The type and shape of spicules is characteristic of the major taxonomic groups and many are specific to genera or species. The more common types are illustrated in Fig. 12.4. Large spicules (megascleres) can be described by reference either to the number of axes (monaxons have a single axis, triaxons have three axes at right angles), or to the number of rays (e.g. diactine, a straight spicule pointed at both ends, and triactine, a spicule having three rays). Descriptive names are also given (e.g. oxea, pointed at both ends, strongyle, rounded at the ends, tylostyle, a knobbed, pointed peg, etc.). Some Demospongiae have minute spicules of unusual beauty and strange shape called microscleres (to distinguish them from the much larger megascleres). They are very useful when identifying sponges as certain shapes are confined to particular families; for example, asterose microscleres are found only in the Tetractinomorpha, whereas sigmas and chelate microscleres are confined to certain families in the Ceractinomorpha.

The spicules of the Calcarea are composed of calcium carbonate with a small amount of magnesium carbonate, crystallizing as a form of calcite. They are ensheathed in a thin layer of collagen which can be revealed by staining. Spicules are formed extracellularly by sclerocytes, that is, their secretion crystallizes between the cells and not within their cytoplasm. Monaxons are formed by two scleroblasts which move apart as the spicule increases in length: triactine spicules are secreted by six scleroblasts. Growth of spicules is rapid. Jones (1964b) measured their rate of growth in *Leucosolenia* and found that triactine spicules increased by 1–2.25 $\mu m\ h^{-1}$, and monaxons increased by up to 3.78 $\mu m\ h^{-1}$, which means that spicules can grow to full size within 24 h. The fate of the sclerocytes is not known but it is assumed that they move into the mesohyl and secrete more spicules.

The siliceous spicules of the Demospongiae are laid down intracellularly. First of all an axial rod of organic material appears in the cytoplasm of the sclerocyte; later, silica is deposited round this axial rod to form the spicule (Simpson, 1984). In most cases siliceous spicules are bound together by collagen or spongin. Spongin is a form of collagen peculiar to sponges that is laid down extracellularly by spongocytes as a fused, three-dimensional framework to form a resilient skeleton. In the family Spongiidae, siliceous spicules are absent and spongin is the only skeletal material.

Spicules are not scattered haphazardly through the sponge; they are arranged in an orderly manner and often with a distinctive architecture. In *Clathrina cortica*, the Y-shaped triactine spicules are orientated tangential to the surface and with one ray pointing away from the vent (Fig. 12.1(b)). In *Scypha ciliata*, at least four main types of spicule are present, each confined to a specific part of the body (Fig. 12.2(c)). Tetraxon spicules lining the spongocoel and triactine spicules lying in the walls of the flagellated chambers form the skeletal framework supporting the sponge, while large, slightly curved, oxeas project outwards; oxeas filter from the inhalent water larger particles and debris that might block the prosopyles, and possibly prevent small animals entering the sponge. The palisade of long fine spicules round the osculum will help to separate the inhalent and exhalent currents and prevent them merging, but they may also prevent small amphipods, syllid worms and polyclads from entering the spongocoel.

The beautiful symmetry and complex architecture of these little sponges prompts us to enquire what controls or regulates the growth processes that lead to a particular-shaped spine being formed at a particular place and orientated in a particular direction with respect to the axis of the sponge. This problem has received little attention, perhaps because of the technical difficulties of investigating it. However, Jones (1958) investigated the influence that direction of water flow had on the orientation of spicules in *Leucosolenia* by reversing the current. When a piece of sponge was removed and water was forced through the osculum in the opposite direction, a new osculum was formed at the cut end (Fig. 12.2(d)). Newly formed spicules were orientated with one ray pointing away from the new vent; however, new spicules forming near the old vent were still correctly orientated to that structure, and not reversed as would be expected if they were orientating to the new direction of flow. In the controls, the cut end healed over without the formation of a vent. Jones concluded that while current direction might influence the formation of a new vent, it did not influence the orientation of spicules. Other observations showed that the rate of growth of the outer pinacocytes was slightly greater than that of the internal choanocyte layer, thus causing shear in the mesohyl. Jones (1965) investigated the effect of shear on the orientation of spicules, but came to the conclusion that it was not responsible for their orientation (Jones, 1970).

Teragawa (1986) found that sand particles increased the growth rate of experimental pieces of the sponge *Dysidea etheria*, possibly by acting as a filler and allowing a given quantity of spongin to produce more fibres. About 20% of the particles settling on the surface of the sponge were incorporated into the primary and secondary spongin fibres.

12.3 FEEDING CURRENTS AND DIGESTION

The choanocytes

Choanocytes are a universal feature of sponges; they are responsible for trapping and ingesting food particles as well as creating the water current that brings food to the sponge. Generally, choanocytes of calcareous sponges are larger (8–12 μm) than those of the Demospongiae (2–3 μm). Their cell body may be rounded or irregular in shape, and lateral projections may interlock with similar irregularities on adjacent cells, but they are not tightly packed or bound together as cells in a normal epithelium (Fig. 12.6(a)), and scanning electron micrographs show spaces between them through which water can pass (Figs 12.6(c), 12.7(a)). As the choanocytes are not bound to a basal membrane, it is possible for them to withdraw their flagellum and pass into the mesohyl by amoeboid movement. The flagellum arises from a basal body and may be plain or have lateral flimmer filaments that form a vane (Fig. 12.6(a)). The base of the flagellum is surrounded by a collar consisting of a palisade of some 20 to 40 microvilli, each 0.1–0.2 μm in diameter and evenly spaced by an interval of about the same dimension. The regularity of spacing shown on all electron micrographs is noticeable and indicates that the microvilli must be relatively stiff structures (Fjerdingstad, 1961; Brill, 1973; Johnston and Hildemann, 1982). An electron micrograph by Willenz (1980) shows a microtubule running inside each microvillus and passing down into the cytoplasm of the cell (Fig. 12.6(b)), which may act as a stiffening rod. Microfilaments have been described which form a net-like mesh between the microvilli (Fig. 12.6(b)) (Fjerdingstad, 1961; Brill, 1973).

Various structures may be found in the cytoplasm. The nucleus lacks a nucleolus and may be situated either basally or next to the basal body of the flagellum; mitochondria and a Golgi apparatus are always present (Fig. 12.6(a)). One or more contractile vacuoles are present in the cytoplasm of

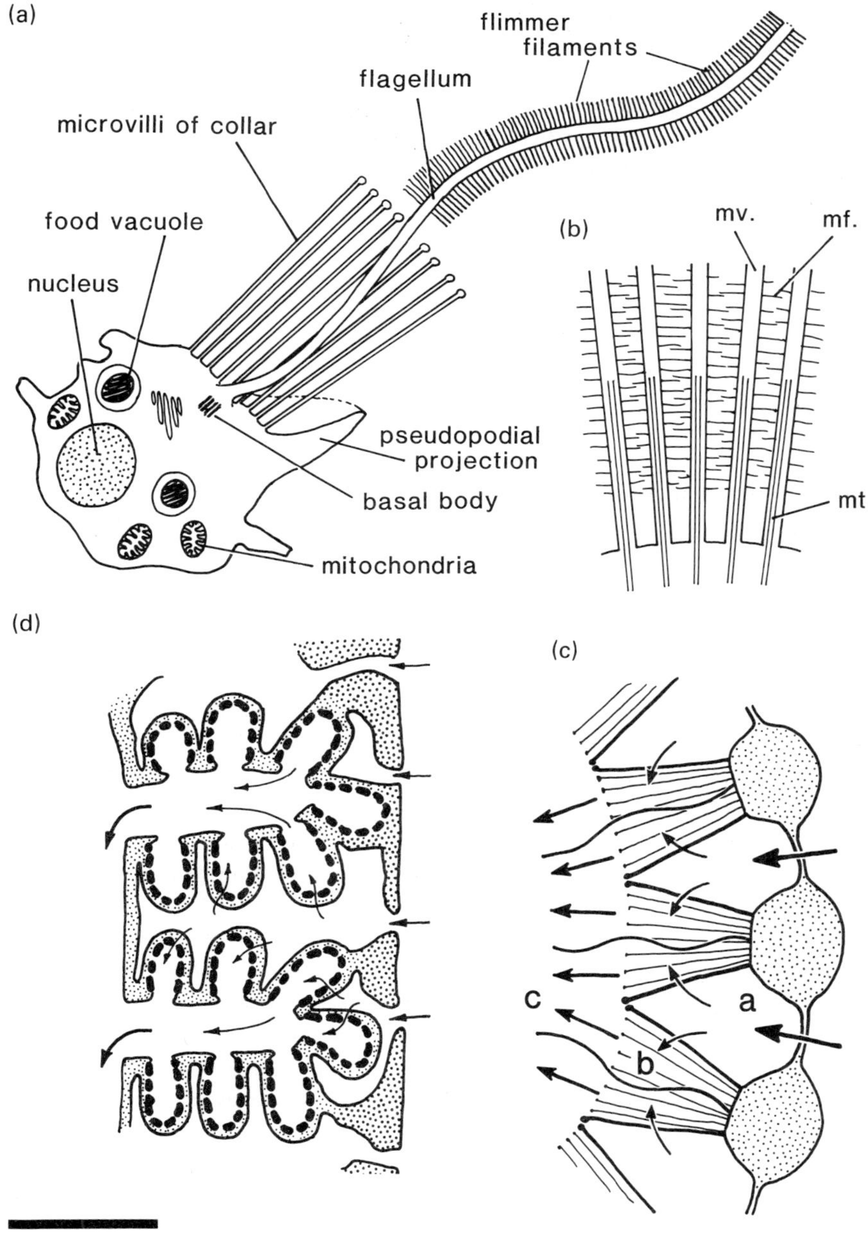

Fig. 12.6 Structure of a choanocyte. (b) Microtubules and filaments (based on electron micrographs, Willenz, 1980). mf., microfilaments; mt., microtubules; mv., mirovilli. (c) Water flow past choanocytes (see text). (d) Water flow through calcareous sponge of Leucon type. Scale bar: (a) = 4 μm, (b) = 1 μm.

freshwater sponges just below the collar (Brill, 1973), but they are absent in marine species. In the sponge *Suberites*, there is a protoplasmic sheath inside the collar that surrounds the base of the flagellum (Connes *et al.*, 1971). Vacuoles with food particles are a regular feature of choanocytes and their number increases when food is abundant. Pigment granules, basal spherules and root-like protoplasmic filaments are other structures that have been described (Tuzet, 1973).

Water flow and arrangement of chambers

The way in which the choanocytes carry out their three basic roles of creating a current, capturing food and digestion is not as well understood as their structure.

It is easy to observe that without any apparent movement, a sponge can produce a current in absolutely still water. As contractile cells are rarely found in sponges it must be inferred that the flagella create this current flowing through the sponge, but how this is done is not easy to understand.

The flagellum of choanocytes beats in one plane and drives water away from its base, but there is no synchronization of beat with surrounding choanocytes, and adjacent flagella may beat in different planes (Jones, 1964a), therefore there is nothing corresponding to the orderly metachronal beating of cilia which is responsible for pumping water through narrow tubes in other animals. Indeed, it is not clear that flagellate cells lining a tube normal to the surface should cause movement of water along the tube at all. It is more likely that each would cause local eddy currents as already noted in the stolonic tubes of *Thyroscyphus marginatus* (see page 255). In the case of the flagellated chambers, however, water accelerated by flagella to the centre of the chamber will be replaced by water drawn in through the spaces between adjacent choanocytes (Figs 12.6(c), 12.7(b)).

The effect of flagella beating in a tube closed at one end (like the flagellated chambers of *Scypha*) will be to accelerate water to the middle of the lumen. Here the velocity of flow will increase progressively along the tube (to the right of diagram in Fig. 12.8(a)). It will be explained on page 289 that this increase in velocity accompanies a reduction in pressure, hence, the pressure (P_3) or (P_4) at the centre of the tube will be less than the pressure outside (P_1), and water flows into the tube via the openings (prosopyles), (Fig. 12.2(b)).

Beating flagella will have another important effect. Water accelerated towards the middle of the tube will cause an equal and opposite thrust on the choanocytes (albeit a very small one) which will keep the lumen of the tube (flagellated chamber) open.

The volume of water passed through the sponge is increased by increasing the number of flagellated chambers. In *Scypha ciliata*, parallel tubular chambers open into a large spongocoel. In the leuconid type of calcareous

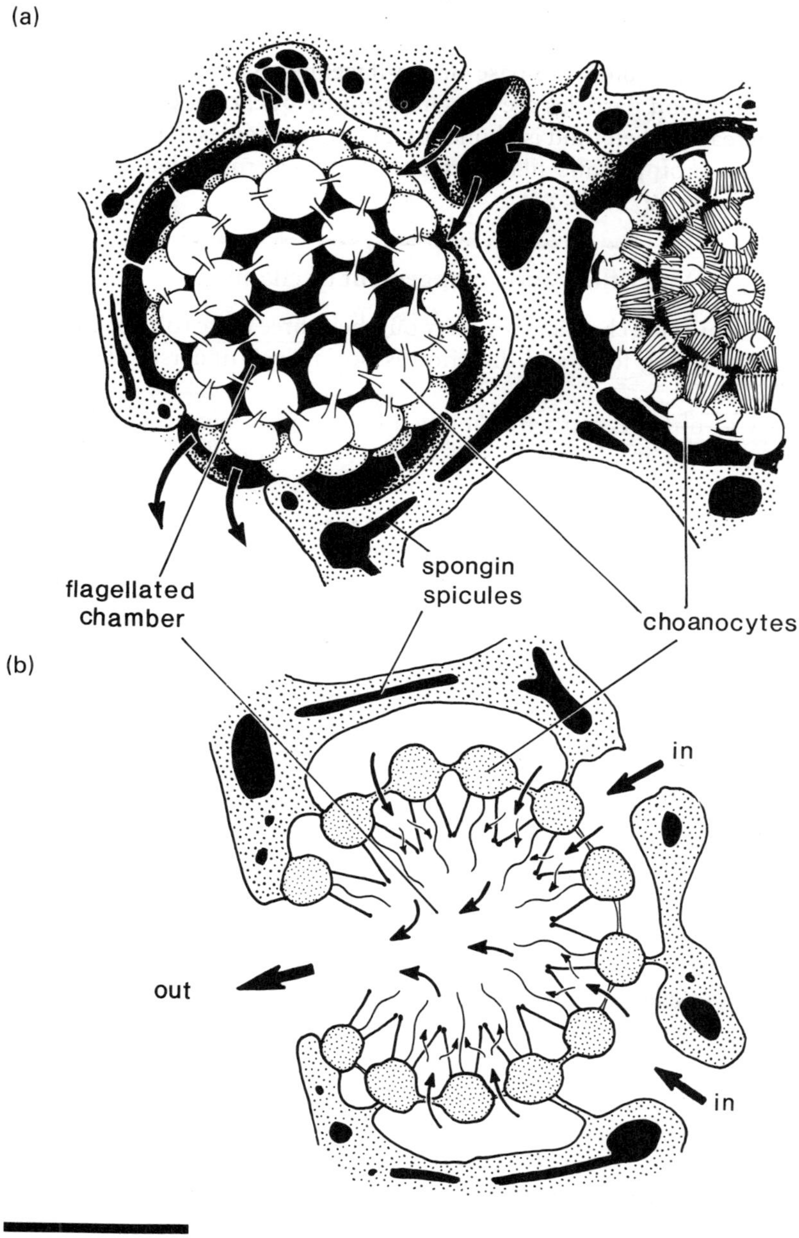

Fig. 12.7 (a) Arrangement of flagellated chambers in a dictyoceratid sponge. (b) Water flow through a chamber. Scale bar = 15 μm.

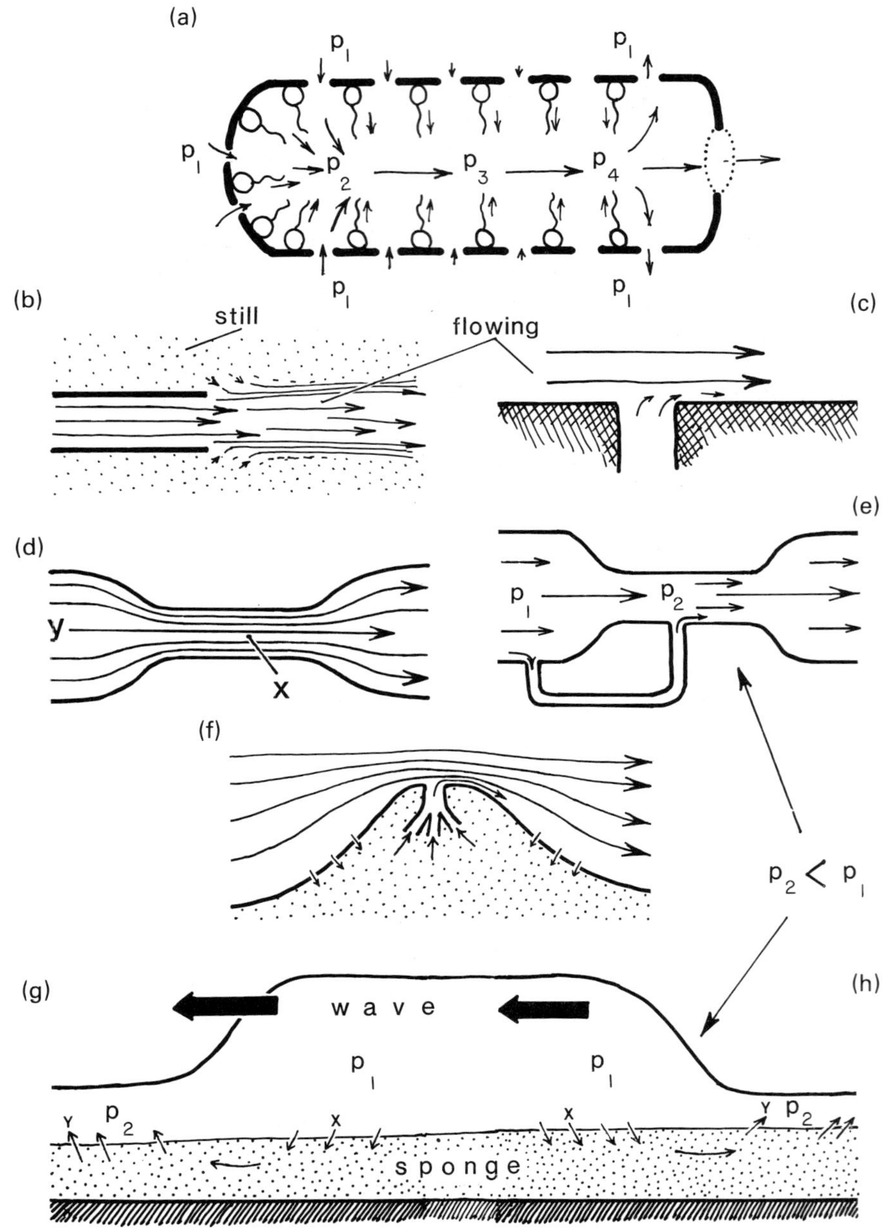

Fig. 12.8 Gneration of water currents in a sponge. See text for explanation. (a) Flagellated chamber. (b) and (c) Viscous entrainment. (d) to (f) Venturi effect. (g) and (h) Pressure changes caused by passing wave.

sponges, the flagellated chambers open into exhalent channels which join together before opening at the vent (Fig. 12.6(d)). Members of the Demospongiae have vast numbers of very small, nearly spherical chambers which connect to small exhalent channels, and these coalesce into progressively larger channels (Fig. 12.7(a)). In *Halichondria panicea*, for example, the average size of chambers is 25 μm and their density is *c.* 1.8×10^7 cm^{-3}.

In the three examples of demosponges chosen here (*Tethya*, *Halichondria* and *Spongia*), there are similarities in the basic arrangement of parts. Each has numerous pores or ostia at the surface, which lead into relatively large cortical or subdermal cavities. From here, inhalent ducts divide repeatedly and convey water to the flagellated chambers through pores called prosopyles. Water exits via apopyles to the exhalent channels, which converge and coalesce to form the major exhalent trunks leading to the oscula. All these internal passageways are lined by endopinacocytes.

When the size of the osculum (or oscula) is compared with the total area of the ostia through which water enters the sponge, there is a marked discrepancy. In *Halichondria panicea*, for example, the cross-sectional area of the oscula is less than 1% of the area of the ostia. Since the velocity of flow in a tube is inversely proportional to the cross-sectional area, it follows that there must be a marked difference in velocity between water entering and water leaving the sponge.

Reiswig (1975) has determined the effective rate of flow out of the osculum of the sponge *Haliclona permollis* as 51 mm s^{-1}, whereas the calculated velocity of water entering the ostia was only 0.5 mm s^{-1}, i.e., the exit velocity is 100 times greater than the entry velocity. This has important implications.

A high exit velocity carries the effluent water further away from the sponge, thereby reducing the chances of its being returned to the intake area. This is desirable, for effluent water will have been depleted of food and oxygen and may also contain waste products or gametes. On the other hand, a very low intake velocity is desirable since it will reduce the likelihood of drawing in silt particles and other foreign matter that might block the ostia and the narrow inhalent passageways. Sponges have not evolved the cilia and mucus-secreting glands that are used by other animals to trap and carry away sand particles that might block or interfere with their feeding mechanisms. This might account for the fact that few sponges live in very sheltered waters where silt and mud accumulate, or in places where there is much sand. Their habit of growing on rocky surfaces where there is vigorous water movement and on the underside of rocks and ledges means that denser inorganic particles will not settle on them. It is known that both exo- and endopinacocytes engulf particles that settle on their surface and carry them to the exhalent channels, where they are voided (Bergquist, 1978).

Table 12.1 Current velocities in *Haliclona permollis**

Site	*Velocity* (mm s^{-1})
Ostia	0.57
Subdermal cavity	0.18
At base of choanocyte collar	0.005
Apopyles (exits from chambers)	0.04
Osculum	51.3

* Source: Reiswig (1975).

The velocity of flow is not constant throughout a sponge (Table 12.1). These figures show that the passage of water through the sponge is extremely slow. The drop in velocity that occurs in the subdermal cavity will allow particles that have passed through the ostia, but are too large to pass through the finer channels to the flagellated chambers (i.e. from *c.* 5 to 20 μm), to settle and be phagocytosed by the endopinacocytes. The water passing into the flagellated chambers will contain only the smallest of particles, bacteria and fine organic material less than 5 μm. Velocity increases again as more exhalent channels meet. This will help to remove waste particles being discharged by pinacocytes into the exhalent stream.

In very small tubes, frictional resistance to flow is extremely high, thus a high percentage of the choanocyte's energy output will be expended in overcoming viscous forces rather than accelerating water. It is safe to conclude that the architecture of each species of sponge is adapted to creating certain velocity characteristics at points where these velocities maximize the efficiency of collecting food, removing inorganic matter or disposing of wastes.

Passive flow

Although there is no doubt that the choanocytes can produce a current, the shape of most encrusting sponges is such that movements of the surrounding water will draw water passively through the canal system. Thus the volume of water passing through a sponge (and hence the amount of food available) can be enhanced by a sponge's shape and position on the rocks. This has been graphically demonstrated by Vogel (1974) on the sponge *Halichondria bowerbanki* from Woods Hole, Massachusetts. After pieces of sponge were immersed in fresh water to stop the flagella of choanocytes beating, they were placed in flowing seawater. Dye was drawn in passively through the ostia and flowed out through the osculum.

There are three ways in which passive flow can be induced. The first is viscous entrainment (Fig. 12.8(b) and (c). If a jet of water is passed through still water, shear forces develop between the moving and static water. Viscous forces will oppose this shear and some of the static water will be carried along with the jet. By a similar principle, water flowing over the surface of a sponge would draw water out of the oscula.

In encrusting species where the oscula are raised up like volcanoes, water will be sucked out of the oscula by water flowing over the surface, as a result of the Venturi effect (Fig. 12.8(d) and (e)). This phenomenon can be explained as follows: the relationship between velocity (u) and pressure (p) of a fluid flowing through a tube is given by Bernoulli's equation:

$$P + 0.5\,\varrho U^2 = \mathrm{K} \qquad (12.1)$$

where ϱ is the density of the fluid and K is a constant. For water, ϱ is unity, and so $P + U^2$ is proportional to K. The velocity of water flowing through a tube is inversely proportional to the diameter of the tube, and so velocity will increase where the tube narrows (indicated in Fig. 12.8(d) by drawing the streamlines closer together). From equation 12.1, pressure must drop if velocity increases, consequently pressure at X will be less than at Y and beyond the constriction. This pressure difference would draw water through a side tube connecting Y and X (Fig. 12.8(e) where $P_1 > P_2)_5$ this is called the Venturi effect and only applies to fluids in motion. The raised osculum of a sponge resembles the narrowed tube and causes a local increase in velocity of water flowing over it. Consequently pressure at the mouth of the osculum will be slightly less than inside the sponge and so water will be drawn through the sponge and out of the osculum (Fig. 12.8(f)).

A large sponge subjected to wave action will experience pressure differences in different parts of its body as a wave passes over it (Fig. 12.8(g) and (h)). At X, water would enter the sponge, and at Y leave it. As the wave passes, the conditions would reverse. Thus wave action would tend to cause an oscillation of water in the sponge rather than a flow.

It has been noted by many observers that sponges grow much larger in certain places, where they are subject to external water currents. This may reflect the extra food brought to the sponge by passive flow. The important observation to be made here is that a sponge's food supply may be increased, without any extra effort or energy expenditure, by virtue of its shape and where it lives. Of all modes of food collection this would appear to be the most energy efficient, and this might help to explain the extreme conservatism of the Porifera. It is likely that the method of nutrition we see in sponges today is the same as that used in sponges from Cambrian times, for it is difficult to imagine what selective advantage could be obtained by adopting a less efficient mode of food collection.

Control of flow

There are three ways in which a sponge might regulate the water flowing through its body: by varying the aperture of the porocytes and ostia, by closing the diaphragm at the osculum, or by stopping the flagella beating.

If the flagella stop beating, it is possible that the choanocyte chambers might collapse and block the passage of water. In *Tethya*, myocytes in the medulla can cause a reduction in volume of the sponge, which would have the same effect. Regulation of flagellar beating is uncertain. It is known that flagella may be stopped by changes in salinity, but flagella are not known to be under the same close nervous or neuroid control as the cilia of other invertebrates, which can stop, reverse or synchronize to a different phase of beat, as already mentioned in mussels and ascidians.

The porocytes found in calcareous sponges can open or close the intracellular aperture leading into the flagellated chamber. Reiswig (1975) has demonstrated the presence of porocytes round the ostia of certain Demospongiae, presumably with the ability to control the aperture. Ostia in the Demospongiae are formed by several cells and can be closed by myocytes.

The delicate diaphragm across the osculum, already noted in *Scypha ciliata*, can regulate water flow by opening or closing. When the diaphragm is stimulated (for example, by a jet of water from a pipette) it closes slowly. The pressure of the water still flowing from the cloaca pushes it up into a cone (Fig. 12.1(g)).

A number of sponges, including *Halichondria panicea*, are known to have a large, curiously shaped amoeboid cell in the centre of the flagellated chamber. At times it appears to block the apopyle, and it is thought that this might regulate water flow through the chamber (Connes *et al.*, 1971).

The question of nervous co-ordination in sponges has been the subject of considerable debate. Although nerve-like cells have been reported from histological studies (Pavans de Ceccatty, 1966), neurophysiological studies have failed to detect any form of action potential from sponges (Lawn, 1982). A nervous system would require some degree of permanence in the specific relationship between nerve cells and the effector organs they control. In view of the instability of sponge structure and the continual movement or transformation of cells, it is unlikely that a nervous system as understood in the Metazoa could exist in sponges.

Food capture and digestion

In calcareous sponges the porocytes will place an upper limit on the size of particle that can enter the flagellated chamber. A stained specimen of *Clathrina cortica* was found to have porocyte apertures ranging from 3 to 14 μm in diameter with a mean of 7 μm, thus few organisms larger than

10 μm would be able to enter the flagellated chambers; however, it is doubtful that the choanocytes could ingest such large particles.

In the Demospongiae the situation is slightly different. It will be the ostia in the first instance that limit the size of particles that may enter. Reiswig (1975) found that the ostia of *Halichondria panicea* ranged from 6.5 to 32 μm with a mean of 17.9 μm. Presumably organisms up to 20 or 25 μm will be taken in, but as already pointed out, these are likely to be taken up by pinacocytes lining the inhalent channels. Only particles smaller than 2–3 μm will reach the flagellated chambers and, since the choanocytes themselves are only 3–4 μm, it is unlikely that particles larger than 1 μm will be ingested.

The exact functioning of the microvilli forming the collar has not been satisfactorily explained. They appear to be ideally suited to act as a sieve to collect particles in the 0.1–0.5 μm range. However, all observations on feeding, both with nutritive particles and with inert material such as India ink, carmine or latex beads, shows that phagocytosis takes place at a point on the cell surface below the collar. In many electron micrographs, protoplasmic processes extend up on the outside of the microvilli; these might be pseudopodia responsible for phagocytosis. Unfortunately the very small size of these structures (near the limit of resolution of the light microscope) makes it difficult to observe their activity. Theoretically there are two possibilities. The microvilli might be like the filopodia of Foraminifera, and particles adhering to the plasmalemma of a microvillus might be carried back to the cell in a manner similar to that of filopodia. Alternatively, the collar 'net' might be passive, and particles trapped there might be removed by pseudopodia which extend up from the cell and ingest the particle.

The universal presence of choanocytes in sponges suggests that the collar is an essential structure and without it the cells would not be able to collect food. Electron micrographs of flagellated chambers (Willenz, 1980; Johnston and Hildemann, 1982) suggest that the microvilli touch those of adjacent cells so that the chamber is separated into two regions: a central lumen where the rate of water flow might be higher, and a 'sub-collar' space in which the water moves very slowly (respectively c and a in Fig. 12.6(c)). This region of slow movement would favour collection of particles by pseudopodia and reduce the likelihood of their being washed away before being engulfed. The collar will be much more effective as a net in the very small chambers of the Demospongiae than it is in the relatively large flagellated chambers of species such as *Scypha ciliata*, where choanocytes are less than one-twentieth the width of the chamber.

Whatever interpretation is put on the way choanocytes function, it is clear that they capture food in a very different way from all other particle feeders.

It is generally believed that sponges are not selective feeders, for both choanocytes and pinacocytes phagocytose inert material, such as carbon particles or polystyrene beads, as readily as bacteria or other microscopic organisms that would provide nourishment. Choanocytes may pass inert material to the underlying archaeocytes, or they may retain it in vacuoles for up to 12 h before eventually discharging it into the mesohyl (Willenz, 1980). In calcareous sponges, digestion of food particles takes place within the choanocyte. The products of digestion are taken into the cytoplasm and any undigested material remaining in the vacuole is discharged from the cell. Thus the choanocyte both captures and digests food material.

In those Demospongiae that have been investigated, such as the freshwater sponges, the choanocytes do not appear to have a digestive role. Within an hour of capture, particles in vacuoles are passed from the choanocytes to the underlying archaeocytes where digestion takes place. It is clear that these choanocytes are specialized for food capture and that the process of digestion is relegated to other cells. The reason for this trend has not been explained. If the choanocytes have lost their ability to digest food, we are faced with the problem of explaining how they obtain their energy for constant flagellar activity. When food particles are abundant, choanocytes soon become replete with numerous food vacuoles, far beyond their capacity (one would imagine) to digest, unless further collecting activity was shut down for a long period. But by passing food to the larger amoebocytes, which are capable of digesting and storing food reserves, the choanocytes are able to collect food continuously.

The origin of the choanocytes that replace those leaving the chambers or in the growth of new flagellated chambers is uncertain. Choanocytes multiply by mitotic division and this will replace those that migrate into the mesohyl, but there is the possibility that they could arise from the archaeocytes as they do in the earliest developmental stages of gemmules (see page 293). The problem is of some theoretical interest as it raises the question whether sponge cells are totipotent (i.e., capable of changing to any other type of cell), or whether there are cell lines derived from the early blastomere stages.

Many sponges harbour symbionts in the mesohyl which may contribute to the host sponge's nutrition, either by releasing nutrients or by being phagocytosed by the amoeboid archaeocytes. In some members of the order Verongida, symbiotic blue-green algae (Cyanophyceae) and bacteria may be present in vast numbers. Symbiotic bacteria are also found in many Dictyoceratid sponges. The boring sponge, *Cliona*, harbours symbiotic zoochlorellae (Bergquist, 1978).

12.4 GROWTH AND REPRODUCTION OF SPONGES

Neither flagellated chambers nor oscula arise by budding, therefore there is nothing in the structure of calcareous sponges and demosponges that can be equated to the zooid of compound ascidians or the modules of bryozoans and hydroids. The gradual increase in complexity of a sponge is the outcome of normal growth processes and the rearrangement of water channels to meet changes in hydrodynamic pressures resulting from increased size. Flagellated chambers can have no other status than trophic organs or units serving the sponge as a whole. Moreover, the number and arrangement of oscula is entirely dependent upon hydrodynamic requirements. The sponge that develops from a single parenchymella larva must be considered as an individual organism, not a colony. In the words of Bergquist (1978), "Individuality in sponges is recognizable only at the level of entire specimens invested by a continuous pinacoderm."

Asexual reproduction has already been mentioned in connection with *Tethya corticata*. It allows a sponge, once established in a favourable place, to exploit the surrounding available surface. Another method of asexual reproduction is by gemmule formation. Gemmules are best known from freshwater sponges, but some marine species also produce them. They consist of a spongin capsule, strengthened with spicules, which is packed with archaeocytes containing food reserves (glycogen, vitelline platelets and lipids).

When favourable conditions occur, the first archaeocytes to emerge from the capsule transform into pinacocytes, collencytes and sclerocytes, which produce an outer covering, the mesohyl, and spicules, respectively. At first there are no choanocytes, but these eventually appear and form the first flagellated chamber. Since the gemmule originally had only one type of cell, the archaeocyte, it follows that every functional cell type in the sponge must be derived from it (Fig. 12.9).

Dispersal in the majority of sponges is by a motile embryo (coeloblastula, amphiblastula or parenchymella) which has been produced by sexual reproduction. This allows dispersal and the potential to colonize a wider area

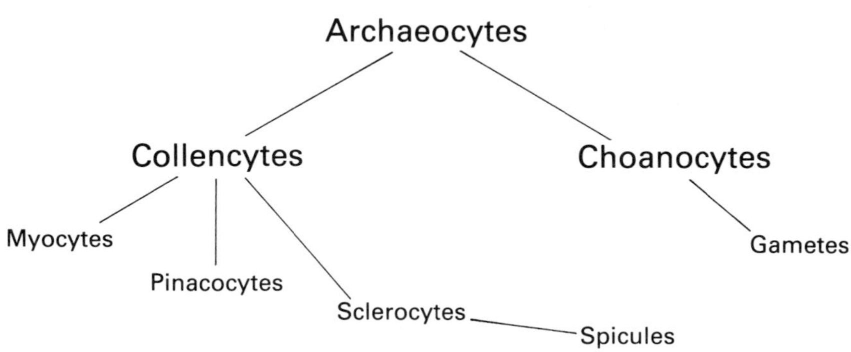

Fig. 12.9 Derivation of functional cell types from archaeocytes (see text).

than could be achieved by asexual means. Although these dispersive embryos are simple in structure and lack special sensory organs, a combination of photonegative and geonegative behaviour tends to bring them to settlement sites where the probability of successful adult life is relatively high. For species living on or near the shore, where there are few suitable settlement sites, this behaviour is of prime importance.

The details of sexual reproduction are poorly known for most sponges, but Duboscq and Tuzet (1937) revealed a number of unique features for the calcareous sponge *Scypha ciliata* (= *Sycon raphanus*).

Both sperm and ova are derived ultimately from choanocytes. In sponges producing sperm, choanocytes withdraw their flagellum and migrate into the mesohyl, where they undergo cell division to produce a spermatic cyst in which spermatogenesis takes place. When sperm is released it passes out through the osculum with the exhalent current and is dispersed in the sea. Ova are derived from choanocytes that take up a position immediately under the choanocyte layer and extend long protoplasmic processes to the choanocytes above (Fig. 12.10(a)). They accumulate nutrients and grow to a large size.

Duboscq and Tuzet described a unique method of fertilization in this sponge. Spermatozoa entering via the water current were picked up by the choanocytes and retained in a vacuole. The choanocytes then withdrew their flagellum and migrated through the mesohyl to the ova, with which they made protoplasmic contact. The spermatozoon (and sometimes the whole carrying cell) entered the cytoplasm of the ovum (Fig. 12.10(b)). Maturation division took place before male and female pronuclei fused to form the zygotic nucleus. This method of sperm transport is without parallel in the animal kingdom and calls for some comment on its significance and possible evolution.

There are clear advantages in retaining a fertilized egg within the body of the parent: it can be protected, surrounded by a favourable environment and nourished during development, but a prerequisite for incubation is internal fertilization. Spermatozoa are only equipped for locomotion in an aqueous medium. The ovum lies in the mesohyl and to reach it a spermatozoon would have to traverse either the pinacocytes or the choanocyte layer. This migration could be achieved only if it was able to transform into an amoeboid form with pseudopodia. However, there is a far more serious problem to be negotiated. Choanocytes are highly specialized for, and extremely efficient at, removing food particles from the water passing through the sponge. Spermatozoa fall within the size range of such particles, and there can be little doubt that they would be taken up into the cytoplasm of the choanocytes by phagocytosis before they could find a way through to the mesohyl. Attempts to penetrate the pinacocytes are unlikely to meet with more success in view of

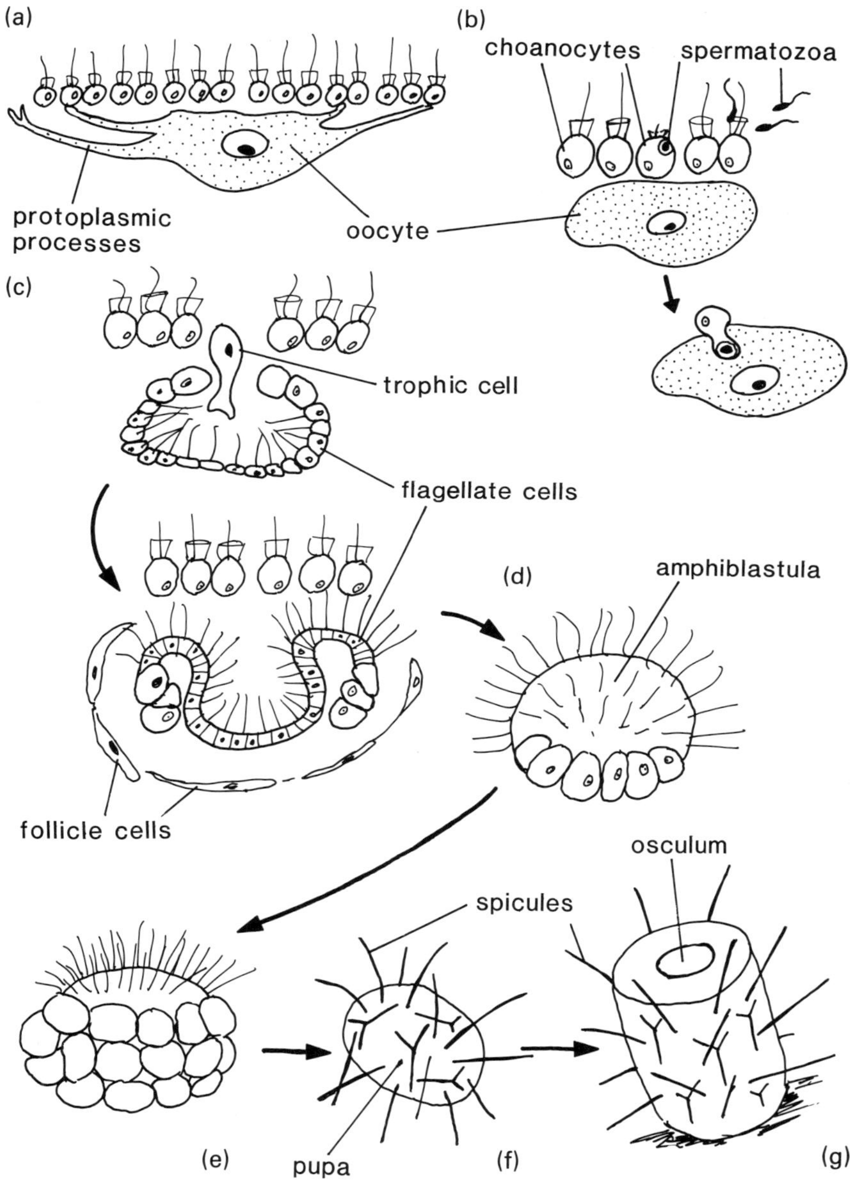

Fig. 12.10 Development of calcareous sponge. (a) Oocyte with protoplasmic extensions to choanocytes. (b) Fertilization. (c) Trophic phase of blastula (following cleavage). (d) Formation of amphiblastula. (e) Amphiblastula. (f) Pupa. (g) Young sponge. ((b) to (g) based on DuBosq and Tuzet, 1937.)

their known phagocytotic ability. One would have to postulate that the choanocytes can discriminate between spermatozoa and food particles and not ingest them, but in view of the known inability of choanocytes to feed selectively, this seems unlikely.

It has already been pointed out that choanocytes pass food vacuoles to archaeocytes and ova, and that they can withdraw their flagellum and become amoeboid wanderers in the mesohyl. Thus a parsimonious hypothesis requiring no special conditions is that spermatozoa are drawn into the flagellated chambers, where they are picked up by the choanocytes, which in turn migrate towards the ova as they would do with food and pass the spermatozoa to the ova.

Finally it might be pointed out that this process greatly enhances the probability of spermatozoa coming into contact with an ovum. Due to the very large volume of water passing through the sponge and the high efficiency in capturing particles, the sponge is in effect concentrating the spermatozoa.

Duboscq and Tuzet (1937) also described a number of remarkable features in development of the larva, which takes place in the parent sponge. Cleavage gives rise to a hollow blastula composed of micromeres with their flagella directed into the blastocoel, and larger macromeres adjacent to the choanocyte layer of the parent sponge. The sponge continues to nourish the late blastula by means of trophic cells containing nutrients which pass into the blastula through an opening between the macromeres (Fig. 12.10(e)). Follicle cells from the mesohyl form a capsule round the blastula, and a transformation takes place in which the blastula turns inside out so that the flagella now project outwards. The macromeres regroup ventrally and the embryo assumes the form of an amphiblastula (Fig. 12.10(d)). At this stage the larva is set free to swim in the sea. Before it settles, the macromeres mutliply and grow over the flagellate micromeres, which now line the shallow cavity (Fig. 12.10(e)). After settling on the substratum, the larva undergoes another unusual transformation or metamorphosis. The flagella are withdrawn and the cells form a central mass obliterating the original cavity. Some of these cells degenerate whilst others have the appearance of archaeocytes. The external cells flatten and secrete a number of spicules. This stage has been called a 'pupa' (Fig. 12.10(f)). Eventually a cavity forms in the central mass of cells which is destined to be the first flagellated chamber and later the spongocoel. Some of the outer cells, which can now be called pinacocytes, develop into porocytes and establish a connection with the central cavity, but flagellate cells have not yet appeared. The juvenile sponge now takes on a cylindrical appearance. At one end it is attached to the substratum, and at the opposite end a central aperture forms the vent. Many spicules project from the surface, and flagellate collar cells make their appearance lining the central cavity. It is now a little sponge (Fig. 12.10(g)).

This sequence of events is unusual for several reasons. The development of flagella in the blastocoel is not known in the Metazoa and may represent a trophic specialization that has evolved in *Scypha*, The turning inside out of the blastula is only known elsewhere in the Volvocales. The process results in flagella coming to lie on the outer surface of the embryo, where they can play a locomotory role later in larval life. The larger blastomeres (macromeres) give rise to the pinacocytes or external covering of the juvenile sponge, as well as the spicules and other cells of the mesohyl. The choanocytes that line the internal cavity or spongocoel are derived from the smaller flagellate cells (micromeres). This is the direct opposite to what happens in the Metazoa, where micromeres give rise to the ectoderm and the macromeres give rise to the gut. Bergquist (1978) points out that "this type of derivation of adult regions cannot be reconciled with conventional conceptions of germ layers and their derivatives in other Metazoa."

In ceractinomorph sponges each ovum is surrounded by follicle cells. Hundreds of amoeboid 'nurse' cells migrate towards the ova and are incorporated into its cytoplasm, resulting in a tenfold increase in diameter. The egg now appears multinucleate. Fertilization takes place within the parent sponge and the remains of the nurse cells become distributed among the blastomeres. Embryos remain in the sponge until they have developed into the dispersive parenchymella (Bergquist, 1978).

SUMMARY

Many sponges live on the lower shore in places where they will experience regular emersion during spring tides. The large volume of water retained in their body on emersion, and their habit of colonizing shaded places under rocks and seaweed, prevents water loss or solar radiations affecting them. The tough elastic texture of keratose sponges and the presence of spicules prevent damage by wave action and possibly predators, however, many sponges produce organic compounds (allelomones) as defence against predators, attack by micro-organisms and competition for space by other species.

Their mode of feeding is unique, but can only take place whilst they are submerged. Very small particles of detritus and bacteria are captured and ingested by the choanocytes lining the chambers. Mucus is not involved.

Most sponges on the lower shore can regenerate or regrow when parts are damaged by rock movements, eaten or killed by abnormal exposure.

13

Colonization of the sea shore

In previous chapters, adaptations have been described which enable particular animals to endure regular submersion and emersion, but they tell nothing of the way in which a species is established on the shore or the relationships it has with other species living in the same community. These aspects will now be considered.

Because adult sessile animals are fixed and unable to select a place to live, they are entirely dependent upon a free-living dispersive phase in their life history to colonize new areas or maintain an existing population. For littoral species, suitable substrata will only occur in a very narrow band along the shore, and it is imperative that larvae settle in this zone. Adults are forced to live wherever a larva finally attaches, therefore successful colonization is contingent upon the larva settling in a place suited to the needs of adult life. In nearly all cases the motile phase is a sexually produced larva or embryo, although asexually produced hydroid medusae and planuloids are a notable exception. This suggests that genetic variability, introduced through gene recombination, offers some selective advantage to colonization where the habitat itself is heterogeneous (patchy).

13.1 REPRODUCTIVE STRATEGIES

It might be thought that shore-living animals, all subjected to the same environmental fluctuations, would adopt similar reproductive strategies, but this is not the case. A wide spectrum of reproductive tactics is found with a tendency towards two extremes: those species that produce vast numbers of offspring which feed in the plankton for a period (planktotrophic larvae), and those that produce relatively few offspring which depend mainly on the yolk reserves of the egg for growth (lecithotrophic larvae). The reproductive trends of animals dealt with in previous chapters is set out in Table 13.1.

Table 13.1 Egg production and larval types of some sessile littoral animals

Type of larval phase	*Magnitude of egg production per reproductive episode* $>10^6$	$<10^6$	$<10^4$	$<10^2$	<10
Solitary species and sponges					
Planktonic (with planktotrophic larva)					
Eggs shed	*Crassostrea* *Mytilus*				
Eggs retained		*Ostrea*	Barnacles *Semibalanus*		
Short planktonic phase with lecithotrophic larva					
Eggs shed	*Pyura*		*Galeolaria* *Tethya*		
Eggs incubated		Ceractinomorph		*Spirorbis*	*Janua*
No planktonic phase		sponges		*Actinia*	
*Modular and colonial animals**					
Planktonic larva†					
Eggs shed			MO‡ ←	[*Obelia*]‡ (medusa)	
			MO ←		[*Membranipora*]
			MO ←	[*Silicularia*] (eumedusa)	
Lecithotrophic eggs incubated			MO ←		[*Beania*]
			Clone ←		[*Eleutheria*] (medusa)
			Colony ←		[*Botryllus*]
			Colony ←		[*Aplidium*]
Parental nourishment of embryo			MO ←		[*Laomedea*] (gonophore)
			MO ←		[*Bugula*]

* Only blastozooids produce eggs which develop into oozoids. Asexual reproduction of oozoid (and blastozooids) replicates genome and increases number of eggs produced per oozoid.
† Entries in this section of the table show a progressive suppression of the planktonic phase (reading downwards, i.e. from *Obelia* to *Bugula*).
‡ MO, modular organism; [ ] denotes sexually reproducing blastozooid.

There are six parameters regulating the reproductive potential of sexually reproducing animals and each may be subject to variation by environmental influences.

First, the number of eggs produced at each reproductive episode may be influenced by availability of food or temperature, but tends to be characteristic of each species. Older animals produce more eggs; for example, Moore (1935) found that the mean output of nauplii from *Semibalanus balanoides* increased annually from less than 500 in the second year to between 3000 and 4000 in the sixth and seventh year. Some old specimens produced up to 13 000 nauplii in a year. Consequently older animals will tend to contribute a higher proportion of their genes to the population.

Second, the number of reproductive episodes per breeding season varies widely among species. Some species have a single spawning period (they are termed semelparous), while others spawn several times during the breeding season (iteroparous). In an unpredictable environment or where attachment depends upon the chance appearance of newly exposed rock, iteroparous species are at an advantage, as they can 'sample' a series of conditions at different times of the year.

A third parameter is reproductive span, the number of reproductive episodes from the age of first reproduction to death of the parent organism. Long-lived species will contribute to population recruitment over many years and so average out year-to-year fluctuations in colonization success. This offers a significant advantage to species where recruitment is very low because of scarcity of attachment sites or intense interspecific competition. It also overcomes the problem of recruitment if breeding fails in any particular year.

Fourth, in many sessile species asexual multiplication of reproductive units (gonophores or zooids) increases the number of gametes produced from a specific genome. Here the tendency is towards the production of one or a very few lecithotrophic eggs per reproductive unit, and the organism's fecundity is this number multiplied by the number of gonophores in a modular organism or zooids in a colony.

Fifth, the amount of yolk provided for each egg determines how far development can proceed before the larva has to obtain its own food. Eggs produced in vast numbers are normally small and represent a low investment per egg by the parent animal. They develop into small planktotrophic larvae which, by means of their own feeding and digestive system, feed upon phytoplankton for further nourishment and growth. When phytoplankton is abundant, as during the spring 'bloom', growth will be rapid and large numbers of potential recruits are produced, but under conditions of food shortage there could be a complete failure of planktotrophic larvae. The production of large numbers of larvae is advantageous where attachment sites are sparsely distributed or where settlement is difficult owing to strong wave action. Planktotrophic larvae may spend weeks or months in the plankton and be distributed widely, but mortality is likely to be high and proportional to the length of time spent there.

Other species produce relatively few eggs, which are well provided with yolk. This allows development to advanced planktotrophic larvae or, if there is an abundance of yolk, will carry them through a non-feeding larval stage to metamorphosis and juvenile development before it is necessary for them to collect their own food. This represents a higher investment per egg, lower mortality and relatively higher chances of recruitment to the adult population. Suppression of the planktotrophic phase eliminates dependence on seasonal changes in the abundance of phytoplankton.

Production of a few yolk-laden eggs has led to the evolution of incubation or retention of embryos during their development to larvae. This eliminates most of the pre-larval mortality suffered by eggs shed directly into the sea. A radically different approach has evolved in *Bugula*, where nourishment is provided after fertilization by the parent zooid through a placenta.

Finally, the time of year at which reproduction takes place (or larvae are released) may be critical, particularly for planktotrophic larvae, and may be initiated by environmental changes such as seasonal ambient temperature or photoperiod.

The wide range of reproductive patterns outlined in previous chapters can be looked upon as the outcome of selective pressures exerted by the environment. These fall into two categories: density-dependent factors, which are affected by the degree of crowding and hence the level of competition for vital resources (food, sunlight, shade, attachment area, feeding space), and density-independent factors, which are not influenced in any way by the number of animals present, but are the result of an unstable environment in which there occur wide fluctuations in predation, accidental destruction, lethal fluctuations in temperature, salinity, radiation etc., and catastrophic events that cause mass mortality.

When competition for a vital resource adversely affects reproductive potential or increases mortality, organisms are said to be subjected to K-selection. This selective pressure favours individuals with enhanced competitive ability. It will occur when the size of the population exceeds the capacity of the environment to suppy its need for food or space, and is therefore density dependent. On the other hand, when reproductive potential is reduced or mortality is increased by instability of the environment (density-independent factors), the organism is said to be subjected to r-selection. Here natural selection favours accelerated growth rates and high reproductive output.

Density-dependent and density-independent factors rarely operate in isolation, and most natural populations will be subjected to varying degrees of both. The concept is a comparative one. An organism is said to be more or less K- or r-selected than some other organism (Gadgil and Solbrig, 1972; MacArthur, 1972).

Theory predicts that K-selected animals will tend to delay the time of first reproduction, produce relatively few eggs, have several reproductive episodes during the breeding season, and show a lower reproductive effort (produce fewer lecithotrophic larvae and possibly retain eggs during development). Such animals would tend to be long-lived and to form stable communities. On the other hand, it is predicted that the life span of adult r-selected animals will be short (less than 1 y) and that they will start breeding at an early age, producing numerous eggs in one breeding episode. Reproductive effort is large, with the production of vast numbers of planktotrophic larvae. It is predicted that r-selected animals should be good colonizing species, but would not produce stable communities.

Evidence to support these theories has been sought in birds and lizards, where the number of offspring, survivorship, resource allocation and degree of competition can be quantified, or in dandelion plants, rotifers, cladocerans and copepods, where self-fertilization or parthenogenesis eliminate the complications of genetic variability (Cody, 1966; Gadgil and Solbrig, 1972; Allan, 1976), but little attention has been paid to marine littoral species, and even less to sessile animals.

Many sessile animals of the marine littoral zone appear to contradict these theoretical predictions. Interpretation is made difficult by the lack of data on growth rates, fecundity, resource allocation etc., and by uncertainty about which selective forces are acting on a particular species. Barnacles living on the upper shore, for example, are subjected to severe density-independent pressures, but tidal movements will limit the time available for feeding, and this acts in the same way as competition for food, which is a density-dependent factor. Moreover, barnacles produce large numbers of planktonic larvae with wide dispersal and colonizing potential – all characteristics of r-selection, but at times of heavy spat fall, competition for space will be intense, and this would exert a K-type selective pressure. Colonial ascidians like *Aplidium* show the opposite extreme; they produce one or two larvae with massive yolk reserves and poor dispersal characteristics. Here K-selection is indicated, but it is far from clear what vital resource is in short supply as a result of competition. It is unlikely to be food. Possibly space for development of a new colony is the limiting factor.

A critical review of the theoretical aspects of r- and K-selection and an alternative theory, 'bet-hedging', has been given by Stearns (1976). The theoretical predictions for bet-hedging are the reverse of r- and K-selection, and appear to fit observation on sessile littoral animals better.

Bet-hedging predicts that, where fluctuations of the environment have their main impact upon juvenile mortality, selection will favour reduced reproductive effort, smaller numbers of well-endowed eggs, and longevity of adults. On the other hand, where adult mortality is affected more by fluctuations in the environment than juvenile mortality, shorter life span, increased reproductive effort and large numbers of eggs appear to be favoured.

13.2 LARVAL MORTALITY, DISPERSAL AND SETTLEMENT

Many factors affect larval mortality. Fertilization of an ovum may fail through lack of synchronization between production of eggs and sperm, extreme dilution of sperm, short period of sperm activity, entry of a spermatozoon from a different species or incompatibility between sperm and ovum derived from the same parent (self-fertilization). All of these events will reduce the number of eggs that develop. Most would be overcome by internal fertilization, but this is a rare phenomenon among sessile animals other than barnacles.

Even when fertilization has been successful, abnormalities may occur in the early stages of development and reduce the number of viable larvae. Planktotrophic larvae may not find sufficient food to complete growth, although the absorption of dissolved organic matter could be a valuable supplement (Manahan and Crisp, 1982). The longer that larvae remain in the plankton, the greater is the risk of being consumed by planktonic predators such as fish and decapod crustacean larvae, predatory copepods like *Corycaeus*, siphonophores and species of *Sagitta*. They may be carried by currents away from the littoral zone and out to sea, where future development and adult life is impossible.

Even greater dangers confront larvae at the time of settlement, for they enter a region where filter-feeding barnacles and passive raptorial cnidarians efficiently remove planktonic larvae. Larvae attempting to settle may be captured by predators such as amphipods or terebellid polychaetes before they have time to attach. Newly metamorphosed barnacle spat may be destroyed by the abrasive action of waves, desiccation during emersion, or crushing by browsing limpets before they can secrete a protective shell. Survival after metamorphosis may be reduced for other reasons. Some larvae settle in places where adult life is impossible. Others select a suitable site, but are confronted with competition for space by other organisms or are killed by predators.

The two extremes of reproduction previously discussed overcome this mortality in different ways. The probability of survival to reproductive age could be raised by increasing the number of offspring produced, provided that density-dependent mortality did not increase. The release of vast numbers of eggs or larvae simultaneously might overwhelm predators and allow more offspring to survive. At the other extreme, incubation of a few lecithotrophic eggs and shortening the free-swimming dispersal phase will reduce mortality prior to settlement, and give those larvae that do find a space to settle a better chance of establishing themselves against competitors.

In still water, free-swimming larvae would not move very far from the point of release using their own locomotory organs, but near the shore, tidal movements and currents carry them further afield, depending on the duration of the planktonic phase and the velocity of the current.

Very little is known about the distances or directions larvae are transported. Within the littoral zone, tidal water movements do little more than move larvae up and down the shore. Further out, there may be rip currents parallel with the shore, or deeper offshore currents at right angles to the coastline; both will carry larvae far from their place of origin, and usually to regions where adult life is impossible. This is clearly demonstrated when man-made structures, such as buoys and oil rigs some distance from the shore, are colonized by littoral species. Wilson (1982) found *Semibalanus balanoides* larvae in samples taken from the deep offshore current 4.8 km from the Northumberland coastline. This barnacle does not occur on the North Sea gas platforms (West Sole) 75 km from the coast, indicating that the larvae cannot survive transportation over such long distances, but *Mytilus edulis* does colonize the platforms (Lane *et al.*, 1982). The planktonic plantigrade stage of this mussel secretes a long, fine thread which enables it to drift in the water column for periods up to several weeks. The thread differs from the byssus attachment threads in not having an attachment plaque and appears to be an adaptation to planktonic life (Lane *et al.*, 1985). Plantigrades may settle temporarily and disperse again on many occasions before they make a final attachment.

Along-shore currents would carry larvae over far wider areas than is possible with their own locomotory powers, and this would greatly increase their chances of finding a favourable settlement site. There is the possibility that larvae may encounter new areas to colonize and so avoid competition with the parental stock for food and space. Strathmann (1974) suggested that wide dispersal is an advantage to sibling larvae as it averages-out variations in success of colonization under variable environmental conditions. Far more important, however, it the increased gene flow between populations that resuls from wide dispersal.

Wide spacing of individuals could be a disadvantage to sessile species if it caused failure of fertilization. Close settlement is essential to barnacles because of their method of sperm exchange, but it will be beneficial to all animals that shed their sperm into the sea.

Before settlement can take place, larvae must migrate back from coastal waters to the shore, but little is known about how this is achieved. It seems unlikely that the larvae can do this by their own locomotory powers, and onshore winds must be ruled out as cyprids may settle heavily during periods of offshore winds. Shanks (1986) found that cyprids of *Chthamalus* showed a peak of settlement during rising tides two or three days before maximum spring tide, and suggested that they are transported passively by onshore currents and surface slicks.

The question now arises, is the place of larval attachment arrived at by chance or choice? By analogy with wind-dispersed seeds, larvae of sessile animals might be distributed at random over the shore. Some larvae would settle in unfavourable places and die, while others would, by chance, settle in areas favourable for growth. According to this hypothesis, adult distribution may have no relation to the pattern of settlement; it merely reflects a pattern of survival. Zonation of species would then be attributed to environmental factors.

There is some evidence to support this view. Connell (1961) found that cyprids of *Semibalanus balanoides* settled as much as 1.5 m above adults during spring tides, but died during the long period of exposure during the following neap tides. The upper limit of this barnacle is about mean high water neap tides. Recently-settled *Chthamalus stellatus*, however, only suffered a 26–40% mortality after 9 d exposure, and this enabled them to colonize the rocks as high as mean high water spring tides.

The overwhelming majority of investigations support the view that attachment is a response to a specific set of environmental conditions encountered by the larva during an exploratory phase. The place of attachment appears to be 'chosen', but this may not require any decision-making on the part of the larva.

The free-swimming larval stage of sponges, cnidarians and tunicates rarely lasts more than a few hours and may occupy only a matter of minutes before settlement takes place, consequently dispersal is seldom far from the point of release. The parenchymella of sponges usually crawls about over the substratum before attachment, but ascidian tadpoles do not appear to explore the substratum; they attach immediately they contact a solid surface.

Gregarious settlement is not common among sponges or ascidians (Bergquist, 1978; Millar, 1971), however, the larvae of both groups usually show a response to light and gravity. For the sponge parenchymella this is remarkable, as it does not possess complex light or gravity receptors like those of ascidian tadpoles. In most cases larvae are positively phototactic when first released, and become negatively phototactic or indifferent to light at a later stage in larval development. A combination of simple negative phototactic and negative geotactic behaviour leads them to settle on the underside of rocks or on the roof of caverns where they are shielded from direct sunlight. The tadpoles of *Botryllus schlosseri* metamorphose almost as soon as they are released and do not disperse very far. Grosberg and Quinn (1986) found that they settled significantly closer to their parental colony (with which they share one of the histocompatibility genes) than to other colonies. A few species, like *Pyura stolonifera*, are indifferent to light and settle on the upper surface of rocks exposed to the sun.

Less is known about the settlement of cnidarians. The planula larva is of simple structure without any special sensory organs, and so complex exploratory behaviour would not be expected. Despite this, many hydroids show a specific association with particular seaweeds, which suggests that they are reacting to some specific stimulus, probably a chemical, that induces attachment. For example, *Silicularia bilabiata* is commonly found on *Phyllospora comosa*, *Clava multicornis* on *Ascophyllum nodosum*, and *Obelia geniculata* on *Laminaria digitata* round the British Isles or *Macrocystis pyrifera* in New Zealand (Ralph, 1956). These associations are not exclusive and occasionally these hydroids are found on other seaweeds. A more specialized association is formed between *Hydractinia*, a hydroid, and hermit crabs. The hydroid only lives on shells occupied by a hermit crab and not on the shell of live gastropods of the same species.

Settlement behaviour of barnacle, oyster, tube worm and bryozoan larvae has been the subject of considerable experimental analysis, and in all species investigated it has been found that they pass through a searching or exploratory phase prior to settlement, giving the impression that they are actively seeking a particular combination of environmental conditions before attachment (reviews: Crisp, 1974, 1976).

Although larvae give the impression that they choose a site for attachment, it is more likely that their simple sensory and effector systems are programmed to respond to certain specific environmental stimuli with a stereotyped activity. Several diverse stimuli (the stimulus situation) may be necessary to evoke the full sequence of settlement and fixation. Initially the stimulus threshold is high, and fixation only takes place when all aspects of the stimulus situation are present, but many larvae are able to delay settlement for several days. As time passes, the stimulus threshold diminishes and attachment can be evoked by only part of the stimulus situation. This frequently leads to attachment in unsuitable places.

A pattern of behaviour common to most species is that the larva alights on a surface and then moves rapidly over it in a more or less straight line (the wide searching phase of Crisp, 1976). If it does not encounter favourable stimuli (or receives unfavourable stimuli) it will swim away and explore some other surface, but if it encounters appropriate stimuli it passes into a new phase of slower, more intense searching of a particular area by traversing back and forth over it (close searching). The impression is given that the larva is in some way testing its suitability as an attachment site. This may be followed by rotation or motion to and fro within its own length (inspection). Provided that the larva does not receive any adverse stimuli (like proximity to another animal), it will fix itself at this spot by a special secretion and then proceed to metamorphosis.

Several cues may lead to attachment. Most larvae settle more readily on surfaces that have a bacterial film. Ryland (1959) found that *Alcyonidium hirsutum*, *A. gelatinosum* and *Flustrellidra hispida* all settled more heavily on *Fucus serratus* than on 16 other species of seaweed when offered the choice. Presumably they respond to some qualitative difference between the weeds, such as the nature of chemical substances diffusing from the surface.

With barnacles, oysters and *Spirorbis*, a far greater inducement to settling is the presence of adults of the same species or recently metamorphosed larvae. This is not surprising, for the best indication that a particular site is a suitable place to live is the presence of adults. Cypris larvae of *Semibalanus balanoides* settle readily near adults or young barnacles of their own species, but not on them. They can delay settlement up to 2 weeks if they do not encounter favourable stimuli for fixation, thus increasing their chances of finding a suitable place to metamorphose.

Cyprids can detect and respond to the protein complex called arthropodin in the cuticle of adult barnacles, particularly the base left behind on the rock when the shell had been removed (Knight-Jones, 1953), but they can only do so if their antennules came into contact with it whilst exploring the surface. Proteins in solution were ineffective in inducing attachment. Larkman and Gabbott (1975) have shown that a number of acidic proteins and protein–carbohydrate complexes extracted from the shell of barnacles, mussels and oysters would induce attachment of *Semibalanus balanoides*, provided they were absorbed onto a solid surface. However, Wethey (1984a) found that cyprids of *Semibalanus balanoides* settled gregariously in response to short-lived chemical cues from other cyprids, and to a lesser extent from metamorphosed spat. However, there was no evidence for preferential settlement on the cuticular bases of adults which had been killed and dislodged by storms 2 months previously.

Cyprids walk over the surface on their antennules during the exploratory phase. The antennules have a terminal disc with numerous sensory structures and gland cells which secrete a proteinaceous substance for temporary adhesion at each step (Nott and Foster, 1969; Walker and Yule, 1984). Cyprids of *Semibalanus balanoides* can adhere to roughened surfaces treated with barnacle cuticle extract with a force of 2×10^5 N m^{-2}, but they can easily detach themselves. Yule and Walker (1984b) have shown that colour of the substratum influences adhesion (cyprids hold on to dark surfaces more tightly than light ones), thus temporary adhesion is at least in part under control of the cyprid, and is not due entirely to cohesive molecular forces in the protein secretion.

Arthropodin is virtually insoluble and so detection by chemo-receptors is unlikely. Crisp (1974) suggested a mechanism analogous to antibody–antigen recognition. Parts of the antennular disc of the cyprid may possess properties that cause adhesion to specific parts of the arthropodin molecule, and sensory proprio-receptors may be able to detect this force of adhesion. The antennules are well supplied with sensory receptors of several types.

The effectiveness of barnacle cuticle to induce settlement reflects systematic position. Knight-Jones (1955) showed convincingly that while some members of the Balanidae (for example *Balanus crenatus*) could induce settlement of *Semibalanus balanoides*, they did not induce settlement of *Chthamalus stellatus*, which belongs to a different family. Cyprids of *Elminius modestus* responded as readily to the presence of *S. balanoides* as to their own species (Larkman and Gabbott, 1975). Because of the strong attractiveness of conspecifics to cyprids, they appear to be gregarious and congregate in areas where adults occur; however, Wethey (1984a) found that the old bases of detached barnacles lost their attractiveness to settling cyprids, and settlement did not necessarily occur where adults had previously been attached. Nevertheless, cyprids tend to aggregate. Yule and Walker (1985) suggested that the 'footprints' of the temporary protein adhesive are attractive to other cyprids and may induce gregarious settlement.

Pediveliger larvae of oysters carry out exploratory behaviour similar to that of the cyprid, but locomotion is by the ciliated foot. Larvae of *Crassostrea virginica* settle more readily on oyster shells with the periostracum intact; when it is removed, the shell becomes less attractive. Both oyster and tube worm larvae produce a mucous thread whilst exploring the surface, and it is possible that these threads may induce other larvae to settle in the same area when they encounter them.

A number of other factors, such as light intensity, current velocity, topography and texture of the surface, may influence attachment. Many species of bryozoan tend to settle on smooth surfaces (Ryland, 1976), but *Semibalanus balanoides* shows a marked preference for settlement in grooves, pits, depressions and crevices in the rock, and avoids smooth surfaces. Many cyprids settle inside the empty shells of dead barnacles (Crisp and Barnes, 1954). Young barnacles are most vulnerable during the period of metamorphosis and the first few days of growth when the protective shell is being formed, thus it is likely that the slight shelter the pits and crevices offer against wave action is important at this time. The larvae of *Spirorbis spirorbis* show a tendency to settle in the slight concavity on either side of the midrib of *Fucus serratus*, but what advantage this may give to the adult is obscure.

Larval settlement is subject to two conflicting tendencies, a centripetal tendency (gregariousness) drawing individuals close together, and a centrifugal tendency (territoriality) which forces them apart. The density or spacing out achieved will be determined by the relative strength of the two tendencies. The advantages of living close to conspecifics have already been noted, but too dense a settlement may be detrimental and reduce the percentage of larvae that survive to adulthood. Animals that are closely packed together will prevent each other from growing normally, weaken their attachment to the substratum and possibly interfere with feeding. In dense settlement of *Semibalanus balanoides*, young barnacles cannot grow laterally to form the typical conical shell; they are forced to grow upwards as a tightly packed bunch of tubular shells that form mounds or hummocks (Fig. 5.1(f)). Each individual has a small base, and lateral pressure from neighbours weakens attachment, so the mass is easily broken up and detached by heavy seas (Barnes and Powell, 1950; Gutmann, 1960). In consequence the probability of any individual surviving to reproductive status is greatly reduced. Although *S. balanoides* rarely settles on the shell of living adults of its own species, it readily settles on other species of barnacle and mussels, thus in years of very heavy spat fall, the other species may be smothered and their feeding obstructed with fatal consequences.

Larval movements just prior to attachment suggest that they are directed at spacing out the larvae, but have little to do with adult requirements for space. They are very similar in both cyprids of barnacles and larvae of *Spirorbis*. The larvae make frequent changes in direction so that they move back and forth over a relatively small area. If they encounter another individual that has already settled, they will move away and closely search another area. Cyprids carry out close searching by walking on their antennules. At each step they stand on one antennule and twist the body round in an arc with the caudal setae extended. It has been suggested that this would give information to the cyprid about surrounding objects and conspecifics, thus enabling it to select depressions and avoid settling too closely (Crisp and Barnes, 1954; Barnes, 1955; Crisp, 1974, 1976).

The average distance between settled larvae of *Spirorbis spirorbis* is 1–2 mm, which would allow for growth to adult size (Wisley, 1960). A similar average distance was found for *Semibalanus balanoides* (Knight-Jones and Moyse, 1961), but this is far too close to allow growth to adult size (20–50 cyprids could settle at this density on an area covered by the base of a single adult *c.* 1 cm in diameter). Wethey (1984a) considered that the very close settlement of *S. balanoides* cyprids indicated the absence of any form of territorial behaviour and spacing out to optimal distances. Such close settlement would appear to be a disadvantage. However, it might prevent other species from settling in the same area where they might smother them, interfere with feeding, or reduce chances of sperm exchange. It is clear that the majority of cyprids that attach and metamorphose are later displaced. The first cyprids to settle are likely to find the most favourable sites for growth and will have a size advantage over later arrivals.

SUMMARY

Sessile animals depend upon a sexually-produced motile phase for dispersal and recolonization. It is imperative that larvae settle in situations and at levels on the shore that will be optimal for adult life, for the adult is compelled to live wherever the larva settles.

Many reproductive strategies are found among sessile animals. Continuous or repeated reproduction allows offspring to be produced under different conditions at different times of the year. Some seasons may be more favourable to survival than others. Long adult life, spanning many years, enables an animal to survive years of complete reproductive failure and exploit good years. Asexual multiplication increases the number of gametes that can be produced from one genome; this is particularly advantageous in production of large yolky (lecithotrophic) eggs by small animals.

Lecithotrophic eggs allow a short larval period and settlement near the adult population. Eggs with little yolk develop into planktonic larvae which feed on phytoplankton. Such larvae are at great risk from predation and may be dispersed over a very wide area; some will be transported away from suitable substrata for settlement. Planktotrophic larvae are dependent upon favourable plankton growth for survival.

The larvae of some species settle in favourable places by chance, but in the majority of cases studied, larvae pass through an exploratory phase in which they search for a particular set or pattern of environmental features before settling. Larval sense organs and patterns of behaviour are tuned to achieving settlement in certain places. Texture and topography of the substratum, water flow, light, gravitational orientation, presence of a bacterial surface film and certain specific chemical stimuli have all been identified as influencing settlement in one species or another. However, the presence of adults or newly settled larvae of the same species is a powerful stimulus and leads to gregarious settlement. Overcrowding occurs when settlement is heavy and few animals will survive to adulthood.

14

Community structure and interactions

Space for attachment on the narrow intertidal zone is finite and so competition must occur if plants and sessile animals increase in number or grow in size. Milne (1961) critically reviewed the diverse meanings of the term competition and the following definition, based on Milne's, has been adopted in the following discussion: competition is the endeavour of organisms to gain sufficient of a common resource when the supply of that resource is insufficient for all. For sessile animals on rocky shores, space for attachment and growth is a resource as vital as food.

Competition may take the form of exploitation, in which the resource is more or less equally divided among all competitors, and interference, where some individuals gain what they require while others are denied or excluded, as in territorial competition. Exploitation tends to be more severe between members of the same species because all share the same requirements. With interference it is unlikely that any two animals will be equally matched, and so one dominates the other. This may result in one species displacing the other and taking over all the available resource, or one of the competitors may be reduced in numbers or capacity to breed, but may still coexist with the dominant species. Among sessile littoral animals, competition is not an obvious phenomenon that can be witnessed by casual observation; it takes place slowly and surreptitiously, depending on growth and not on overt activity. In the natural habitat, a dynamic coexistence is more common than exclusion because several species may interact simultaneously, preventing any one species from displacing the others.

The middle and upper shore is dominated by solitary animals (such as barnacles, mussels and tube worms), each of which tends to monopolize a particular zone. On the lower shore and sublittoral zone, colonial animals, modular organisms and sponges become the dominant life forms. Solitary animals are excluded or overgrown unless they grow large or protect themselves with spines (e.g., *Pyura stolonifera* and *Trichomya hirsuta*).

Jackson (1977) considered colonial species superior competitors for space, but their inability to tolerate the physical stress of the environment prevented them from colonizing the middle shore. This competitive superiority stems from five characteristics of colonial organisms.

1. Growth tends to be exponential and continues throughout life (i.e., it is indeterminate). This results in lateral expansion into the available space;
2. Predators or physical forces rarely kill the whole organism or colony, and rapid regeneration can take place from the stolon or remaining undamaged zooids. This means that recolonization does not have to wait for settlement and growth of new recruits;
3. There is a tendency towards continuous sexual reproduction. This means that larvae will be ready to colonize vacant space at any time
4. Colonial species are more successful at preventing overgrowth and settlement of larvae on their surface than solitary species;
5. Growth in certain directions may bring colonial organisms into more favourable locations; *Didemnum molle* colonies, for example, can migrate up vertical rock faces towards the light (Olson, 1986).

When a sessile animal grows or spreads out over the surrounding rock it will come into contact with other animals. The nature of the interaction and its eventual outcome depends upon whether the other animal is of the same or a different species; the outcome in the two cases is so different that they will be considered separately.

14.1 INTERACTION BETWEEN DIFFERENT SPECIES

When two different species of sessile animal come into contact, several things may happen. Both animals may stop growing along the line of contact and divert their growth in some other direction, or one animal may grow over the other. In the latter case, the animal underneath may survive and live normally or it may be deprived of food and eventually die. Sometimes one animal causes necrosis or even death of the animal with which it comes into contact. Finally, one animal may succeed in physically displacing the other. The eventual outcome of such events may be determined by relative size of the interacting species, by their rate and pattern of growth, or by some other factor, such as the production of toxic substances. Some species react to a competitor by altering their pattern of growth so that they are not overgrown.

The interaction among four species of Bryozoa living on *Fucus serratus* has been analysed by Stebbing (1973). In most cases when members of different species came into contact, growth stopped. However, when *Alcyonidium hirsutum* came into contact with *Electra pilosa*, it grew over the latter in 95.3% of cases. Another species, *A. gelatinosum*, was less successful in growing over *E. pilosa*. This was attributed to the development of longer spines by *Electra* along the border where it came into contact with the other bryozoan. Growth of these spines can be regarded as a defensive response.

It has been questioned whether growth actually stops when the two species come into contact. Osborne (1984) found that over longer periods of time, considerable growth and interaction took place and the boundary was by no means static. Various types of stolon formation and limited overgrowth were observed in interactions between bryozoan species and between bryozoans and compound ascidians. When *Celleporaria* and *Stylopoma* came into contact, both produced terminal stolons; those of the latter species were longer and overgrew the *Celleporaria*. The *Celleporaria* reacted by forming a new secondary growth edge and growing over the *Stylopoma* which, in response, started another round of stolonic overgrowth. In time this resulted in the growth of a ridge along the boundary between the two species.

A different response has been described when *Alcyonidium hirsutum* or *A. gelatinosum* came into contact with *Spirorbis spirorbis* (Stebbing, 1973). Although the bryozoans grew over the tube, the tube worm responded by extending its tube above the bryozoan, and continued to live normally. This phenomenon is frequently encountered where encrusting sponges, the anthozoan *Clavularia novaezelandiae*, and compound ascidians, such as *Didemnum moseleyi*, spread over an area already occupied by tube worms and hydroids, both of which can grow above the encroaching organism.

Sometimes a sessile animal is killed by a larger species that grows over it. For example, when the cyprids of *Megabalanus nigrescens* or the larvae of *Pyura stolonifera* settle on or near *Galeolaria caespitosa* and *Tesseropora rosea*, they may rapidly grow over and kill the latter, although in other cases *G. caespitosa* may overgrow or displace young barnacles (Denley and Underwood, 1979).

The jewel anemone, *Corynactis californica*, of Pacific North American shores can kill other sea anemones (*Anthopleura elegantissima*, *Metridium senile*) with its nematocysts. It does this by extruding the gastric filaments and bringing them into contact with the other sea anemone, which is killed within 3 weeks (Chadwick, 1987).

The position of interacting species may determine the outcome of competition. Walters and Wethey (1986) have shown that on a level surface, *Alcyonidium* grows over *Electra*, but *Electra* may overgrow *Alcyonidium* if it has a height advantage; evidently topography and position are important factors of competition.

Physical factors play an equally important role in competition. *Chthamalus fragilis* and *Semibalanus balanoides* compete for space throughout the intertidal zone on the New England coast of North America. In shade, *S. balanoides* overgrows *C. fragilis* and is dominant, but on surfaces exposed to direct sunlight, it is killed by heat or desiccation, and *C. fragilis* becomes the dominant species (Wethey, 1984b). *Mytilus californianus* is the dominant mussel on exposed rocky shores. It has a stronger shell and byssal attachment than *M. edulis* and can survive heavy wave impact and storms. The edible mussel (*M. edulis*), however, can survive in very sheltered places because it is able to detach its byssus and escape being buried by silt. Between these extremes the two mussels may coexist owing to the heterogeneity of the environment and variations in predation pressures (Harger, 1972; Seed, 1976).

Complex organic compounds called allelochemicals produced by certain sponges are believed to be responsible for preventing overgrowth by other animals or close settlement by their larvae. It is possible that the production of allelochemicals as a defence against interspecific interaction may be a widespread phenomenon among sessile animals, for there is evidence that it occurs in some compound ascidians, bryozoans and cnidarians (Stoecker, 1980; Branch, 1984).

Physical displacement has been noted among barnacles. When *Semibalanus balanoides* grows up against *Chthamalus stellatus* it can force the edge of its shell under the latter's shell plates. This weakens attachment to the rock and eventually *Chthamalus* is displaced (Connell, 1961). Paine (1974) has recorded the effect of competition between *Pollicipes polymerus* and *Mytilus californianus*. An area of bare rock was successfully colonized by the barnacles, which grew in size, but after 5 y they were completely excluded by lateral pressure from the surrounding mussels.

Sometimes it is possible for two or more species to make use of the same area for attachment. This is seen where the stolons of *Thyroscyphus marginatus* and *Amphisbetia minima* intertwine. The former grows five to ten times higher than the latter, allowing the two species to feed at different levels in the water column.

14.2 INTERACTION BETWEEN MEMBERS OF THE SAME SPECIES

Members of the same species usually stop growth when they come into contact, but in some cases they fuse. In nearly all cases where fusion has been investigated, it has been shown that both animals were of the same genotype (or isogeneic, i.e., they were clonemates or distant parts of the same branched organism). Fusion of isogeneic stolons is a widespread phenomenon among sponges, hydroids, bryozoans and compound ascidians. This is clearly illustrated by the hydroids.

If the growing tip of the main stolon of *Silicularia bilabiata* comes into contact with any part of a different animal (i.e., one derived from a different planula larva and therefore genetically different or allogeneic), it will stop growing. After a while, one of the lateral branches will take over the role of leading stolon and will start a new main stolon at right angles to the first. Because of its pattern of growth, it frequently happens that one of the stolonic side branches comes into contact with some other part of the same stolonic system. In this case fusion takes place. Upon fusion the coenosarc joins up, so that hydroplasmic flow can take place through the new connection. Fusion of this type leads to a reticulate stolon. In *Obelia geniculata*, the leading stolon does not stop when it encounters the stolon of another individual, but grows over it and continues in a straight line.

Under certain conditions the distal end of the vertical hydrocaulus of *Thyroscyphus marginatus* starts to grow rapidly to form a stolon-like outgrowth without lateral hydranths. If the club-shaped growing tip comes into contact with any other part of the hydrocaulus or stolon of the same organism, it will fuse to it as described above. In one instance, one of these branches came into contact with a hydrotheca and eventually fused with the hydranth inside. This observation is significant, for it shows that fusion is a property of the growing tip of a stolon. Fusion never occurs when hydranths or hydrocauli lie in contact with one another.

Fusion in hydroids raises a number of questions that have not been satisfactorily answered. It is clear that the response to contact involves recognition of self and non-self (i.e., discrimination of isogeneic and allogeneic individuals), but recognition cannot involve sense organs, nor can it involve direct cellular contact, for the cells of the coenosarc are isolated by the perisarc. Initial contact is between the very thin pellicle at the growing tip (which is continuous with the perisarc) and the perisarc of some other region. Recognition must be at the molecular level, but whether this is due to the molecular structure of the perisarc surfaces, or whether it depends upon some substance diffusing across the perisarc from the coenosarc, is not known.

Another problem arises in understanding how the coenosarc links up. It is assumed that the perisarc is dissolved by enzymes (chitinase or a proteinase) secreted at the tip of the stolon. The simplest hypothesis is that these are produced by the stolon tip. For enzymes to be produced by some other part of the coenosarc, one would have to postulate that a substance produced by the tip of the stolon diffuses through the thick perisarc to stimulate certain localized ectodermal cells to synthesize the enzyme. This seems unlikely, as breakdown of the perisarc is not a common phenomenon and only occurs at a few places (such as at the distal end of the gonotheca in *Obelia geniculata* prior to release of medusae). The phenomenon of self-recognition seen in hydroids is not isolated and occurs during conjugation of filamentous algae (for example *Spirogyra*) or moulds, where the cells are enveloped in cellulose and chitin respectively. Both sex and species specificity in gamete fusion of the unicellular alga, *Chlamydomonas*, have been shown to depend upon specific glycoprotein substances on the surface of the cells (Wiese, 1974). It is probable that recognition of self in both hydroids and ascidians depends upon glycoproteins on the outer surface of their perisarc and tunic respectively.

Fusion and rejection occur between colonies of the compound ascidians *Botryllus*, *Botrylloides* and *Perophora* (Taneda *et al.*, 1985). Oka (1970) has described the phenomenon in *Botryllus primigenus* and has shown that fusibility between colonies is controlled by a single multiallelic co-dominant histocompatibility gene (H). Fusion only occurred between colonies possessing a common allele. When the tunic of such colonies came into contact, the ampullae extended and made contact tip-to-side (Fig. 14.1(b) and (c)) and eventually fused. Within 48 h a common vascular system was established and the two colonies functioned as a single unit in which reproduction was synchronized.

In contrast, when allogeneic colonies came into contact, the ampullae extended but did not fuse. Watanabe and Taneda (1982) have shown that blood cells accumulated in the ampullae, which were eventually cut off from the rest of the vascular system, and new tunic walls were formed, separating the two colonies (Fig. 14.1(d) to (f)). They also demonstrated that a humoral factor present in the blood, tunic matrix, and morula blood cells was responsible for the destruction of the tunic and the formation of new tunic. Interactions somewhat similar to those of *Botryllus* occur between the stolons of *Perophora sagamiensis* (Koyama and Watanabe, 1982).

Fusion of two histocompatible colonies will give rise to a chimera, or compound colony, with a common vascular system circulating blood and pigment cells derived from both of the original colonies, but the zooids will be of different genotypes. Colony fusion is of common occurrence both in nature and in the laboratory, and a number of selective advantages have been suggested on the assumption that the chimera is a functional entity. However, the experiments of Rinkevich and Weissman (1987a) call this to question. They found that fusion invariably resulted in zooids of one genotype dominating those of the other genotype. Growth of the latter was suppressed and the zooids were eventually resorbed. The time course varied from weeks to months for complete resorption, depending upon the size of fusing colonies. The surviving colony presumably benefits nutritionally from resorbing the other zooids, but there can be no genetic advantage. There was a tendency for large colonies to suppress the zooids of a smaller colony, but this was not always the case; certain genotypes invariably suppressed certain other genotypes irrespective of size, suggesting that dominance is dependent upon a specific gene. It is not linked with the histocompatibility gene or genes controlling colour.

Because the tadpoles of *Botryllus schlosseri* tend to settle on or near parental colonies, they are likely to come into contact with the latter as they grow. Rinkevich and Weissman (1987b) found that offspring (oozooids and young colonies) were killed and their zooids were resorbed if they fused with,

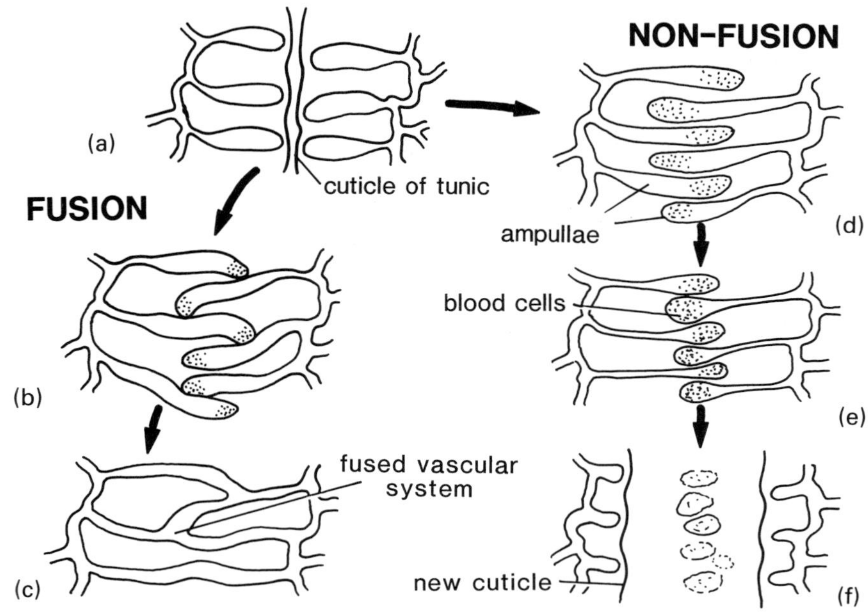

Fig. 14.1 Histocompatibility in *Botryllus primigenus*. (a) Two colonies may come into contact. If they share a common allele, they fuse ((b) and (c)); if they do not share a common allele, a non-fusion reaction ((d) to (f)) takes place. (Based on Oka, 1970; Watanabe and Taneda, 1982.)

or remained in contact with, their parental colony. It is difficult to suggest any selective advantage that could result from fusion or the tendency for siblings to settle near their parents in the light of these experiments.

The histocompatibility gene also controls fertilization in *Botryllus primigenus*, but in this case fertilization only takes place between allogeneic gametes (i.e., when the spermatozoon and the chorion cells round the ovum carry different alleles). Since all the spermatozoa produced by a particular colony will have one or other of the two alleles present in the somatic cells of the chorion, self-fertilization cannot occur and outbreeding is obligatory. All individuals in a natural population of this species are heterozygous for the H gene. Self-fertilization is possible under experimental conditions in *Botryllus schlosseri* although, as pointed out previously, this leads to inbreeding depression. Sabbadin (1978) considered that the mechanism of self and non-self recognition in the latter species was a means of preventing highly inbred colonies being formed with the inevitable impoverishment of the gene pool.

Under experimental conditions, Fuke (1980) has shown that when the blood of two allogeneic or xenogeneic (different species) solitary ascidians is mixed, a characteristic 'contact reaction' takes place between the blood cells which results in their lysis. In *Halocynthia roretzi* this contact reaction of blood cells is controlled by two multiallelic genes (Fuke and Nakamura, 1985). These experiments demonstrate the ability of solitary ascidians to recognize self and non-self, but what significance this can have to the sea squirts is not clear, for under natural conditions the blood of two individuals would never come into contact. In solitary ascidians histocompatibility may be important in preventing self-fertilization.

Another type of interaction involving recognition of self has been investigated in *Hydractinia echinata*, a hydroid that grows on shells inhabited by hermit crabs. The form of the colony is variable: it may be mat-like, or it may consist of an extensively branched network of stolons from which polyps resembling *Clava* arise. Isogeneic stolons fuse when they meet, but a dense mass of intertwining hyperplastic stolons is formed when allogeneic stolons from two different colonies come into contact. Necrosis and damage to one or both colonies may result from the discharge of nematocysts (Buss *et al.*, 1984). This response is remarkable for two reasons. It appears that a special type of nematocyst, the basitrichous isorhiza, is produced in the hyperplastic stolons from interstitial cells, in response to contact with allogeneic tissue. The formation of nematocysts in the stolon is normal for hydroids (except for the hydras, which lack this structure), but the discharge of nematocysts from the stolon is unique. In other hydroids they are carried by the nematocytes to some definitive position (usually on the tentacles or nematophores) and will only discharge when correctly orientated. Normally the perisarc would preclude their discharge from the stolon, but outgrowths from the hyperplastic tissue containing the nematocysts are not enclosed in perisarc. This example illustrates the highly specific nature of the nematocyst discharge, although the stimulus that causes it is unknown.

Behaviour of sea anemones

Apotreptic behaviour (i.e., activities tending to cause withdrawal of conspecifics; Barnett, 1981) is common among sea anemones. It involves a number of interactions between adult animals which have been variously described as aggression or territorial behaviour, but neither of these terms is satisfactory as both carry implications that do not apply to cnidarians. The general term of apotrepsis is preferred. In sea anemones it is believed that all instances of apotreptic behaviour are in response to allogeneic interaction, although no genetic analysis comparable with that of *Botryllus* has been carried out. Many species of sea anemone have several different colour forms and, as asexually produced clonemates are always identical in colour, colour differences have been taken to indicate allogeneic relationship. Analysis of isozyme distribution in natural populations of *Actinia equina* supports this interpretation (Quicke *et al.*, 1983; Quicke and Brace, 1984). It must be assumed that histocompatibility is tested when tentacles of two adjacent anemones come into contact. If they are clonemates (isogeneic), no further reaction takes place, but if they are allogeneic or xenogeneic, an elaborate pattern of apotreptic behaviour ensues, in which specially produced nematocysts are discharged into the opponent's tissues. This may cause necrosis and damage of the tissues, and usually results in the two animals moving apart; under laboratory conditions it may result in death of one of the individuals (Francis, 1973b). The special nematocysts are confined to the acrorhagi of *Actinia equina*, *Actinia tenebrosa* and *Anthopleura elegantissima*, or the very large fighting tentacles (formerly called catch tentacles) of *Metridium senile* and other acontiate anemones (Francis, 1973b; Bigger, 1980).

The details of the various patterns of behaviour may vary from species to species, but all show the following sequence of events:

1. Initial contact between the tentacles of two individuals. At this time histocompatibility must be determined, for contact with allogeneic or xenogeneic tissue (or more probably the mucopolysaccharides produced by the ectodermal cells) appears to be the stimulus that initiates apotreptic behaviour. Many contacts in succession may be necessary before a response is elicited. There is evidence that animals may become habituated to the presence of allogeneic individuals (Purcell and Kitting, 1982);
2. Preparation. The acrorhagi or fight tentacles are inflated on the side of stimulation, and this is often accompanied by partial withdrawal of the feeding tentacles or elongation of the column;
3. Fairly rapid movement is made towards the source of stimulation so that the acrorhagi or fighting tentacles come into contact with the other anemone;
4. Nematocysts discharge on contact and the other anemone's tentacles may be rapidly withdrawn. The ectoderm of the acrorhagi or of the tip of fighting tentacles is torn off as they retract and it remains in contact with the victim;
5. The victim may contract or bend away. After a time one of the anemones moves away. Necrosis of the victim's tissues occurs during the next few days.

The sequence of events has been described for *Anthopleura elegantissima* (Francis, 1973b), *Metridium senile* (Purcell, 1977) and *Actinia equina* (Brace *et al.*, 1979) (Fig. 14.2).

Young anemones do not possess the special nematocysts, acrorhagi or fighting tentacles found in adults and cannot inflict any injury; nevertheless they respond to contact by writhing movements of the feeding tentacles and

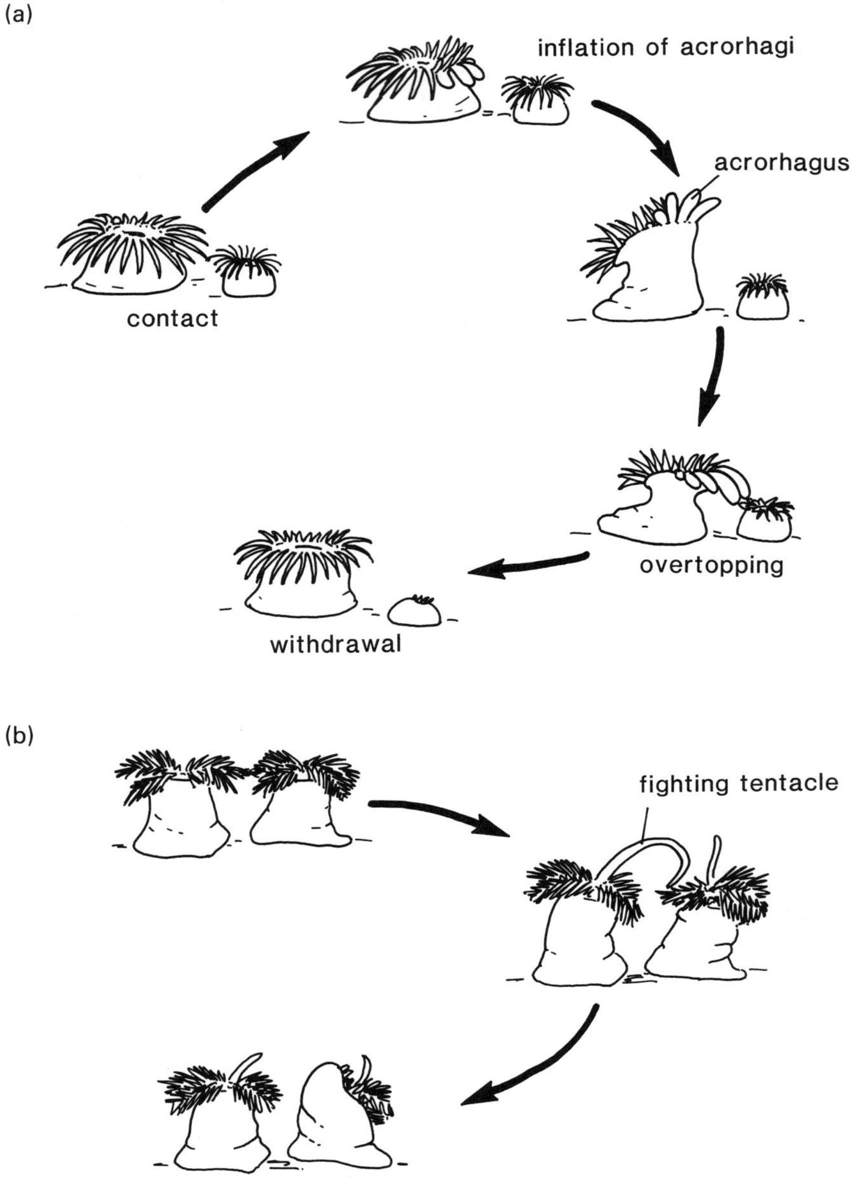

Fig. 14.2 Apotreptic behaviour of sea anemones. (a) Use of acrorhagi (*Actinia, Anthopleura*). (b) Use of fighting tentacles (*Metridium*). (Based on Francis, 1973b, and Purcell, 1977.)

move away from each other. This shows that the ability to detect histocompatibility is present at an early age.

Purcell (1977) has shown that fighting tentacles in *Metridium senile* are formed by modification of feeding tentacles near the mouth. This transformation with the production of a completely new cnidom takes *c.* 9 weeks and is brought about after repeated contact with allogeneic individuals. Fighting tentacles are more frequently found on individuals at the periphery of a clone adjacent to a different clone.

The function of this agonistic behaviour is difficult to understand, although it is clear that it will tend to space out young individuals produced by sexual reproduction and therefore of different genotype. It does not inhibit clustering of clonemates. Under natural conditions, different clones of *Anthopleura elegantissima* are separated by a strip of 'no man's land' believed to be maintained by apotreptic interaction between individuals along the border (Francis, 1973a). Similar clustering of clonemates was found in *Metridium senile* (Purcell, 1977).

Apotreptic behaviour is not evoked in response to predators (Francis, 1973b), and so it cannot be a defence mechanism. However, this behaviour is often evoked between different species of sea anemone (xenogeneic reaction) and could be interpreted as a mechanism for defending living space from encroachment. The fact that it also occurs as a response to allogeneic and not isogeneic interaction has led to the suggestion that it reserves space for the spread of the genotype (i.e., for new clonemates). Large closely packed clones may offer certain advantages; collectively they might be more efficient at preventing predation, desiccation or displacement by wave action and in competing against other species for space. Brace *et al.* (1979) have postulated that 'aggression' between dissimilar adult genotypes would promote isogeneic clustering and help to maintain space for settlement of autogeneic offspring. Observations on *Actinia tenebrosa* by Ayre (1983) seem to support this idea; however, isogeneic clustering would be detrimental to sexual reproduction, and the idea of reserving space for clonemates has been questioned. Purcell and Kitting (1982) found that some aggregations of *Metridium senile* were of mixed genotype, and apotreptic behaviour was suppressed by habituation to being touched by allogeneic individuals. Kaplan (1983), studying the same species, found that the sex of an individual influenced its reaction to conspecifics. Male–male and female–female encounters of allogeneic individuals resulted in apotreptic behaviour, but the interaction of males with females was non-aggressive. This would allow members of the opposite sex to mingle with a clone even though they were allogeneic, and thus increase the probability that successful cross-fertilization will occur. In *Metridium senile*, where spawning may not be synchronous, the presence of an allogeneic male anemone among a clone of females has obvious advantages in promoting outbreeding. The apotreptic behaviour of sea anemones, although based upon histocompatibility, is clearly a different phenomenon from the interaction of compound ascidians and their mechanisms of fertilization; nevertheless, in both groups allogeneic reaction may prevent inbreeding and promote gene flow.

14.3 PREDATORS

A very different type of interaction between species is that of predator and prey. Our knowledge of the feeding habits of predatory animals in the littoral zone is far from complete, but there is no doubt about the effects they have on community structure and stability. Besides affecting population densities, predators may increase community diversity by preventing competitive exclusion (Branch, 1984).

Many sessile animals will be consumed accidentally. Fish feeding on amphipods and small shrimps inevitably take fragments of the hydroids or bryozoans in its diet, and Newman (1960) attributed the paucity of intertidal particularly vulnerable to being scraped off the rock surface by limpets and chitons that feed on diatoms and algal sporelings. However, there are several groups of predators specialized for feeding on particular sessile animals. They will be discussed according to the feeding technique they adopt.

Certain regular sea urchins chisel barnacles and bryozoans from the rock surface with their powerful teeth. Some specimens of *Echinus esculentus* collected from Plymouth, England, were found to have fed exclusively on barnacles (species not determined). This urchin is known to feed on the sublittoral barnacle, *Verruca* (Foster, 1959) and various bryozoans (Ryland, 1976). The blenny, *Blennius pholis*, may take a significant proportion of bryozoans in its diet, and Newman (1960) attributed the paucity of intertidal barnacles (*Chthamalus stellatus* and *Tesseropora* sp.) on tropical reefs in the Western Pacific to the feeding habits of coral fish (*Acanthurus*, *Chaetodon* and *Balistes*), which feed upon young barnacles.

The common European sea star, *Asterias rubens*, is a predator of mussels. It may become a pest reaching plague proportions. Densities of 200–450 m^{-2} have been recorded on mussel beds in Morecambe Bay, England, that are capable of consuming several thousand tonnes of mussels in a season (Seed, 1976). After opening the mussel, it everts its stomach and slowly digests the soft tissues. Contrary to the belief that this sea star opens the mussel by the sustained pull of its tube feet, it has been shown that mussels can remain tightly closed against a constant pull for a period far exceeding the time taken by the sea star to gain entry. Lavioe (1956) found that the sea star exerts a series of short pulls with periods of relaxation, and within 5–10 min the valves of the mussel may be opened sufficiently (0.1 mm) for part of the stomach to be inserted and digestive enzymes secreted, or alternatively for the enzymes to diffuse into the mantle cavity. The mussel may become habituated to a series of pulls and fail to respond by tight closure. No evidence of any substance causing relaxation of the mussel's adductor muscles could be found in extracts from the sea star's stomach and other tissues.

Connell (1972) observed that the sea star, *Pisaster ochraceus*, on San Juan Island, Washington, moved up into the intertidal zone during high tide, where it was able to wrench large barnacles, *Semibalanus cariosus*, from the rock and return to the sublittoral to digest them.

Undoubtedly the most important predators between tide marks are the carnivorous prosobranch molluscs, particularly those belonging to the family Thaididae.

On British shores the common species is *Thais* (= *Nucella*) *lapillus*, the dog whelk, which preys upon *Semibalanus balanoides* and *Mytilus edulis*. When feeding on the former, it waits until the operculum is partly opened, and then thrusts its proboscis into the mantle cavity. The radula at the end of the proboscis is then used to rasp away the barnacle's flesh (Moore, 1938). Mussels are eaten in a different way. A small hole is bored through the valve through which the proboscis can be inserted. This difference in behaviour may reflect the fact that mussels can keep their valves closed for a much longer period than barnacles.

The American oyster drill, *Urosalpinx cinerea*, is the major predator of commercial oysters on the Atlantic Coast of North America and has been introduced into the Pacific Coast of North America and to Britain. It is particularly destructive to young oysters (Galtsoff, 1964). The mechanism used by *U. cinerea* to drill through oyster shells has been elucidated by Carricker (1961). A gland (the accessory boring gland) in the foot is applied to the shell for a period and then withdrawn. The radula is used to rasp off a thin layer of shell and then the gland is applied again. This cycle is repeated many times until the shell is penetrated. Carriker *et al.* (1967) found that carbonic anhydrase, which was produced in the gland, increased the concentration of bicarbonate and hydrogen ions; the latter, passing to the exterior, dissolved the calcium carbonate of the oyster shell. The concentration of calcium ions increased in the accessory boring gland during the drilling process.

Carnivorous prosobranchs may exert a profound effect upon the vertical distribution of intertidal animals. On the coast of Washington, Connell (1970, 1972) found that the lower limit of *Balanus glandula* and *Semibalanus cariosus* was determined mainly by the predatory activity of *Thais emarginata*. The niche occupied by *Thais lapillus* on British shores is taken by the mulberry shell, *Morula marginalba* on New South Wales shores, and by *Lepsiella scorbina* in New Zealand (Morton and Miller, 1968). These gastropods are the major predator of *Chamaesipho* and other small barnacles of the lower middle shore. Many species of larger carnivorous prosobranchs are found in the infralittoral fringe. *Cabestana spengleri* is of interest because it feeds on *Pyura stolonifera* by boring through the thinner flexible tunic near the siphons. Penetration may be aided by enzymes, for when a whelk is pulled away, a patch *c.* 1 cm in diameter of frayed whitish tunic is revealed, from which the sand grains have been removed. Cowries belonging to the Cypraeidae and Lamellariidae feed on hydroids as well as on compound and solitary ascidians. Many show a marked preference for a single prey species. For example, the cowrie, *Erato voluta*, feeds on *Botryllus schlosseri* by inserting its proboscis through the oral siphon to extract the zooid (Fretter, 1951).

Various species of the opisthobranch gastropod, *Pleurobranchus*, feed on sessile animals. Thompson and Slinn (1959) found that *P. membranaceus* readily ate both solitary and compound ascidians by boring through the tunic with the radula. They also have a large acid gland, capable of secreting sulphuric acid, that opens by a duct into the mouth. This suggests that they are capable of boring through calcareous shells and may feed on bivalves or even barnacles. These gastropods do not have an external shell and protect themselves from predators by sulphuric acid-secreting glands in the fleshy mantle.

A number of the nudibranchs have a relatively broad radula and may browse on bryozoans. A member of the Arminacea found in New Zealand feeds on *Beania magellanica* by devouring the zooids one by one (Morton and Miller, 1968), and a number of unspecialized members of the Doridacea feed on sponges; for example, *Archidoris* feeds on *Haliclíona*, *Rostangia* on *Microciona*, and *Glossodoris* on various sponges (Morton, 1963).

Several families of the more specialized aeolid nudibranchs feed on hydroids, sea anemones and soft corals. They are unaffected by the nematocysts of their prey, which pass through their gut and accumulate in the cnidosacs at the end of the club-shaped cerata on the dorsal surface of the body. Presumably the nudibranchs can eat cnidarians because they do not provide the appropriate stimulus for nematocyst discharge. It is claimed that the nematocysts give protection to the nudibranchs, but it is not clear under what condition they would discharge in the cnidosac, nor are they orientated as in the hydroid or sea anemone. Isolated nematocysts can be made to discharge under abnormal conditions of pH or osmotic stress, and so it is possible that they could be triggered to discharge in the stomach of teleost fish. Fish might learn to avoid aeolid nudibranchs if this affected them in some way. It is significant that many species of the Aeoliidae and Caryophylliidae found on Australasian shores are among the most colourful animals in the sea. The colours may be aposomatic and indicate distastefulness, but this makes the assumption that the predator can discriminate different colours. In the littoral zone only certain fish and birds fall into this category. Very small (2–3 mm) members of the family Dotidae are common on hydroids such as *Silicularia*, *Amphisbetia* and *Plumularia*. They differ from the aeolids in only having from six to twelve short branched cerata.

Many members of the family Dorididae are highly specialized as suctorial feeders on bryozoans. There has been a tendency for reduction of the radula and the development of a muscular pharynx that acts as a pump to suck out the polypide or body fluids. On British shores *Acanthodoris pilosa* and *Goniodoris nodosa* are found on *Fucus*, where they feed on *Alcyonidium*, *Flustrellidra* and *Botryllus schlosseri* (Millar, 1971; Ryland, 1976). *Onchidoris fusca* is unusual in that it may feed on *Balanus crenatus* by sucking out the more fluid parts of the body (Barnes and Powell, 1954). Species of *Dendrodoris*, feeding on solitary ascidians, are the most highly specialized of all nudibranchs, for the radula and buccal mass have been lost. Enzymes are used to penetrate the tunic of their prey, and the muscular pharynx sucks out the fluid remains of the ascidian after digestion (Morton and Miller, 1968).

The suctorial method of feeding is used by another group of predators, the Pycnogonida. Some species feed on algae, others on detritus, but a large proportion feed on sponges, hydroids, sea anemones and bryozoans. The pycnogonids are peculiar for the relatively large size of their proboscis, which is larger than the body in some species. A large muscular pharynx occupies most of the proboscis and acts either as a suction pump to suck up liquid food or, by its contractions, to macerate food that has been bitten off by the horny jaws surrounding the mouth. Fine setae at the back of the pharynx act as a filter preventing solid food from entering the gut.

Nymphon gracile, common on the British coast, feeds on sea anemones (such as *Actinia equina*) by rasping off pieces of the body wall with its jaws, but when it feeds on hydroids it uses its chelifores to pick off and macerate pieces of the hydroid, which are then brought to the mouth and seized by the jaws. *Pycnogonum littorale*, which lacks chelifores, feeds on *Actinia equina* and *Metridium senile* by thrusting its proboscis into the wall of the anemone and biting off tissue, which is sucked up by the pharynx. *Achelia echinata* feeds on the bryozoan *Flustra foliacea* and various hydroids. It aligns its proboscis so that the tip is in front of the operculum and then waits for the polypide to emerge. It then thrusts the proboscis into the aperture and moves forward so that the proboscis ruptures the frontal membrane; this allows it to eat the zooid (King, 1973).

Certain vertebrates are predators on mussels and oysters. When the tide is out, ducks can destroy large numbers of young oysters, while oyster-catchers and gulls may feed extensively on the edible mussel. When the tide is in, bottom-living fish may eat mussels.

Defence against predators

Many tactics have evolved among sessile animals to deter or counteract predation. These are usually more effective against generalist than specialist predators, for the latter have evolved counter strategies that overcome the prey's defences. A shell or operculum that can be tightly closed may isolate the prey and stop a predator detecting its presence. The heavy calcareous shells of the larger barnacles, adult mussels and oysters may be sufficient protection against the smaller carnivorous gastropods, but larger predators such as *Thais orbita* and *Cabestana spengleri* will attack adult *Megabalanus nigrescens* and *Pyura stolonifera*. Predators take prey within a certain size range, thus animals that are too small or have grown beyond this range will be ignored.

Many sessile animals rely upon behavioural defences. The sudden withdrawal of serpulid worms and closure of the tube by the operculum would fall into this category, along with closure of siphons (sea squirts), opercula (Bryozoa and barnacles), or the closure of bivalves. Other activities include rocking of the operculum in balanomorph barnacles, squirting of sea squirts, and the activity of avicularia in Bryozoa.

Perhaps the most effective defence against predators and competitors is the production of allelochemicals and toxins. Toxic compounds present in the nematocysts of cnidarians are not only effective in capturing prey, but they are believed to act as a defence against predators, although the evidence for this is somewhat circumstantial.

Many species of sea squirts possess vesicles in their tunic which, when damaged, release sulphuric acid. This could be a deterrent to both predators and browsers. A high correlation was found between refusal to eat pieces of ascidian and their vanadium content. Fish and crabs when tested rarely ate species with vanadium, but readily ate other species of ascidian, suggesting that the presence of vanadium renders them unpalatable (Stoecker, 1980).

A large number of complex organic compounds, called allelochemicals or allelomones, have been extracted from sessile animals, particularly sponges, and may affect the growth, health, behaviour or reproductive success of other organisms (Whittaker and Feeny, 1971). Little is known about the biological significance of these compounds to the animal producing them or their allelopathic effects on other organisms. Amade *et al.* (1987) found that all of 28 species of tetractinomorph and ceractinomorph sponge from the Mediterranean Sea showed some antibacterial or antifungal activity. Species that normally contained symbiotic bacteria showed lower antibacterial activity, and it has been suggested that the allelochemicals function to control the symbiotic population. Many sponges of the Dictyoceratida produce a variety of terpenoid compounds which may have an antibiotic effect or may render the sponge unpalatable to predators. The terpinols give sponges an unpleasant odour resembling carbon disulphide, and will kill any other animals (polychaete worms, crustaceans etc.) placed in the same container or aquarium. Many alcyonarians possess ichthyotoxic compounds in their tissues that deter predatory fish. *Heteroxenia fuscescens*, for example, appears to lack nematocysts but produces terpenols (Schlichter, 1982) which are toxic to fish. However, Barre *et al.* (1986) found that other alcyonarians, which lacked ichthyotoxic compounds, were not eaten, possibly because of the general deterrent action of nematocysts. The sponge, *Suberites domuncula*, possesses a neurotoxin, suberitine, which may likewise deter predators or browsers (Cariello *et al.*, 1980).

A number of ascidians, bryozoans, and certain anthozoans as well as sponges are known to produce substances that act as antibiotics or inhibit settlement of larvae of other species on or near them. They also inhibit the growth of surrounding organisms competing for space (Branch, 1984).

A most unusual antipredator strategy has been described for *Mytilus edulis*. Petraitis (1987) found that mussels could immobilize their predator, *Thais lapillus*, with byssus threads. Although whelks were not common on natural mussel beds, 30% of those that were found there had been trapped by byssus threads.

14.4 DISTRIBUTION OF ANIMALS ON THE SHORE

The structure and physiological capacities of an animal determine where it is possible for it to live, i.e., its potential range. A species may have several physiological races with slightly different tolerances to the environment, consequently the geographical range of the species will be greater than the potential range of individuals. Stephenson and Stephenson (1972) compared the physical features of coasts in Africa, Australasia, British Isles and North American and gave an account of the major features of the various plant and animal communities. Their work shows a conspicuous similarity in the vertical zonation of the shore in different parts of the world, and emphasizes the fact that, although species may differ, animals of a similar type occupy similar positions on the shore in other places. A more detailed analysis of the physical environment and community structure on rocky shores round the British Isles has been made by Lewis (1964). In both these works the emphasis has been on the physical factors that influence distribution. However, distribution cannot be explained by physical factors alone: biotic factors are often of far greater significance in determining where an animal is permitted to live. The actual area occupied by a species at any instant of time will be less than its potential range, owing to interactions with other species (competition and predation), and the chance event of a larva successfully establishing itself in that area.

The communities of animals and plants living on rocky shores are diverse and complex. In any one place or zone, a surprisingly large number of species make up the community, with one or two of the larger species dominating the area and occupying most of the space. A much larger number of subdominant and occasional species occupy the remainder. On the Pacific Coast of Washington State, Paine (1974) found that 30–40% of the middle shore was occupied by barnacles (*Semibalanus cariosus*, *Balanus glandula*, *Chthamalus fissus*), and 36–43% was occupied by algae (*Endocladia*, *Corallina*, and *Lithothamnium*). Mussels (*Mytilus californianus*) and a sponge (*Halichondria panacea*) occupied a lesser area (*c.* 5% each), and 9–14% was unoccupied. Thus eight species occupied 80% or more of the available space. Another ten species, including the major predators of barnacles and mussels (*Thais emarginata* and *Pisaster ochraceus*), were present as subdominant or minor components. However, when occasional species and small animals are taken into account, the number of species in this community is likely to be three or four times the number censused by Paine.

14.5 THE DYNAMIC NATURE OF SHORE COMMUNITIES

Occasional visits to the coast give the impression of great community stability over periods of time, and this tends to be borne out by photographs taken from time to time. Paine (1974), for example, was able to show from photographs taken over a period of 9 y that the upper and lower boundaries of mussel bands changed very little during that period, but when individual mussels were marked in a small area, a considerable turnover of the population was revealed. Some individuals were killed and the vacated space was taken by movement of surrounding mussels, recruitment of juveniles or the invasion of other species.

Seasonal changes in composition also occur. In the same study, Paine showed that *Pisaster ochraceus* tends to migrate into deeper water during the winter, but moves up shore during the summer. Consequently predation pressure on barnacles and mussels is significantly higher during the summer. A similar situation has been noted for *Thais lapillus* in Scotland (Connell, 1961). Seasonal changes are often more obvious among the seaweeds. During winter the mid-tidal region of SE Australian shores may be a bright green because of a carpet of *Ulva*, but this dies back in spring, revealing the grey-green *Hormosira banksii* and a mat-like growth of pink coralline algae or brown *Lobophora*. Perhaps the most significant changes to temperate shores occur in autumn and winter, when storms at sea cause extensive damage to populations of sessile animals and seaweeds. Dead barnacles, the gaping shells of dead mussels, and dense aggregates of barnacles or mussels forming hummocks are points of weakness which allow heavy wave action to strip larger areas of these animals from the rocks. Similarly, the mat-like growth of stunted coralline seaweed consolidated with sand or silt may be swept away from the rocks, leaving a clean surface ready to be recolonized in the spring.

Among animal species, the two most important biotic factors influencing community composition are competition and predation. The ecological and evolutionary implications of competition between marine littoral organisms have been extensively reviewed by Branch (1984), who points out that far less attention has been paid to marine than terrestrial animals.

The final outcome of competition is very different from that of predation. If no other complication is involved, one competitor will displace or eliminate another and so occupy its space, whereas a predator is not likely to eliminate its prey species, for this would be to its own detriment. Non-specialized predators will switch to a new source of food before their normal prey is eliminated. Competition or predation rarely occur in isolation; they frequently interact, and other members of the community may modify their effect. This is clearly seen in the vertical zonation of species on the shore as well as in the composition and diversity of littoral communities.

The upper limit of barnacle and mussel distribution is determined by the tolerance of juveniles and adults to the physical conditions of the environment, but the lower boundary results from competition and predation (Connell, 1972). In those cases where the lower limit of distribution on the shore has been tested experimentally, it has been found that barnacles, mussels and tube worms are all capable of living at much lower levels than their observed distribution would suggest. Their lifespan, however, is much shorter at lower levels because of predation. *Semibalanus balanoides* is a stronger competitor for space and displaces *Chthamalus stellatus* when the two species occur in the same area. *C. stellatus*, however, is more tolerant of desiccation and can live higher on the rocks, consequently it tends to occupy a band above *S. balanoides*, its lower boundary being determined by competition with the latter species. In areas where *S. balanoides* is scarce, *C. stellatus* extends lower on the shore (Connell, 1972). On the coast of Washington State, *Mytilus californianus* is prevented from living lower on the shore by its chief predator, *Pisaster ochraceus*, and forms a conspicuous lower boundary. When *P. ochraceus* was removed manually, the mussel extended its range nearly 2 m further down the shore over a period of 3 y (Paine, 1974). Thus the lower boundary of the mussel band coincides with the sea star's upper limit of tolerance to emersion, and this in turn is influenced by tide height.

This interaction between mussel and sea star may be influenced by other members of the community. Mussels were found to increase in density in areas where the jewel anemone, *Corynactis californica*, was abundant, because of the deterrent effect of the latter on sea stars (Wolfson *et al.*, 1979).

It is possible that the lower boundary of distribution of many other shore animals is determined by similar interactions.

When a predator feeds selectively upon the dominant of two competitors, it may modify the intensity of competition to a level where both competitors can coexist. This has been demonstrated in the natural situation by placing cages over groups of animals to exclude predators or control the number of predators. Menge (1976) studied the competitive interaction between *Semibalanus balanoides* and *Mytilus edulis* at mid-tide levels on the coast of Massachusetts and Maine. In exposed places where predators were few, barnacles were the first to colonize bare rock surfaces in the spring and soon occupied most of the area. Juvenile mussels settling among the barnacles during summer soon overgrew and replaced the barnacles to become the dominant species. When mussels were removed experimentally, the barnacles continued to grow and monopolize the area. In less exposed situations the predator, *Thais lapillus*, exerted a marked effect upon community structure. By feeding on young mussels, it reduced the competition between barnacles and mussels to the point where both species could coexist.

The effect of predators upon the composition of the shore community is graphically demonstrated when a whole community is eliminated by a catastrophic event. Red tides caused by astronomical numbers of the dinoflagellate *Gyrodinium* have caused mass mortality of shore animals and algae in Ireland (Cross and Southgate, 1980). During the period of recolonization after a red tide, Southgate *et al.* (1984) found that settlement of barnacles and fucoid seaweeds was exceptionally heavy, owing to the drastic reduction in numbers of browsing and predatory gastropods.

Most rocky shore communities show high species diversity, and interaction between their members is expected to be extremely complex. The experiments of Paine (1974) in removing *Pisaster ochraceus* reveal an unexpected aspect of predation. In the absence of its main predator, *Mytilus californianus* not only expanded its range, but displaced other species in the community to become the single occupant of the available space. Of the 25 macroscopic species present prior to removal of the sea stars, only the mussels remained. It is clear from this result that predation significantly increases species diversity.

Predators usually show preferences for a particular prey species within a certain size range, although this preference might change. An important factor for predatory gastropods is the time needed to bore through the shell of their prey, for this will determine how many food items can be consumed during one feeding period (tide cycle). There is a tendency for predators to take small prey (i.e., juveniles), so once an animal has grown past a certain size it can escape predation. This might reflect the relationship between nutritional benefit and the amount of time spent in obtaining it. In the lower region of the barnacle zone on less exposed shores, predation of young barnacles by thaid whelks may be extremely high, so that few members of that year class survive. On the other hand, events that lead to high mortality of the predators may result in large numbers of young barnacles surviving and becoming the dominant age class in future years.

The interactions of organisms on some shores are far more complex than has been indicated in the examples considered so far, and a particular organism may exert different effects according to its density. In some cases a reciprocal effect is noticed with respect to density. Underwood *et al.* (1983) have studied the interactions of five main members of the *Tesseropora–Cellana* community on moderately exposed shores in New South Wales. Interaction between *T. rosea* and the limpet, *C. tramocerica*, depends upon the density of both limpets and barnacles. When barnacle density is low, limpets 2–3 cm in diameter can browse among them. This reduces recruitment by destroying barnacle spat, but it is beneficial in stopping macroalgae from growing and smothering the barnacles. At high barnacle densities, the limpets cannot feed effectively and leave the area, although the smaller acmaeid limpet, *Patelloida latistrigata*, may become abundant. In the absence of limpets, barnacle recruitment increases. The main predator is the oyster drill, *Morula marginalba*, which preys on younger barnacles, but shows a preference for *Patelloida*. Larger barnacles are not attacked. The presence of young *Patelloida* is, therefore, beneficial to recruitment of *T. rosea*. A tufted greenish-brown seaweed, *Gelidium pusillum* (a red alga), which also grows in the barnacle zone, makes it difficult for *Morula* to feed upon *T. rosea*, but excessive growth of the seaweed will smother and kill both young and old barnacles.

It becomes apparent from this example that neither competition nor predation has one single effect. Any one relationship may have both beneficial and detrimental results, according to the densities of the two species involved. Moreover, the presence or absence of another member of the community may profoundly affect intensity of competition or predation. Community structure on the sea shore is far from static; it must be looked upon as the dynamic interaction between a cyclically changeable physical environment and a highly complex community of plants and animals, both of which change with time and season in a way that can seldom be predicted.

SUMMARY

Space for attachment and growth is a vital resource for sessile animals and competition for it may take place between two species. Colonial and modular organisms are superior to solitary animals as competitors, but they are physiologically incapable of competing for space on the middle shore.

Interaction between two animals of different species may result in one displacing the other or causing its death, but other animals could complicate the outcome. When two members of the same species come into contact they stop growing, or they may fuse together (some compound ascidians, hydroids and sponges). Fusion only occurs between animals that are isogeneic (i.e. clonemates or distant parts of the same branching organism), and may be controlled by a histocompatibility gene.

Some sea anemones display a type of territorial behaviour in which special organs (acrorhagi and fighting tentacles) and special nematocysts are used as weapons of offence. Isogeneic individuals do not interact, but allelogeneic animals interact causing damage and even death with their nematocysts. There is evidence that allelogeneic individuals of opposite sex do not display this behaviour.

Many members of the shore community are predators on sessile animals; some are specialized to avoid or over-run the normal defences of their prey. Examples of predatory species are given. Defence strategies of sessile animals involve protective shells or spines, growing beyond the optimal size for the predator, behavioural reactions, isolation of sensory information to the predator, and the production of toxic compounds or deterrents (allelomones).

The vertical distribution of animals on the shore is determined by both physical and biotic factors. The upper limit for barnacles, mussels and tubeworms is determined by their degree of adaptation to physical conditions, but the lower limit is regulated by competition and predation from animals living on the lower shore.

Within any specified community one or a few species occupy most of the space, but there may be many sub-dominant species together with numerous minor members occupying the remainder.

Seasonal changes in abundance may be caused by rate of recruitment, migration of mobile species and mortality resulting from life-span, climate, competition, predation and disease. Competition could result in elimination of certain species, and thereby reduce species diversity, but this rarely happens because of interactions with other members of the community, particularly predators. Predators increase species diversity by preventing certain species becoming dominant.

Community structure and species distribution is not static; it is constantly changing under the interacting forces of environmental conditions, competition and predation.

Appendix
Classification of sessile shore animals

The following classification indicates the systematic position of groups with sessile genera referred to in the text, but does not give a full classification of mobile animals. Vertebrates, Protozoa and seaweeds have been omitted. Synonymous names are shown in brackets.

PORIFERA (CHAPTER 12)

Class Calcarea

Order Clathrinida	*Clathrina*
Order Leucosolenida	*Leucosolenia*
Order Sycettida	*Scypha* (= *Sycon*)

Class Demospongiae

Subclass Tetractinomorpha

Order Hadromerida	*Tethya*; *Suberites*; *Cliona*

Subclass Ceractinomorpha

Order Halichondrida	*Halichondria*
Order Poecilosclerida	*Myxilla*; *Ophlitaspongia*; *Microciona*
Order Haplosclerida	*Haliclona*
Order Dictioceratida	*Spongia*; *Dysidea*
Order Verongida	*Verongia*

CNIDARIA

Class Anthozoa (Chapter 10)

Subclass Octocorallia (Alcyonaria)

Order Stolonifera	*Clavularia*
Order Alcyonacea	*Heteroxenia*

Subclass Hexacorallia (Zoantharia)

Order Actiniaria	*Anemonia*; *Actinia*; *Anthopleura*; *Urticina* (= *Tealia*); *Oulactis*; *Metridium*; *Calliactis*; *Sagartia*; *Cereus*; *Anthothoë*; *Stomphia*
Order Corallimorpharia	*Corynactis*
Order Scleractinia (Madreporaria)	*Culicia*
Order Ceriantharia	*Cerianthus*

Class Hydrozoa (Chapter 11)

Subclass Hydroidea

Order Hydroida	
Suborder Athecata	*Australomedusa*; *Clava*; *Eleutheria* (= *Clavatella*); *Sarsia* (= *Syncoryne*); *Hydractinia*; *Tubularia*
Suborder Thecata	*Silicularia*; *Obelia*; *Laomedea* (= *Campanularia*); *Hartlaubella*; *Thyroscyphus*; *Phylactotheca*; *Amphisbetia*; *Dynamena*; *Plumularia*; *Aglaophenia*
Order Hydrida	*Hydra*

Subclass Siphonophora *Physalia*

MOLLUSCA (CHAPTER 6)

Class Bivalvia (Chapter 6)

Order Anisomyaria	*Mytilus*; *Austromytilus*; *Perna*; *Trichomya*; *Modiolus*; *Ostrea*; *Crassostrea*; *Saccostrea* (= *Crassostrea*)
Order Heterodonta (not sessile animals)	*Cerastoderma* (= *Cardium*); *Lasaea*; *Chione*

Class Gastropoda

Subclass Prosobranchia (not sessile animals)	*Cellana*; *Patelloida*; *Thais*; *Urosalpinx*; *Morula*; *Lepsiella*; *Cabestana*; *Erato* *Vermetus* (not described in this text)
Subclass Opisthobranchia (not sessile animals)	*Pleurobranchus*; *Archidoris*; *Rostangia*; *Glossodoris*; *Acanthodoris*; *Gonodoris*; *Onchidoris*; *Dendrodoris*

ANNELIDA

Class Polychaeta (Chapter 7)

Order Sabellida	
Family Sabellariidae	*Idanthyrsus*; *Sabellaria*
Family Sabellidae	*Sabella*; *Sabellastarte*
Family Serpulidae	*Pomatoceros*; *Galeolaria*; *Salmacina*; *Mercierella*
Family Spirorbidae	*Spirorbis*; *Janua* (= *Dexiospira*)

CRUSTACEA

Class Copepoda (not sessile animals)

Corycaeus; *Notodelphus*

Class Cirripedia (Chapter 5)

Order Thoracica	
Suborder Lepadomorpha	*Pollicipes*
Suborder Verrucomorpha	*Verruca*
Suborder Balanomorpha	
Superfamily Chthamaloidea	*Catomerus*; *Chthamalus*; *Chamaesipho*
Superfamily Balanomorphoidea	*Tesseropora*; *Tetraclita*; *Tetraclitella*
Superfamily Balanoidea	*Balanus*; *Semibalanus* (= *Balanus*); *Megabalanus* (= *Balanus*); *Elminius*

CHELICERATA

Class Pycnogonida (not sessile animals)

Nymphon; *Pycnogonum*; *Achelia*

BRYOZOA (ECTOPROCT POLYZOA) (CHAPTER 9)

Class Stenolemata

Order Cyclostomata	*Crisia*

Class Gymnolaemata

Order Ctenostomata	*Zoobotryon*; *Alcyonidium*; *Flustrellidra*
Order Cheilostomata	*Membranipora*; *Electra*; *Flustra*; *Beania*; *Bicellariella*; *Bugula*; *Celleporaria*; *Stylopoma*; *Sertella*; *Watersipora*

ENTOPROCT POLYZOA (not described in text)

BRACHIOPODA (not described in text)

PHORONIDA (not described in text)

ECHINODERMATA

Class Echinoidea (not sessile animals)

Echinus

Class Asteroidea (not sesssile animals)

Asterias; *Pisaster*

CHAETOGNATHA

(not sessile animals)
Sagitta

CHORDATA

TUNICATA

Class Ascidiacea (Chapter 8)

Order Enterogona	
Suborder Aplousobranchiata	*Eudistoma*; *Aplidium*; *Didemnum*; *Diplosoma*; *Lissoclinum*; *Parascidia*
Suborder Phlebobranchiata	*Perophora*; *Ciona*; *Phallusia*; *Ascidia*
Order Pleurogona	*Botryllus*; *Botrylloides*; *Dendrodoa*; *Styela*; *Pyura*; *Halocynthia*

References

Aiello, E.L. (1974) Control of ciliary activity in metazoa, in *Cilia and Flagella* (ed. M.A. Sleigh), Academic Press, New York and London, pp. 353–76.

Alexander, R. McN. (1962) Visco-elastic properties of the body-wall of sea anemones. *J. Exp. Biol.*, **38**, 373–86.

Alexander, R. McN. (1979) *The Invertebrates*, Cambridge University Press, Cambridge.

Allan, J.D. (1976) Life history patterns in zooplankton. *Am. Nat.*, **110**, 165–80.

Almada-Villela, P.C. (1984) The effects of reduced salinity on the shell growth of small *Mytilus edulis*. *J. Mar. Biol. Ass. U.K.*, **64**, 171–82.

Amade, P., Charroin, C., Baby, C. and Vacelet, J. (1987) Antimicrobial activities of marine sponges from the Mediterranean Sea. *Mar. Biol.*, **994**, 271–5.

Andersen, S.O. (1976) Cuticular enzymes and sclerotization in insects, in *The Insect Integument* (ed. H.R. Hepburn), Elsevier Scientific Publishing Co., Amsterdam, pp. 121–44.

Anderson, D.T. (1981) Cirral activity and feeding in the barnacle *Balanus perforatus* Bruguière (Balanidae), with comments on the evolution of feeding mechanisms in thoracican cirripedes. *Phil. Trans. R. Soc.*, **B**, **291**, 411–49.

Anderson, D.T. (1983) *Catomerus polymerus* and the evolution of the balanomorph form in barnacles (Cirripedia). *Mem. Aust. Mus.*, **18**, 7–10.

Anderson, D.T. and Anderson, J.T. (1985) Functional morphology of the balanomorph barnacles *Tesseropora rosea* (Krauss) and *Tetraclitella purpurascens* (Wood) (Tetraclitidae). *Aust. J. Mar. Freshwat. Res.*, **36**, 87–113.

Ayre, D.J. (1983) The effect of asexual reproduction and inter-genotypic aggression on the genotypic structure of populations of the sea anemone *Actinea tenebrosa*. *Oecologia*, **57**, 158–65.

Ballantine, W.J. (1961) A biologically-defined exposure scale for the comparative description of rocky shores. *Fld Stud.*, **1**, 1–19.

Banta, W.C., McKinney, F.K. and Zimmer, R.L. (1974) Bryozoan monticules: excurrent water outlets? *Science, N.Y.*, **185**, 783–4.

Barnes, H. (1955) Further observations on rugophilic behaviour of *Balanus balanoides* (L). *Vidensk. Meddr. Dansk. Naturh. Foren.*, **117**, 341–8.

Barnes, H. (1959) Stomach contents and microfeeding of some common cirripedes. *Can. J. Zool.*, **37**, 231–6.

Barnes, H. (1963) Light, temperature and the breeding of *Balanus balanoides*. *J. Mar. Biol. Ass. U.K.*, **43**, 717–27.

Barnes, H. and Barnes, M. (1954) The general biology of *Balanus balanus* (L.) Da Costa. *Oikos*, **5**, 63–76.

Barnes, H. and Barnes, M. (1957) Resistance to desiccation in intertidal barnacles. *Science, N.Y.*, **126**, 358.

Barnes, H. and Barnes, M. (1968) Egg numbers, metabolic efficiency of egg production and fecundity: local and regional variations in a number of common cirripedes. *J. Exp. Mar. Biol. Ecol.*, **2**, 135–53.

Barnes, H. and Barnes, M. (1977a) Studies on the reproduction of cirripedes II. Setting of the lamellae; action of protease and disintegration of the oviducal sac. *J. Exp. Mar. Biol. Ecol.*, **27**, 219–31.

Barnes, H. and Barnes, M. (1977b) Studies on the reproduction of cirripedes III. Ovisacase; properties of the enzyme released by the late stage embryos. *J. Exp. Mar. Biol. Ecol.*, **27**, 233–49.

Barnes, H. and Crisp, D.J. (1956) Evidence of self-fertilization in certain species of barnacles. *J. Mar. Biol. Ass. U.K.*, **35**, 631–9.

Barnes, H. and Powell, H.T. (1950) The development, general morphology and subsequent elimination of barnacle populations, *Balanus crenatus* and *B. balanoides*, after a heavy initial settlement. *J. Anim. Ecol.*, **19**, 175–9 + 3 pl.

Barnes, H. and Powell, H.T. (1954) *Onchidoris fusca* (Muller); a predator of barnacles. *J. Anim. Ecol.*, **23**, 361–3.

Barnes, H., Finlayson, D.M. and Piatigorsky, J. (1963) The effect of desiccation and anaerobic conditions on the behaviour, survival and general metabolism of three common cirripedes. *J. Anim. Ecol.*, **32**, 233–52.

Barnett, S.A. (1981) *Modern Ethology*. Oxford University Press, Oxford.

Barre, S.C., Coll, J.C. and Sammarco, P.W. (1986) Defensive strategies of soft corals (Coelenterata: Octocorallia) of the Great Barrier Reef. II The relationship between toxicity and feeding deterrence. *Biol. Bull. Mar. Biol. Lab., Woods Hole*, **171**, 565–76.

Barrington, E.J.W. (1957) The distribution and significance of organically bound iodine in the ascidian *Ciona intestinalis* L. *J. Mar. Biol. Ass. U.K.*, **36, 1–16.**

Barrington, E.J.W. (1974) Biochemistry of primitive deuterostomians, in *Chemical Zoology*, Vol. VIII (eds M. Florkin and B.T. Scheer), Academic Press Inc., London, 61–95.

Barrington, E.J.W. and Barron, N. (1960) On the organic binding of iodine in the tunic of *Ciona intestinalis* L. *J. Mar. Biol. Ass. U.K.*, **39**, 513–23.

Barrington, E.J.W. and Thorpe, A. (1968) Histochemical and biochemical aspects of iodine binding in the tunic of the ascidian *Dendrodoa grossularia* (van Beneden). *Proc. R. Soc.* **B, 171**, 91–109.

Batham, E.J. and Pantin, C.F.A. (1950a) Muscular and hydrostatic action in the sea-anemone *Metridium senile* (L.). *J. Exp. Biol.*, **27**, 264–89.

Batham, E.J. and Pantin, C.F.A. (1950b) Inherent activity in the sea-anemone *Metridium senile* (L.). *J. Exp. Biol.*, **27**, 290–301.

Batham, E.J. and Pantin, C.F.A. (1950c) Phases of activity in the sea-anemone *Metridium senile* (L.). *J. Exp. Biol.*, **27**, 377–99.

Batham, E.J. and Pantin, C.F.A. (1951) The organization of the muscular system of *Metridium senile* (L.). *Q. J. Microsc. Sci.*, **92**, 27–54 + 2 pl.

Batham, E.J. and Pantin, C.F.A. (1954) Slow contraction and its relation to spontaneous activity in the sea-anemone *Metridium senile* (L.). *J. Exp. Biol.*, **31**, 84–103.

Batham, E.J., Pantin, C.F.A. and Robson, E.A. (1960) The nerve-net of the sea-anemone *Metridium senile*: the mesenteries and the column. *Q. J. Microsc. Sci.*, **101**, 487–510.

Bayne, B.L. (1975) Aspects of physiological condition in *Mytilus edulis* L., with special reference to the effects of oxygen tension and salinity, in *Proceedings of the 9th European Marine Biological Symposium* (ed. H. Barnes), Aberdeen University Press, Aberdeen, pp. 213–38.

Bayne, B.L. (1976) The biology of mussel larvae, in *Marine Mussels* (ed. B.L. Bayne), Cambridge University Press, Cambridge, pp. 81–120.

Bayne, B.L., Thompson, R.J. and Widdows, J. (1976a) Physiology I. Feeding, digestion and respiration, in *Marine Mussels* (ed. B.L. Bayne). Cambridge University Press, Cambridge, pp. 121–206.

Bayne, B.L., Widdows, J. and Thompson, R.J. (1976b) Physiology II. Circulation and Excretion, in *Marine Mussels* (ed. B.L. Bayne). Cambridge University Press, Cambridge, pp. 207–60.

Beedham, G.E. (1958) Observations on the non-calcareous component of the shell of the Lamellibranchia. *Q. J. Microsc. Sci.*, **99**, 341–57.

Beedham, G.E. and Owen, G. (1965) The mantle and shell of *Solemya parkinsoni* (Protobranchia: Bivalvia). *Proc. Zool. Soc. Lond.*, **145**, 405–30.

Beentjes, M.P. and Williams, B.G. (1986) Endogenous circatidal rhythmicity in the New Zealand cockle, *Chione stutchburyi* (Bivalvia, Veneridae). *Mar. Behav. and Physiol.*, **12**, 171–80.

Bergquist, P.R. (1978) *Sponges*, Hutchinson, London.

Berrill, N.J. (1929) Digestion in ascidians and the influence of temperature. *J. Exp. Biol.*, **6**, 275–92.

Berrill, N.J. (1941) The development of the bud in *Botryllus*. *Biol. Bull. Mar. Biol. Lab., Woods Hole*, **80**, 169–84.

Berrill, N.J. (1949) The polymorphic transformations of *Obelia*. *Q. J. Microsc. Sci.*, **90**, 235–64.

Berrill, N.J. (1950) *The Tunicata*, Ray Society, London.

Berrill, N.J. (1961) *Growth, Development and Pattern*, W.H. Freeman, San Francisco.

Bigger, C.H. (1980) Interspecific and intraspecific acrorhagal aggressive behaviour among sea anemones: A, recognition of self and non-self. *Biol. Bull. Mar. Biol. Lab., Woods Hole*, **159**, 117–34.

Black, R. and Johnson, M.S. (1979) Asexual viviparity and population genetics of *Actinia tenebrosa*. *Mar. Biol.*, **53**, 27–31.

Bocquet-Védrine, J. (1965) Étude du tégument et de la mue chez le cirripède operculé *Elminius modestus* Darwin. *Arch. Zool. Exp. Gén.*, **105**, 30–76 + 2 pl.

Bocquet-Védrine, J. (1970) Contribution à l'étude du système cémentaire chez *Balanus crenatus* Brug., (crustacé, cirripède thoracique). *Arch Zool. Exp. Gén.*, **111**, 521–57.

Bouillon, J. (1968) Introduction to coelenterates, in *Chemical Zoology II. Porifera,*

Coelenterata, Platyhelminthes, (eds M. Florkin and B.T. Scheer), Academic Press, New York and London, pp. 81–147.

Bourget, E. and Crisp, D.J. (1975) Factors affecting deposition of shell in *Balanus balanoides. J. Mar. Biol. Ass. U.K.*, **55**, 231–49.

Brace, R.C., Pavey, J. and Quicke, D.L.J. (1979) Intraspecific aggression in the colour morphs of the anemone *Actinia equina*: the 'convention' governing dominance ranking. *Anim. Behav.*, **27**, 553–61.

Branch, G.M. (1984) Competition between marine organisms. Ecology and evolution. *Oceanogr. Mar. Biol. Ann. Rev.*, **22**, 429–593.

Brien, P. (1948) Embranchement des tuniciers, in *Traité de Zoologie*, Tome XI (ed. P.P. Grassé), Masson et Cie, Paris, pp. 553–930.

Brien, P. (1973) Les démosponges. Morphologie et reproduction, in *Traité de Zoologie*, Tome III, fasc. 1 (ed. P.P. Grassé), Masson et Cie, Paris, pp. 133–461.

Brill, B. (1973) Untersuchungen zur Ultrastruktur der Choanocyte von *Ephydatia fluviatilis* L. *Z. Zellforsch. mikrosk. Anat.*, **144**, 231–45.

Brown, C.H. (1950) Keratins in invertebrates. *Nature, Lond.*, **166**, 439.

Bubel, A. (1975) An ultrastructural study of the mantle of the barnacle, *Elminius modestus* Darwin, in relation to shell formation. *J. Exp. Mar. Biol. Ecol.*, **20**, 287–324.

Bullivant, J.S. (1968a) The rate of feeding of the bryozoan, *Zoobotryon verticillatum. N.Z. J. Mar. Freshwat. Res.*, **2**, 111–34.

Bullivant, J.S. (1968b) The method of feeding of lophophorates (Bryozoa, Phoronida, Brachiopoda). *N.Z. J. Mar. Freshwat. Res.*, **2**, 135–46.

Bullock, T.H. (1955) Temperature adaptation in poikilothermic animals. *Biol. Rev.*, **30**, 311–41.

Burighel, P. (1979) Peroxidase absorption in the ascidian gut. *J. Exp. Zool.*, **207**, 131–42.

Burighel, P. and Milanesi, C. (1973) Fine structure of the gastric epithelium of the ascidian *Botryllus schlosseri*, vacuolated and zymogenic cells. *Z. Zellforsch. Mikrosk. Anat.*, **145**, 541–55.

Burnett, B.R. (1977) Blood circulation in the balanomorph barnacle, *Megabalanus californicus* (Pilsbury). *J. Morph.*, **153**, 299–306.

Burnett, A.L., Diehl, N.A. and Diehl, F. (1964) The nervous system of *Hydra*. II Control of growth and regeneration by neurosecretory cells. *J. Exp. Zool.*, **157**, 227–36.

Burton, M. (1963) *A Revision of the Classification of the Calcareous Sponges*, Trustees of the British Museum (Natural History), London.

Buss, L.W., McFadden, C.S. and Keene, D.R. (1984) Biology of hydractiniid hydroids. 2. Histocompatibility effector system/competitive mechanism mediated by nematocyst discharge. *Biol. Bull. Mar. Biol. Lab., Woods Hole*, **167**, 139–58.

Campbell, R.D. (1967) Tissue dynamics of steady state growth in *Hydra littoralis*. II. Patterns of tissue movement. *J. Morph.*, **121**, 19–27.

Campbell, R.D. (1974) Development, in *Coelenterate Biology* (eds. L. Muscatine and H.M. Lenhoff), Academic Press, New York, pp. 1–91.

Campbell, R.D. (1980) Role of muscle processes in *Hydra* morphogenesis, in *Developmental and Cellular Biology of Coelenterates* (eds P. Tardent and R. Tardent), Elsevier/North-Holland Biomedical Press, Amsterdam, pp. 421–28.

Cariello, L., Salvato, B. and Jogi, G. (1980) Partial characterization of suberitine, the neurotoxic protein purified from *Suberites domuncula*. *Comp. Biochem. Physiol.*, **67B**, 337–44.

Carré, D. and Carré, C. (1980) On triggering and control of cnidocyst discharge. *Mar. Behav. Physiol.*, 7, 109–17.

Carriker, M.R. (1961) Comparative functional morphology of boring mechanisms in gastropods. *Am. Zool.*, **1**, 263–6.

Carriker, M.R., Charlton, G. and Zandt, D. van (1967) Gastropod *Urosalpinx*: pH of accessory boring organ while boring. *Science, N.Y.*, **158**, 920–22.

Carter, M.A. and Thorp, C.H. (1979) The reproduction of *Actinia equina* L. var. *mesembrianthemum*. *J. Mar. Biol. Ass. U.K.*, **59**, 989–1001.

Cawthorne, D.F. (1979) A comparative study of the closure response of some cirripede species exposed to falling sea water concentrations. *J. Mar. Biol. Ass. U.K.*, **59**, 811–17.

Cawthorne, D.F. and Davenport, J. (1980) The effects of fluctuating temperature, salinity, and aerial exposure upon larval release in *Balanus balanoides* and *Elminius modestus*. *J. Mar. Biol. Ass. U.K.*, **68**, 367–77.

Chadwick, N.E. (1987) Interspecific aggressive behaviour of the Corallimorpharian *Corynactis californica* (Cnidaria: Anthozoa): effect on sympatric corals and sea anemones. *Biol. Bull. Mar. Biol. Lab., Woods Hole*, **173**, 110–25.

Chapman, D.M. (1969) The nature of cnidarian desmocytes. *Tissue Cell*, **1**, 619–32.

Chapman, D.M. (1974) Cnidarian histology, in *Coelenterate Biology* (eds L. Muscatine and H.M. Lenhoff), Academic Press, New York, pp. 179–210.

Chia, F.S. (1976) Sea anemone reproduction: patterns and adaptive radiations, in *Coelenterate Ecology and Behaviour* (ed. G.O. Mackie), Plenum Press, New York, pp. 261–9.

Clare, A.S., Walker, G., Holland, D.L. and Crisp, D.J. (1985) The hatching substance of the barnacle, *Balanus balanoides* (L). *Proc. R. Soc.* **B**, **224**, 131–47.

Clark, S.D. and Cook, C.B. (1986) Inhibition of nematocyst discharge during feeding in the colonial hydroid *Halocordyle disticha* (= *Pennaria tiarella*): the role of previous prey killing. *Biol. Bull. Mar. Biol. Lab., Woods Hole*, **171**, 405–16.

Clemmesen, B. and Jørgensen, C.B. (1987) Energetic cost and efficiencies of ciliary filter feeding. *Mar. Biol.*, **94**, 445–9.

Cloney, R.A. (1977) Larval adhesive organs and metamorphosis in ascidians. 1. Fine structure of the everting papillae of *Distaplia occidentalis*. *Cell Tissue Res.*, **183**, 423–44.

Cloney, R.A. (1982) Ascidian larvae and the events of metamorphosis. *Am. Zool.*, **22**, 817–26.

Cody, M.L. (1966) A general theory of clutch size. *Evolution*, **20**, 174–84.

Connell, J.H. (1961) Effects of competition, predation by *Thais lapillus*, and other factors on natural populations of the barnacle *Balanus balanoides*. *Ecol. Monogr.*, **31**, 61–104.

Connell, J.H. (1970) A predator–prey system in the marine intertidal region. I. *Balanus glandula* and several predatory species of *Thais*. *Ecol. Monogr.*, **40**, 49–78.

Connell, J.H. (1972) Community interactions on marine rocky intertidal shores. *Ann. Rev. Ecol. Syst.*, **3**, 169–94.

Connes, R. (1967) Structure et développement des burgeons chez l'éponge siliceuse

Tethya lyncurium (Lamarck), recherches expérimentales et cytologiques. *Archs. Zool. Exp. Gén.*, **108**, 157–95.

Connes, R., Diaz, J.P. and Paris, J. (1971) Choanocytes et cellule centrale chez la démosponge *Suberites massa* Nardo. *C. R. Hebd. Séanc. Acad. Sci., Paris*, **273**, 1590–3.

Costlow, J.D. (1956) Shell development in *Balanus improvisus* Darwin. *J. Morph.*, **99**, 359–415 + pls 1–8.

Cox, G. (1983) Engulfment of *Prochloron* cells by cells of the ascidian, *Lissoclinum*. *J. Mar. Biol. Ass. U.K.*, **63**, 195–8.

Cranfield, H.J. (1974) Observations on the morphology of the mantle folds of the pediveliger of *Ostrea edulis* L. and their function during settlement. *J. Mar. Biol. Ass. U.K.*, **54**, 1–12.

Cranfield, H.J. (1975) The ultrastructure and histochemistry of the larval cement of *Ostrea edulis* L. *J. Mar. Biol. Ass. U.K.*, **55**, 497–503.

Crisp, D.J. (1974) Factors influencing the settlement of marine invertebrate larvae, in *Chemoreception in Marine Organisms* (eds P.T. Grant and A.M. Mackie), Academic Press, London and New York, pp. 177–265.

Crisp, D.J. (1976) Settlement responses in marine organisms, in *Adaptation to Environment : Essays on the Physiology of Marine Animals* (ed. R.C. Newell), Butterworths, London and Boston, pp. 83–124.

Crisp, D.J. and Barnes, H. (1954) The orientation and distribution of barnacles at settlement with particular reference to surface contour. *J. Anim. Ecol.*, **23**, 142–62.

Crisp, D.J. and Davies, P.A. (1955) Observations *in vitro* on the breeding of *Elminius modestus* grown on glass slides. *J. Mar. Biol. Ass. U.K.*, **34**, 357–80.

Crisp, D.J. and Lewis, A.H. (1984) Factors controlling cold tolerance and breeding in *Balanus balanoides*. *J. Mar. Biol. Ass. U.K.*, **64**, 125–45.

Crisp, D.J. and Patel, B.S. (1960) The moulting cycle in *Balanus balanoides* L. *Biol. Bull. Mar. Biol. Lab., Woods Hole*, **118**, 31–74.

Crisp, D.J. and Ritz, D.A. (1967) Changes in temperature tolerance of *Balanus balanoides* during its life cycle. *Helgoländer wiss. Meeresunters.*, **15**, 98–115.

Crisp, D.J. and Southward, A.J. (1961) Different types of cirral activity of barnacles. *Phil. Trans. R. Soc.*, **B**, **243**, 271–307.

Crisp, D.J. and Spencer, C.P. (1958) The control of the hatching process in barnacles. *Proc. R. Soc.* **B**, **148**, 278–99.

Cross, T.F. and Southgate, T. (1980) Mortalities of fauna of rocky substrata in South-West Ireland associated with the occurrence of *Gyrodinium aureolum* blooms during Autumn 1979. *J. Mar. Biol. Ass. U.K.*, **60**, 1071–3.

Crowell, S. (1953) The regression–replacement cycle of hydranths of *Obelia* and *Campanularia*. *Physiol. Zool.*, **26**, 319–27.

Crowell, S. (1957) Differential responses of growth zones to nutritive level, age, and temperature in the colonial hydroid *Campanularia*. *J. Exp. Zool.*, **134**, 63–90.

Dales, R.P. (1957) Some qualitative aspects of feeding in sabellid and serpulid fan worms. *J. Mar. Biol. Ass. U.K.*, **36**, 309–16.

Dales, R.P. (1967) *Annelids*, Hutchinson, London.

Darwin, C. (1854) *The Balanidae (or Sessile Cirripedes); the Verrucidae, etc. A Monograph on the Subclass Cirripedia*, Vol. 2, Ray Society, London.

Davenport, D., Ross, D.M. and Sutton, L. (1961) The remote control of nematocyst-discharge in the attachment of *Calliactis parasitica* to shells of hermit crabs. *Vie Milieu*, **12**, 197–209.

Davenport, J. (1976) A comparative study of the behaviour of some balanomorph barnacles exposed to fluctuating sea water concentrations. *J. Mar. Biol. Ass. U.K.*, **56**, 889–907.

Davenport, J. (1979) The isolation response of mussels (*Mytilus edulis* L.) exposed to falling sea-water concentrations. *J. Mar. Biol. Ass. U.K.*, **59**, 123–32.

Davenport, J. (1981) The opening response of mussels (*Mytilus edulis*) exposed to rising sea water concentrations. *J. Mar. Biol. Ass. U.K.*, **61**, 667–78.

Davenport, J. (1985) Osmotic control in marine animals, in *Physiological Adaptations of Marine Animals* (ed. M.S. Laverack), Symposia of the Society for Experimental Biology, 39, Company of Biologists Ltd, Cambridge, pp. 207–44.

Davson, H. (1959) *A Textbook of General Physiology*, J. and A. Churchill, London.

Denley, E.J. and Underwood, A.J. (1979) Experiments on factors influencing settlement, survival and growth of two species of barnacles in New South Wales. *J. Exp. Mar. Biol. Ecol.*, **36**, 267–93.

DeSanto, R.S. and Dudley, R.L. (1969) Ultra microscopic filaments in the ascidian *Botryllus schlosseri* (Pallas) and their possible role in ampulla contractions. *J. Ultrastruc. Res.*, **28**, 259–74.

De Silva, P.H.D.H. (1967) Studies on the biology of Spirorbinae (Polychaeta). *J. Zool., Lond.*, **152**, 269–79.

DeVries, A.L. (1974) Survival at freezing temperatures, in *Biochemical and Biophysical Perspectives in Marine Biology*, Vol. 1 (eds D.C. Malins and J.R. Sargent), Academic Press, London, New York and San Francisco, pp. 289–330.

De Zwaan, A. (1977) Anaerobic energy metabolism in bivalve molluscs. *Oceanogr. Mar. Biol. Ann. Rev.*, **15**, 103–87.

De Zwaan, A. (1983) Carbohydrate metabolism in bivalves, in *The Mollusca*. Vol. I. *Metabolic Biochemistry and Biomechanics* (ed. P.W. Hochachka), Academic Press, New York and London, pp. 137–75.

De Zwaan, A. and Putzer, V. (1985) Metabolic adaptations of intertidal invertebrates to environmental hypoxia. (A comparison of environmental anoxia to exercise anoxia), in *Physiological Adaptations of Marine Animals* (ed. M.S. Laverack), Symposia of the Society of Experimental Biology, 39, Company of Biologists Ltd, Cambridge, pp. 33–62.

De Zwaan, A. and Wijsman, T.C.M. (1976) Anaerobic metabolism in Bivalvia (Mollusca). Characteristics of anaerobic metabolism. *Comp. Biochem. Physiol.*, **54B**, 313–24.

De Zwaan, A., de Bont, A.M.Th., Zurburg, W., Bayne, B.L. and Livingstone, D.R. (1983) On the role of strombine formation in the energy metabolism of adductor muscle of a sessile bivalve. *J. Comp. Physiol.* **B**, **149**, 557–63.

Digby, P.S.B. (1968) The mechanism of calcification in the molluscan shell, in *Studies in the Structure, Physiology and Ecology of Molluscs* (ed. V. Fretter), Symposium 22, Zoological Society of London, Acadmic Press, London, pp. 93–107.

Dixon, D.R. (1980) The energetics of tube production by *Mercierella enigmatica* (Polychaeta: Serpulidae). *J. Mar. Biol. Ass. U.K.*, **60**, 655–9.

Duboscq, O. and Tuzet, O. (1937) L'ovogenèse, la fécondation et les premiers stades du développement des éponges calcaires. *Archs. Zool. Exp. Gén.*, **79**, 157–316.

Dudley, P.L. (1966) *Development and Systematics of Some Pacific Marine Symbiotic Copepods*, University of Washington Press, Seattle and London.

Eggleston, D. (1972) Patterns of reproduction in marine Ectoprocta of the Isle of Man. *J. Nat. Hist.*, **6**, 31–8.

Ellington, W.R. (1980) Partial purification and characterization of a broadly-specific octopine de-hydrogenase from the tissues of the sea anemone *Bunodosoma cavernata* (Bose). *Comp. Biochem. Physiol.*, **67B**, 625–31.

Ellington, W.R. (1982) Metabolic responses of the sea anemone *Bunodosoma cavernata* (Bose) to declining oxygen tensions and anoxia. *Physiol. Zool.*, **55**, 240–49.

Elyakova, L.A. (1972) Distribution of cellulases and chitinases in marine invertebrates. *Comp. Biochem. Physiol.*, **43B**, 67–70.

Endean, R. (1955a) Studies on the test of some Australian ascidians. I. The blood of *Pyura stolonifera* (Heller). *Aust. J. Mar. Freshwat. Res.*, **6**, 157–63, + 3 pls.

Endean, R. (1955b) Studies of the blood and test of some Australian ascidians. II. The test of *Pyura stolonifera* (Heller). *Aust. J. Mar. Freshwat. Res.*, **6**, 139–156.

Endean, R. (1955c) Studies of the blood and test of some Australian ascidians. III. The formation of the test of *Pyura stolonifera* (Heller). *Aust. J. Mar. Freshwat. Res.*, **6**, 157–61.

Endean, R. (1961) The test of the ascidian, *Phallusia mammillata*. *Q. J. Microsc. Sci.*, **102**, 107–17.

Ermak, T.H. (1977) Glycogen deposits in the pyloric gland of the ascidian *Styela clava* (Urochordata). *Cell Tissue Res.*, **176**, 47–55.

Ermak, T.H. (1982) The renewing cell populations of ascidians. *Am. Zool.*, **22**, 795–805.

Ewer, R.F. (1947) On the functions and mode of action of the nematocysts of *Hydra*. *Proc. Zool. Soc. Lond.*, **117**, 365–76.

Fjerdingstad, E.J. (1961) The ultrastructure of choanocyte collars in *Spongilla lacustris* (L). *Z. Zellforsch. mikrosk. Anat.*, **53**, 645–57.

Foster, B.A. (1971) Desiccation as a factor in the intertidal zonation of barnacles. *Mar. Biol.*, **8**, 12–29.

Foster, G.R. (1959) The ecology of *Echinus esculentus* L. Quantitative distribution and rate of feeding. *J. Mar. Biol. Ass. U.K.*, **38**, 361–7 + 2 pl.

Francis, L. (1973a) Clone-specific segregation in the sea anemone *Anthopleura elegantissima*. *Biol. Bull. Mar. Biol. Lab., Woods Hole.*, **144**, 64–72.

Francis, L. (1973b) Intraspecific aggression and its effect on the distribution of *Anthopleura elegantissima* and some related sea anemones. *Biol. Bull. Mar. Biol. Lab., Woods Hole*, **144**, 73–92.

Fretter, V. (1951) Some observations on the British Cypraeids. *Proc. Malac. Soc. Lond.*, **29**, 14–20.

Fuke, M.T. (1980) 'Contact reaction' between xenogeneic and allogeneic coelomic cells of solitary ascidians. *Biol. Bull. Mar. Biol. Lab., Woods Hole*, **158**, 304–15.

Fuke, M.T. and Nakamura, I. (1985) Pattern of cellular alloreactivity of the solitary ascidian, *Halocynthia roretzi*, in relation to genetic control. *Biol. Bull. Mar. Biol. Lab., Woods Hole*, **169**, 631–7.

Furth, A.J. (1980) *Lipids and Polysaccharides in Biology*, Edward Arnold, London.

Fyhn, H.J., Petersen, J.A. and Johansen, K. (1973) Heart activity and high pressure circulation in cirripedes. *Science, N.Y.*, **180**, 513–15.

Gäde, G. (1983) Energy metabolism of arthropods and mollusks during environmental and functional anaerobiosis. *J. Exp. Zool.*, **228**, 415–29.

Gadgil, M. and Solbrig, O.T. (1972) Concept of r- and K-selection: evidence from wild flowers and some theoretical considerations. *Am. Nat.*, **186**, 14–31.

Galtsoff, P.S. (1964) The American oyster *Crassostrea virginica* Gmelin. *Fishery Bull. Fish Wildl. Serv. U.S.*, **64**, 1–480.

Gashout, S.E. and Ormond, R.F.G. (1979) Evidence for parthenogenetic reproduction in the sea anemone, *Actinia equina* L. *J. Mar. Biol. Ass. U.K.*, **59**, 975–87.

Georges, D. (1978) Circadian rhythm in the neural gland of *Ciona intestinalis*, in *8th International Symposium on Comparative Endocrinology* (eds P.J. Guillard and H.H. Boer), Elsevier/North-Holland Biomedical Press, Amsterdam, pp. 161–4.

Gordon, J. and Carriker, M.R. (1980) Sclerotized protein in the shell matrix of a bivalve mollusc. *Mar. Biol.*, **57**, 251–60.

Goreau, T.F. (1959) The physiology of skeleton formation in corals. I. A method for measuring the rate of calcium deposition by corals under different conditions. *Biol. Bull. Mar. Biol. Lab., Woods Hole*, **116**, 59–75.

Goreau, T.F. and Goreau, N.I. (1959) The physiology of skeleton formation in corals. 2. Calcium deposition by hermatypic corals under various conditions in the reef. *Biol. Bull. Mar. Biol. Lab., Woods Hole*, **117**, 239–50.

Gosline, J. (1971) Connective tissue mechanisms in *Metridium senile*. *J. Exp. Biol.*, **55**, 775–95.

Gotelli, N.J. (1987) Spatial and temporal pattern of reproduction, larval settlement and recruitment of the compound ascidian *Aplidium stellatum*. *Mar. Biol.*, **94**, 45–51.

Grainger, F. and Newell, G.E. (1965) Aerial respiration in *Balanus balanoides*. *J. Mar. Biol. Ass. U.K.*, **45**, 469–79.

Grégoire, C. (1972) Structure of the molluscan shell, in *Chemical Zoology VII. Mollusca* (eds M. Florkin and B.T. Scheer), Academic Press, London and New York, pp. 45–102.

Grosberg, R.K. and Quinn, J.F. (1986) The genetic control and consequences of kin recognition by the larvae of a colonial marine invertebrate. *Nature, Lond.*, **322**, 456–9.

Gruffydd, Ll.D., Lane, D.J.W. and Beaumont, A.R. (1975) The glands in the base of the larval foot in *Pecten maximus* and possible homologies in other bivalves. *J. Mar. Biol. Ass. U.K.*, **55**, 463–76.

Gutmann, W.F. (1960) Funktionelle Morphologie von *Balanus balanoides*. *Abh. senckenb. naturforsch. Ges.*, **500**, 1–43.

Hand, C. (1966) On the evolution of the Actiniaria, in *The Cnidaria and their Evolution* (ed. W.J. Rees), Symposium 16, Zoological Society of London, Academic Press, London, pp. 135–46.

Harger, J.R.E. (1970) The effect of wave impact on some aspects of the biology of sea mussels. *Veliger*, **12**, 401–14.

Harger, J.R.E. (1972) Competitive co-existence among intertidal invertebrates. *Am. Scient.*, **60**, 600–7.

Hart, C.E. and Crowe, J.H. (1977) The effect of attached gravel on survival of intertidal anemones. *Trans. Am. Microsc. Soc.*, **96**, 28–41.

Hawkins, C.T., James, G.A., Parry, D.L., Swinehart, J.H. and Wood, A.L. (1983) Intracellular acidity in the ascidian. *Comp. Biochem. Physiol.*, **76B**, 559–65.

Haynes, J.F. (1973) Epithelial–muscle cells, in *The Biology of Hydra* (ed. A.L. Burnett), Academic Press, New York, pp. 233–7.

Hellebust, J.A. (1970) Light: Plants, in *Marine Ecology*. Vol. I, Pt. 1 (ed. O. Kinne), Wiley Interscience, London, New York, Sydney, Toronto, pp. 125–58.

Herdman, E.C. (1924) *Botryllus*. L.M.B.C. Memoir XXVI, Liverpool University Press, Hodder and Stoughton, London.

Hinde, R.A. (1970) *Animal Behaviour. A Synthesis of Ethology and Comparative Psychology*, McGraw-Hill, Kogakusha, Tokyo.

Holstein, T. (1980) An ultrastructural and kinematographic study of the nematocyst morphogenesis in selected coelenterate species, in *Development and Cellular Biology of Coelenterates* (eds. P. Tardent and R. Tardent), Elsevier/North-Holland Biomedical Press, Amsterdam, pp. 313–18.

Hoyle, G. (1952) The response mechanism in ascidians. *J. Mar. Biol. Ass. U.K.*, **31**, 287–305.

Hoyle, G. (1953) Spontaneous squirting of an ascidian, *Phallusia mammillata* Cuvier. *J. Mar. Biol. Ass. U.K.*, **31**, 541–62.

Hoyle, G. (1964) Muscle and neuromuscular physiology, in *Physiology of the Mollusca*. Vol. I (eds K.M. Wilbur and C.M. Yonge), Academic Press, New York, London, pp. 313–51.

Hunt, S. (1972) Scleroprotein and chitin in the exoskeleton of the ectoproct, *Flustra foliacea*. *Comp. Biochem. Physiol.*, **43B**, 571–7.

Istin, M. (1975) The structure and formation of calcified tissue, in *Biochemical and Biophysical Perspectives in Marine Biology* Vol. 2, (eds D.C. Malins and J.R. Sargent), Academic Press, London, New York, San Francisco, pp. 1–68.

Jackson, J.B.C. (1977) Competition on marine hard substrata: the adaptive significance of solitary and colonial strategies. *Am. Nat.*, **111**, 743–67.

Jackson, J.B.C. (1979) Morphological strategies of sessile animals, in *Biology and Systematics of Colonial Organisms* (eds G. Larwood and B.R. Rosen), Academic Press, London and New York, pp. 499–555.

Jah, Raj.K. and Mackie, G.O. (1967) The recognition, distribution and ultrastructure of hydrozoan nerve elements. *J. Morph.*, **123**, 43–62.

Johnston, I.S. and Hildemann, W.H. (1982) Cellular organization in the marine demosponge, *Callyspongia diffusa*. *Mar. Biol.*, **67**, 1–7.

Jones, W.C. (1958) The effect of reversing the internal water-current on the spicule orientation in *Leucosolenia variabilis* and *L. complicata*. *Q. J. Microsc. Sci.*, **99**, 263–78.

Jones, W.C. (1964a) Photographic records of living oscular tubes of *Leucosolenia variabilis*. I. The choanoderm boundary, the choanocytes and the pore arrangement. *J. Mar. Biol. Ass. U.K.*, **44**, 67–85.

Jones, W.C. (1964b) Photographic records of living oscular tubes of *Leucosolenia variabilis*. II. Spicule growth, form and displacement. *J. Mar. Biol. Ass. U.K.*, **44**, 311–31.

Jones, W.C. (1965) Photographic records of living oscular tubes of *Leucosolenia variabilis*. III. Irregular growth of the oscular tube. *J. Mar. Biol. Ass. U.K.*, **45**, 1–28.

Jones, W.C. (1970) The composition, development, form and orientation of calcareous sponge spicules, in *The Biology of the Porifera* (ed. W.G. Fry), Symposium 25, Zoological Society of London, Academic Press, London, pp. 91–123.

Jones, W.C., Pickthall, V.J. and Nesbitt, S.P. (1977) A respiratory rhythm in sea anemones. *J. Exp. Biol.*, **68**, 187–8.

Jørgensen, C.B. (1976) August Putter, August Krogh, and modern ideas on the use of dissolved organic matter in aquatic environments. *Biol. Rev.*, **51**, 291–328.

Jørgensen, C.B. (1981a) A hydromechanical principle for particle retention in *Mytilus edulis* and other ciliary suspension feeders. *Mar. Biol.*, **61**, 277–82.

Jørgensen, C.B. (1981b) Feeding and cleansing mechanism in the suspension-feeding bivalve, *Mytilus edulis. Mar. Biol.*, **65**, 159–63.

Jørgensen, C.B. (1983) Fluid mechanical aspects of suspension feeding. *Mar. Ecol. Progr. Ser.*, **11**, 89–103.

Jørgensen, C.B. and Goldberg, E.D. (1953) Particle filtration in some ascidians and lamellibranchs. *Biol. Bull. Mar. Biol. Lab., Woods Hole*, **105**, 477–89.

Kaplan, S.W. (1983) Intersexual aggression in *Metridium senile*. *Biol. Bull. Mar. Biol. Lab., Woods Hole*, **165**, 416–18.

King, P.E. (1973) *Pycnogonids*, Hutchinson, London.

Kinne, O. (1970) Temperature: invertebrates, in *Marine Ecology*, Vol. I, Pt. 1 (ed. O. Kinne), Wiley Interscience, London, New York, Sydney, Toronto, pp. 407–514.

Knight, D.P. (1968) Cellular basis for quinone tanning of the perisarc in the thecate hydroid *Campanularia* (= *Obelia*) *flexuosa*. *Nature, Lond.*, **218**, 584–6.

Knight, D.P. (1970) Sclerotization of the perisarc of the calyptoblastic hydroid, *Laomedia flexuosa*. 1. The identification and location of dopamine in the hydroid. *Tissue Cell*, **2**, 467–77.

Knight-Jones, E.W. (1953) Laboratory experiments on gregariousness during settlement in *Balanus balanoides* and other barnacles. *J. Exp. Biol.*, **30**, 584–99.

Knight-Jones, E.W. (1955) The gregarious settling reaction of barnacles as a measure of systematic affinity. *Nature, Lond.*, **174**, 266.

Knight-Jones, E.W. and Moyse, J. (1961) Intraspecific competition in sedentary marine animals, in *Mechanisms in Biological Competition* (ed. F.L. Milthorpe), Symposium 15, Society for Experimental Biology, Cambridge University Press, Cambridge, pp. 72–95.

Kott, P. (1963) The ascidians of Australia. IV. Aplousobranchiata Lahille: Polyclinidae Verrill (continued). *Aust. J. Mar. Freshwat. Res.*, **14**, 70–118.

Koyama, H. and Watanabe, H. (1982) Colony specificity in the ascidian, *Perophora sagamiensis*. *Biol. Bull. Mar. Biol. Lab., Woods Hole*, **162**, 171–81.

Kruijf, H.A.M. de (1977) Bursting pacemaker activity in the solitary hydroid *Tubularia solitaria*. *J. Exp. Biol.*, **68**, 19–34.

Lambert, C.C. (1982) The ascidian sperm reaction. *Am. Zool.*, **22**, 841–9.

Lane, D.J.W. and Nott, J.A. (1975) A study of the morphology, fine structure and histochemistry of the foot of the pediveliger of *Mytilus edulis* L. *J. Mar. Biol. Ass. U.K.*, **55**, 477–95.

Lane, D.J.W., Beaumont, A.R. and Hunter, J.R. (1985) Byssus drifting and the drifting threads of the young post-larval mussel, *Mytilus edulis*. *Mar. Biol.*, **84**, 301–8.

Lane, D.J.W., Nott, J.A. and Crisp, D.J. (1982) Enlarged stem glands in the foot of post-larval mussel, *Mytilus edulis*; adaptation for bysso-pelagic migration. *J. Mar. Biol. Ass. U.K.*, **62**, 809–18.

Langton, R.W. (1975) Synchrony in the digestive diverticula of *Mytilus edulis* L. *J. Mar. Biol. Ass. U.K.*, **55**, 221–9.

Langton, R.W. and Gabbott, P.A. (1974) The tidal rhythm of extracellular digestion and the response to feeding in *Ostrea edulis* L. *Mar. Biol.*, **24**, 181–7.

Larkman, V.N. and Gabbott, P.A. (1975) Settlement of cyprid larvae of *Balanus balanoides* and *Elminius modestus* induced by extracts of adult barnacles and other animals. *J. Mar. Biol. Ass. U.K.*, **55**, 183–90.

Lavioe, M.F. (1956) How sea stars open bivalves. *Biol. Bull. Mar. Biol. Lab., Woods Hole*, **111**, 114–22.

Lawn, I.D. (1975) An electrophysiological analysis of chemoreception in the sea anemone *Tealia felina*. *J. Exp. Biol.*, **63**, 525–36.

Lawn, I.D. (1980) A transmesogloeal conduction system in the swimming sea anemone *Stomphia*. *J. Exp. Biol.*, **83**, 45–52.

Lawn, I.D. (1982) Porifera, in *Electrical Conduction and Behaviour in 'Simple' Invertebrates* (ed. G.A.B. Shelton), Clarendon Press, Oxford, pp. 49–72.

Lawn, I.D. and McFarlane, I.D. (1976) Control of shell settling in the swimming sea anemone *Stomphia coccinea*. *J. Exp. Biol.*, **64**, 419–29.

Lentz, T.L. (1968) *Primitive Nervous Systems*, Yale University Press, New Haven, Connecticut.

Lewis, J.B. (1963) Environmental and tissue temperatures of some tropical intertidal marine animals. *Biol. Bull. Mar. Biol. Lab., Woods Hole*, **124**, 277–84.

Lewis, J.R. (1964) *The Ecology of Rocky Shores*, Hodder and Stoughton, London.

Lutaud, G. (1969) Le (plexus) pariétal de Hiller et la coloration du système nerveux par le bleu de méthylène chez quelques Bryozoaires Chilostomes. *Z. Zellforsch. mikrosk. Anat.*, **99**, 302–14.

Lutaud, G. (1973) L'innervation du lophophore chez le Bryozoaire Chilostome, *Electra pilosa* (L). *Z. Zellforsch. mikrosk. Anat.*, **140**, 217–34.

Lutz, R.A. and Rhoads, D.C. (1977) Anaerobiosis and a theory of growth line formation. *Science, N.Y.*, **198**, 1222–7.

MacArthur, R.H. (1972) *Geographical Ecology*, Harper and Row, New York.

McFarlane, I.D. (1969a) Two slow conduction systems in the sea anemone, *Calliactis parasitica*. *J. Exp. Biol.*, **51**, 377–85.

McFarlane, I.D. (1969b) Co-ordination of pedal-disc detachment in the sea anemone, *Tealia felina*. *J. Exp. Biol.*, **51**, 387–96.

McFarlane, I.D. (1970) Control of preparatory feeding behaviour in the sea anemone, *Tealia felina*. *J. Exp. Biol.*, **53**, 211–22.

McFarlane, I.D. (1974) Excitatory and inhibitory control of inherent contractions in the sea anemone *Calliactis parasitica*. *J. Exp. Biol.*, **60**, 397–422.

McFarlane, I.D. (1975) Control of mouth opening and pharynx protrusion during feeding in the sea anemone, *Calliactis parasitica*. *J. Exp. Biol.*, **63**, 615–26.

Mackie, G.O. and Singla, C.L. (1983) Co-ordination of compound ascidians by epithelial conduction in the colonial blood vessels. *Biol. Bull. Mar. Biol. Lab., Woods Hole*, **165**, 209–20.

Mackie, G.O. and Singla, C.L. (1987) Impulse propagation and contraction in the tunic of a compound ascidian. *Biol. Bull. Mar. Biol. Lab., Woods Hole*, **173**, 188–204.

Mackie, G.O., Anderson, P.A.V. and Singla, C.L. (1984) Apparent absence of gap junctions in two classes of cnidarians. *Biol. Bull. Mar. Biol. Lab., Woods Hole*, **167**, 120–23.

Mackie, G.O., Paul, D.H., Singla, C.L., Sleigh, M.A. and Williams, D.E. (1974) Branchial innervation and ciliary control in the ascidian *Corella*. *Proc. R. Soc.* **B**, **187**, 1–35.

Manahan, D.T. and Crisp, D.J. (1982) The role of dissolved organic matter in the nutrition of pelagic larvae: amino acid uptake by bivalve veligers. *Am. Zool.*, **22**, 635–46.

Mariscal, R.N. (1972) The nature of the adhesion to shells of the symbiotic sea anemone *Calliactis tricolor* (Leseur). *J. Exp. Mar. Biol. Ecol.*, **8**, 217–24 + 5 pl.

Mariscal, R.N. (1974) Nematocysts, in *Coelenterate Biology* (eds L. Muscatine and H.M. Lenhoff), Academic Press, New York, pp. 129–178.

Mariscal, R.N. and McLean, R.B. (1976) The form and function of cnidarian spirocysts. 2. Ultrastructure of the capsule tip and wall, and mechanism of discharge. *Cell Tissue Res.*, **169**, 313–21.

Mariscal, R.N., Conklin, E.J. and Bigger, C.H. (1977a). The ptychocyst, a major new category of cnida used in tube construction by a cerianthid anemone. *Biol. Bull. Mar. Biol. Lab., Woods Hole*, **152**, 392–405.

Mariscal, R.N., McLean, R.B. and Hand, C. (1977b) The form and function of cnidarian spirocysts. 3. Ultrastructure of the thread and the function of spirocysts. *Cell Tissue Res.*, **178**, 427–33.

Markham, J.B. and Ryland, J.S. (1987) Function of the gizzard in Bryozoa. *J. Exp. Mar. Biol. Ecol.*, **107**, 21–37.

Matricon, I. (1973) Quelques données ultrastructurelles sur un myoépithélium: le pharynx d'un Bryozoaire. *Z. Zellforsch. mikrosk. Anat.*, **136**, 569–78.

Menge, B.A. (1976) Organization of the New England rocky intertidal community: role of predation, competition and environmental heterogeneity. *Ecol. Monogr.*, **46**, 355–93.

Menon, N.R. (1974) Clearance rates of food suspension and food passage rates as a function of temperature in two North Sea bryozoans. *Mar. Biol.*, **24**, 65–7.

Mettam, C. (1969) Peristaltic waves of tubicolous worms and the problem of irrigation in *Sabella pavonina*. *J. Zool., Lond.*, **158**, 341–56.

Milkman, R. (1967) Genetic and developmental studies on *Botryllus schlosseri*. *Biol. Bull. Mar. Biol. Lab., Woods Hole.*, **132**, 229–43.

Millar, R.H. (1971) The biology of ascidians. *Adv. Mar. Biol.*, **9**, 1–100.

Miller, R.L. (1975) Chemotaxis of sperm of *Ciona intestinalis*. *Nature, Lond.*, **254**, 244–5.

Miller, R.L. (1982) Sperm chemotaxis in ascidians. *Am. Zool.*, **22**, 827–40.

Milne, A. (1961) Definition of competition among animals, in *Mechanisms in Biological Competition* (ed. F.L. Milthorpe), Symposium 15, Society for Experimental Biology. Cambridge University Press, Cambridge, pp. 40–61.

Moore, H.B. (1935) The biology of *Balanus balanoides*. IV. Relation to environmental factors. *J. Mar. Biol. Ass. U.K.*, **20**, 279–305.

Moore, H.B. (1938) The Biology of *Purpura lapillus*. Part III. Life history and relation to environmental factors. *J. Mar. Biol. Ass. U.K.*, **23**, 67–74 + 1 pl.

Morton, B. (1970) The tidal rhythm and rhythm of feeding and digestion in *Cardium edule*. *J. Mar. Biol. Ass. U.K.*, **50**, 499–512.

Morton, J.E. (1956) The tidal rhythm and action of the digestive system of the lamellibranch, *Lasaea rubra*. *J. Mar. Biol. Ass. U.K.*, **35**, 563–86.

Morton, J.E. (1960) The function of the gut in ciliary feeders. *Biol. Rev.*, **35**, 92–140.

Morton, J.E. (1963) *Molluscs*, Hutchinson University Library, London.

Morton, J.E. and Miller, M. (1968) *The New Zealand Sea Shore*, Collins, London.

Muscatine, L. (1974) Endosymbiosis of cnidarians and algae, in *Coelenterate Biology* (eds L. Muscatine and H.M. Lenhoff), Academic Press, New York, San Francisco and London, pp. 359–95.

Muscatine, L. and Lenhoff, H.M. (1965) Symbiosis of hydra and algae. II. Effects of limited food and starvation on growth of symbiotic and aposymbiotic hydra. *Biol. Bull. Mar. Biol. Lab., Woods Hole*, **129**, 316–28.

Nakauchi, M. (1982) Asexual development of ascidians: its biological significance, diversity and morphogenesis. *Am. Zool.*, **22**, 753–63.

Nakauchi, M. (1986) Oozoid development and budding in the polyclinid ascidian, *Parascidia flemingii* (Urochordata). *J. Zool., Lond.*, **208**, 255–67.

Nakauchi, M. and Kawamura, K. (1978) Additional experiments on the behaviour of buds in the ascidian, *Aplidium multiplicatum*. *Biol. Bull. Mar. Biol. Lab., Woods Hole*, **154**, 453–62.

Naylor, E. (1976) Rhythmic behaviour and reproduction in marine animals, in *Adaptation to Environment: Essays on the Physiology of Marine Animals* (ed. R.C. Newell), Butterworths, London and Boston, pp. 393–429.

Naylor, E. (1982) Tidal and lunar rhythms in animals and plants, in *Biological Timekeeping* (ed. J. Brady), Cambridge University Press, Cambridge, pp. 33–48.

Naylor, E. (1985) Tidally rhythmic behaviour of marine animals, in *Physiological Adaptations of Marine Animals* (ed. M.S. Laverack), Symposium of the Society for Experimental Biology, 39, Company of Biologists Ltd, Cambridge, pp. 63–93.

Naylor, E. and Williams, B.G. (1984) Environmental entrainment of tidally rhythmic behaviour in marine animals. *Zool. J. Linn. Soc.*, **80**, 201–8.

Neville, A.C. (1975) *Biology of the Arthropod Cuticle*, Springer-Verlag, New York, Heidelberg, Berlin.

Newell, R.C. (1965) The role of detritus in the nutrition of two marine deposit feeders, the prosobranch *Hydrobia ulvae* and the bivalve *Macoma balthica*. *Proc. Zool. Soc. Lond.*, **114**, 25–45.

Newell, R.C. (1979) *Biology of Intertidal Animals*, 3rd edn, Marine Ecological Surveys Ltd, Faversham, Kent.

Newman, W.A. (1960) On the paucity of intertidal barnacles in the tropical western Pacific. *Veliger*, **2**, 89–94 + 2 pl.

Newman, W.A. and Ross, A. (1976) Revision of the balanomorph barnacles; including a catalogue of the species. *Mem. S. Diego Soc. Nat. Hist.*, **9**, 1–108.

Nicol, E.A.T. (1930) The feeding mechanism, formation of the tube and physiology of digestion in *Sabella pavonina. Trans. R. Soc. Edinb.*, **56**, 537–98 + 2 pl.

North, B.B. (1975) Primary amines in California coastal waters: utilization by phytoplankton. *Limnol. Oceanogr.*, **20**, 20–27.

Nott, J.A. and Foster, B.A. (1969) On the structure of the antennular attachment organ of the cypris larva of *Balanus balanoides* (L.). *Phil. Trans. R. Soc.*, **B, 256**, 115–34.

Nott, J.A. and Parkes, K.R. (1975) Calcium accumulation and secretion in the serpulid polychaete *Spirorbis spirorbis* L. at settlement. *J. Mar. Biol. Ass. U.K.*, **55**, 911–23.

Oka, H. (1970) Colony specificity in compound ascidians. The genetic control of fusibility, in *Profiles of Japanese Science and Scientists* (ed. H. Yukawa), Kodansha, Tokyo, pp. 196–206.

Oka, H. and Watanabe, H. (1959) Vascular budding in *Botrylloides. Biol. Bull. Mar. Biol. Lab., Woods Hole*, **117**, 340–6.

Olson, R.R. (1986) Light enhanced growth of the ascidian *Didemnum molle*/*Procloron* sp., symbiosis. *Mar. Biol.*, **93**, 437–42.

Osborne, S. (1984) Bryozoan interactions: observations on stolonal outgrowths. *Aust. J. Mar. Freshwat. Res.*, **35**, 453–62.

Ottway, J.R. and Kirby, G.C. (1975) Genetic relationship between brooding and brooded *Actinia tenebrosa. Nature, Lond.*, **255**, 221–3.

Owen, G. (1974) Studies on the gill of *Mytilus edulis*: the eu-latero-frontal cirri. *Proc. R. Soc.* **B, 187**, 83–91.

Paine, R.T. (1974) Intertidal community structure. Experimental studies on the relationship between a dominant competitor and its principal predators. *Oecologia*, **15**, 93–120.

Pantin, C.F.A. (1935a) The nerve net of the Actinozoa. I. Facilitation. *J. Exp. Biol.*, **12**, 119–38.

Pantin, C.F.A. (1935b) The nerve net of the Actinozoa. II. Plan of the nerve net. *J. Exp. Biol.*, **12**, 139–55.

Pantin, C.F.A. (1935c) The nerve net of the Actinozoa. III. Polarity and after-discharge. *J. Exp. Biol.*, **12**, 156–64.

Pantin, C.F.A. (1942) The excitation of nematocysts. *J. Exp. Biol.*, **19**, 294–310.

Paparo, A.A. (1972) Innervation of the lateral cilia in the mussel, *Mytilus edulis* L. *Biol. Bull. Mar. Biol. Lab., Woods Hole*, **143**, 592–604.

Paparo, A.A. (1985) Morphological and physiological changes in the lamellibranch *Mytilus edulis*, after 6-OH DOPA administration. *Mar. Behav. and Physiol.*, **11**, 293–300.

Papi, F. (1960) Orientation by night: the moon. *Cold Spring Harb. Symp. Quant. Biol.*, **25**, 475–80.

Pardy, R. and Lewin, R.A. (1981) Colonial ascidians with prochlorophyte symbionts: evidence for translocation of metabolites from alga to host. *Bull. Mar. Sci.*, **31**, 817–23.

Parker, G.H. (1917) Actinian behaviour. *J. Exp. Zool.*, **22**, 193–229.

Patel, B.S. and Crisp, D.J. (1960) The influence of temperature on the breeding and moulting activities of some warm-water species of operculate barnacles. *J. Mar. Biol. Ass. U.K.*, **39**, 667–80.

Pavans de Ceccatty, M. (1966) Connexions cellulaires et jonctions polarisées du réseau intramesenchymateux, chez l'éponge *Hippospongia communis* Lamarck. *C. R. Hebd. Séanc. Acad. Sci., Paris*, **263D**, 145–7 + 2 pl.

Péquignat, E. (1973) The kinetic and autoradiographic study of the direct assimilation of amino acids and glucose by the organs of the mussel *Mytilus edulis*. *Mar. Biol.*, **19**, 227–44.

Petraitis, P.S. (1987) Immobilization of the predatory gastropod, *Nucella lapillus*, by its prey, *Mytilus edulis*. *Biol. Bull. Mar. Biol. Lab., Woods Hole*, **172**, 307–14.

Picken, L.E.R. (1953) A note on the nematocysts of *Corynactis viridis*. *Q. J. Microsc. Sci.*, **94**, 203–27 + 2 pl.

Picken, L.E.R. and Skaer, R.J. (1966) A review of researches on nematocysts, in *The Cnidaria and Their Evolution* (ed. W.J. Rees), Symposium 16, Zoological Society of London, Academic Press, London, pp. 19–50.

Price, H.A. (1980) Seasonal variation in the strength of byssal attachment of the common mussel *Mytilus edulis* L. *J. Mar. Biol. Ass. U.K.*, **60**, 1035–7.

Prosser, C.L. and Brown, F.A. (1961) *Comparative Animal Physiology*, Saunders, Philadelphia.

Purcell, J.E. (1977) Aggressive function and induced development of catch tentacles in the sea anemone *Metridium senile* (Coelenterata, Actiniaria). *Biol. Bull. Mar. Biol. Lab., Woods Hole*, **153**, 355–68.

Purcell, J.E. and Kitting, C.L. (1982) Intraspecific aggression and population distributions of the sea anemone *Metridium senile*. *Biol. Bull. Mar. Biol. Lab., Woods Hole*, **162**, 345–59.

Purchon, R.D. (1968) *The Biology of the Mollusca*, Pergamon Press, Oxford.

Quicke, D.L.J. and Brace, R.C. (1983) Phenotype and genotype spacing within an aggregation of the anemone, *Actinia equina*. *J. Mar. Biol. Ass. U.K.*, **63**, 493–515.

Quicke, D.L.J. and Brace, R.C. (1984) Evidence for the existence of a third, ecologically distinct morph of the anemone, *Actinia equina*. *J. Mar. Biol. Ass. U.K.*, **64**, 531–4.

Quicke, D.L.J., Donoghue, A.M. and Brace, R.C. (1983) Biochemical-genetic and ecological evidence that red/brown individuals of the anemone *Actinia equina* comprise two morphs in Britain. *Mar. Biol.*, **77**, 29–37.

Ralph, P.M. (1956) Variation in *Obelia geniculata* (Linnaeus, 1758) and *Silicularia bilabiata* (Coughtrey, 1875) (Hydroida, F. Campanulariidae). *Trans. R. Soc. N.Z.*, **84**, 279–96.

Ralph, P.M. and Thompson, H.G. (1968) Seasonal changes in growth of the erect stem of *Obelia geniculata* in Wellington Harbour, New Zealand. *Zoology Publs. Vict. Univ. Wellington*, **44**, 1–21.

Raymont, J.E.G. (1963) *Plankton and Productivity in the Oceans*, Pergamon Press, Oxford.

Rees, T.A.V. (1987) The green hydra symbiosis and ammonium. I. The role of the host in ammonium assimilation and its possible regulatory significance. *Proc. R. Soc.* **B, 229**, 299–314.

Reiswig, H.M. (1975) The aquiferous system of three marine Demospongiae. *J. Morph.*, **145**, 493–502.

Rinkevich, B. and Weissman, I.L. (1987a) A long term study of fused subclones in the ascidian *Botryllus schlosseri*: the resorption phenomenon (Protochordata: Tunicata). *J. Zool., Lond.*, **213**, 717–33.

Rinkevich, B. and Weissman, I.L. (1987b) The fate of *Botryllus* (Ascidiacea) larvae cosettled with parental colonies; beneficial or deleterious consequences? *Biol. Bull. Mar. Biol. Lab., Woods Hole*, **173**, 474–88.

Ritz, D.A. and Crisp, D.J. (1970) Seasonal changes in feeding rate in *Balanus balanoides*. *J. Mar. Biol. Ass. U.K.*, **50**, 223–40.

Robson, E.A. (1953) Nematocysts of *Corynactis*: the activity of the filament during discharge. *J. Microsc. Sci.*, **94**, 229–35.

Robson, E.A. (1957) The structure and hydromechanics of the musculo-epithelium in *Metridium*. *Q. J. Microsc. Sci.*, **98**, 265–78.

Ross, D.M. and Sutton, L. (1964) Inhibition of the swimming response by food and of nematocyst discharge during swimming in the sea anemone, *Stomphia coccinea*. *J. Exp. Biol.*, **41**, 751–7.

Rushforth, N.B. and Hofman, F. (1972) Behavioral and electrophysiological studies of *Hydra*. 3. Components of feeding behavior. *Biol. Bull. Mar. Biol. Lab., Woods Hole*, **142**, 110–31.

Russell, F.S. (1970) On a new species of medusa from an inland salt lake in South Australia. *J. Zool.*, Lond., **162**, 449–52.

Russell, F.S. (1970) On a new species of medusa from an inland salt lake in South Australia. *J. Zool., Lond.*, **162**, 449–52.

Ryland, J.S. (1959) Experiments on the selection of algal substrates by polyzoan larvae. *J. Exp. Biol.*, **36**, 613–31.

Ryland, J.S. (1970) *Bryozoans*, Hutchinson University Library, London.

Ryland, J.S. (1976) Physiology and ecology of marine bryozoans. *Adv. Mar. Biol.*, **14**, 285–443.

Sabbadin, A. (1971) Self- and cross-fertilization in the compound ascidian *Botryllus schlosseri*. *Devl Biol.*, **24**, 379–91.

Sabbadin, A. (1978) Genetics of the colonial ascidian, *Botryllus schlosseri*, in *Marine Organisms, Genetics, Ecology and Evolution* (eds B. Battaglia and J.A. Beardmore), Plenum Press, New York, pp. 195–209.

Saleuddin, A.S.M. (1979) Shell formation in molluscs with special reference to periostracum formation and shell regeneration, in *Pathways in Malacology* (eds S. Van der Spoel, A.C. Van Bruggen and J. Lever), Bohn, Scheltema and Holkema, Utrecht, pp. 47–81.

Saunders, D.S. (1977) *An Introduction to Biological Rhythms*, Blackie, Glasgow and London.

Schlichter, D. (1975) The importance of dissolved organic compounds in sea water for the nutrition of *Anemonia sulcata* Pennant (Coelenterata), in *Proceedings of the 9th European Marine Biological Symposium, 1975* (ed. H. Barnes) Aberdeen University Press, Aberdeen, pp. 395–405.

Schlichter, D. (1982) Nutritional strategies of cnidarians: the absorption, translocation and utilization of dissolved nutrients by *Heteroxenia fuscescens*. *Am. Zool.*, **22**, 659–69.

Seed, R. (1976) Ecology of marine mussels, in *Marine Mussels: Their Ecology and Physiology* (ed. B.L. Bayne), Cambridge University Press, Cambridge, pp. 13–65.

Segal, E. (1970) Light: invertebrate animals, in *Marine Ecology*. Vol. 1, Pt. 1 (ed. O. Kinne), Wiley Interscience, London, New York, Sydney, Toronto, pp. 159–211.

Shah, N.M. and Fogg, G.E. (1973) The determination of glycollic acid in sea water. *J. Mar. Biol. Ass. U.K.*, **53**, 321–4.

Shanks, A.L. (1986) Tidal periodicity in the daily settlement of intertidal barnacle larvae and an hypothesized mechanism for the cross-shelf transport of cyprids. *Biol. Bull. Mar. Biol. Lab., Woods Hole*, **170**, 429–40.

Shick, J.M. and Dykens, J.A. (1984) Photobiology of the symbiotic sea anemone *Anthopleura elegantissima*: photosynthesis, respiration, and behavior under intertidal conditions. *Biol. Bull. Mar. Biol. Lab., Woods Hole*, **166**, 608–19.

Silen, L. (1972) Fertilization in the Bryozoa. *Ophelia*, **10**, 27–34.

Silva: *see* De Silva

Silvester, N.R. and Sleigh, M.A. (1984) Hydrodynamic aspects of particle capture by *Mytilus*. *J. Mar. Biol. Ass. U.K.*, **64**, 859–79.

Simpson, T.L. (1984) *The Cell Biology of Sponges*, Springer-Verlag, New York.

Slutterbuck, D.B. (1961) Nematocyst development, in *The Biology of Hydra* (eds H.M. Lenhoff and W.F. Loomis), University of Miami Press, Coral Gables, Florida, pp. 77–129.

Smeathers, J.E. and Vincent, J.F.V. (1979) Mechanical properties of mussel byssus threads. *J. Moll. Stud.*, **45**, 219–30.

Smith, M.J. (1970) The blood cells and tunic of the ascidian, *Halocynthia aurantium* (Pallas). I. Hematology, tunic morphology and partitioning of cells between blood and tunic. *Biol. Bull. Mar. Biol. Lab., Woods Hole*, **138**, 354–78.

Smith, M.J. and Dehnel, P.A. (1970) The chemical and enzymatic analysis of the tunic of the ascidian *Halocynthia aurantium* (Pallas). *Comp. Biochem. Physiol.*, **35**, 17–30.

Somero, G.N. (1978) Interacting effects of temperature and pressure in enzyme function and evolution in marine organisms, in *Biochemical and Biophysical Perspectives in Marine Biology*. *4* (eds D.C. Malins and J.R. Sargent), Academic Press, London, New York and San Francisco, pp. 1–27.

Somero, G.N. and Hochachka, P.W. (1976) Biochemical adaptations to temperature, in *Adaptation to Environment: Essays on the Physiology of Marine Animals* (ed. R.C. Newell), Butterworth, London and Boston, pp. 125–90.

Southgate, T., Wilson, K., Cross, T.F. and Myers, A.A. (1984) Recolonization of a rocky shore in S.W. Ireland following a toxic bloom of the dinoflagellate, *Gyrodinium aureolum*. *J. Mar. Biol. Ass. U.K.*, **64**, 485–92.

Southward, A.J. (1958) Notes on the temperature tolerance of some intertidal animals in relation to environmental temperatures and geographical distribution. *J. Mar. Biol. Ass. U.K.*, **37**, 49–66.

Southward, A.J. (1964) The relationship between temperature and rhythmical cirral activity in some Cirripedia considered in connection with their geographical distribution. *Helgoländer wiss, Meeresunters.*, **10**, 391–403.

Stearns, S.C. (1976) Life-history tactics: a review of the ideas. *Q. Rev. Biol.*, **51**, 3–47.

Stebbing, A.R.D. (1971) The epizoic fauna of *Flustra foliacea* (Bryozoa). *J. Mar. Biol. Ass. U.K.*, **51**, 283–300.

Stebbing, A.R.D. (1973) Competition for space between epiphytes of *Fucus serratus* L. *J. Mar. Biol. Ass. U.K.*, **53**, 247–61.

Stephenson, T.A. and Stephenson, A. (1972) *Life Between Tidemarks on Rocky Shores*, W.H. Freeman and Co., San Francisco.

Stevenson, J.R. (1985) Dynamics of integument, in *Biology of the Crustacea*, Vol. 9, *Integument, Pigments and Hormonal Processes* (eds D.E. Bliss and L.H. Mantel), Academic Press, Orlando, Florida, pp. 1–42.

Stiévenart, J. (1971) Recherches sur la morphologie et étude histochimique de la tunique d'*Halocynthia pappilosa* Gun. (Ascidie Stolidobranche). *Annls Soc. R. Zool. Belg.*, **101**, 25–56.

Stoecker, D. (1980) Chemical defence of ascidians against predators. *Ecology*, **61**, 1327–34.

Storey, K.B. and Storey, J.M. (1983) Carbohydrate metabolism in cephalopod molluscs, in *The Mollusca*, Vol. 1, *Metabolic Biochemistry and Molecular Biomechanisms* (ed. P. Hochachka), Academic Press, New York and London, pp. 91–136.

Strathmann, R.R. (1973) Function of lateral cilia in suspension feeding of lophophorates (Brachiopoda, Phoronida, Ectoprocta). *Mar. Biol.*, **23**, 129–36.

Strathmann, R.R. (1974) The spread of sibling larvae of sedentary marine invertebrates. *Am. Nat.*, **108**, 29–44.

Stubbings, H.G. (1975) *Balanus balanoides*, L.M.B.C. Memoir 37, Liverpool University Press, Liverpool.

Tait, J. and Emmons, W.F. (1925) Experiments and observations on crustacea. Part VI. The mechanism of massive movement of the operculum of *Balanus nubilis*. *Proc. R. Soc. Edinb.*, XLV, 42–47.

Tamarin, A. (1975) An ultrastructural study of byssus stem formation in *Mytilus californianus*. *J. Morph.*, **145**, 151–78.

Taneda, Y., Saito, Y. and Watanabe, H. (1985) Self and non-self discrimination in ascidians. *Zoological Sci.*, **2**, 433–42.

Teragawa, C.K. (1986) Particle transport and incorporation during skeleton formation in a keratose sponge *Dysidea etheria*. *Biol. Bull. Mar. Biol. Lab., Woods Hole*, **170**, 321–34.

Thomas, N.W. (1970) Morphology of cell types from the gastric epithelium of *Ciona intestinalis*. *J. Mar. Biol. Ass. U.K.*, **50**, 737–46.

Thompson, T.E. and Slinn, D.J. (1959) On the biology of the opisthobranch *Pleurobranchus membranaceus*. *J. Mar. Biol. Ass. U.K.*, **38**, 507–24.

Thorndyke, M.C. (1977) Observations on the gastric epithelium of ascidians with special reference to *Styela clava*. *Cell Tissue Res.*, **184**, 539–50.

Thorpe, J.P., Shelton, G.A.B. and Laverack, M.S. (1975) Electrophysiology and co-ordinated behavioural responses in the colonial bryozoan, *Membranipora membranacea* (L). *J. Exp. Biol.*, **62**, 389–404.

Tighe-Ford, D.J. and Vaile, D.C. (1972) The action of crustecdysone on the cirripede *Balanus balanoides* (L). *J. Exp. Mar. Biol. Ecol.*, **9**, 12–28.

Trench, R.K. (1971) The physiology and biochemistry of zooxanthellae symbiotic with marine coelenterates. III. The effects of homogenates of host tissue on the excretion of photosynthetic products *in vitro* by zooxanthellae from two marine coelenterates. *Proc. R. Soc.* **B**, **177**, 251–64.

Tuzet, O. (1973) Introduction et place des Spongiaires dans la Classification. Éponges calcaires. in *Traité de Zoologie*. Tome III, fasc. 1 (ed. P.P. Grassé), Masson et Cie, Paris, pp. 27–132.

Underwood, A.J., Denley, E.J. and Moran, M.J. (1983) Experimental analyses of the structure and dynamics of mid-shore rocky intertidal communities in New South Wales. *Oecologia*, **56**, 202–19.

Vahl, O. (1972) Efficiency of particle retention in *Mytilus edulis* L. *Ophelia*, **10**, 17–25.

Van-Praët, M. (1985) Nutrition of sea anemones. *Adv. Mar. Biol.*, **22**, 65–99.

Vogel, S. (1974) Current-induced flow through the sponge, *Halichondria*. *Biol. Bull. Mar. Biol. Lab., Woods Hole*, **147**, 443–56.

Vries: *see* De Vries

Vyver: *see* Van de Vyver

Waite, J.H. (1983a) Adhesion in byssally attached bivalves. *Biol. Rev.*, **58**, 209–31.

Waite, J.H. (1983b) Quinone-tanned scleroproteins, in *The Mollusca*. Vol. 1, *Metabolic Biochemistry and Molecular Biomechanics* (ed. P.W. Hochachka), Academic Press, New York and London, pp. 467–504.

Walker, G. (1972) The biochemical composition of the cement of two barnacle species, *Balanus hameri* and *Balanus crenatus*. *J. Mar. Biol. Ass. U.K.*, **52**, 429–35.

Walker, G. (1980) A study of the oviducal glands and ovisacs of *Balanus balanoides* (L.), together with comparative observations on the ovisacs of *Balanus hameri* (Ascanius) and the reproductive biology of the two species. *Phil. Trans. R. Soc.*, **B**, **291**, 147–62.

Walker, G. and Yule, A.B. (1984) Temporary adhesion of the barnacle cyprid: the existence of an antennular adhesive secretion. *J. Mar. Biol. Ass. U.K.*, **64**, 679–686.

Walley, L.J., White, F. and Brander, K.M. (1971) Sperm activity and fertilization in *Balanus balanoides*. *J. Mar. Biol. Ass. U.K.*, **51**, 489–94.

Walters, L.T. and Wethey, D.S. (1986) Surface topography influences competitive hierarchies on marine hard substrate: a field experiment. *Biol. Bull. Mar. Biol. Lab., Woods Hole*, **170**, 441–9.

Watanabe, H. and Lambert, C.C. (1973) Larval release in response to light by the compound ascidians *Distaplia occidentalis* and *Metandrocarpa taylori*. *Biol. Bull. Mar. Biol. Lab., Woods Hole*, **144**, 556–66.

Watanabe, H. and Taneda, Y. (1982) Self and non-self recognition in compound ascidians. *Am. Zool.*, **22**, 775–82.

Webb, D.A. (1939) Observations on the blood of certain ascidians, with special reference to the biochemistry of vanadium. *J. Exp. Biol.*, **16**, 499–522.

Weill, R. (1934) Contribution à l'étude des cnidaires et de leurs nematocystes. *Trav. Stn Zool. Wimereux*, **10,11**, 1–701.

Westfall, J.A. (1970) Ultrastructure of synapses in a primitive coelenterate. *J. Ultrastruc. Res.*, **32**, 237–46.

Westfall, J.A., Yamataka, S. and Enos, P.D. (1970) Ultrastructural evidence of polarized synapses in the nerve net of *Hydra*. *J. Cell. Biol.*, **51**, 318–23 + 7 pl.

Wethey, D.S. (1984a) Spatial pattern of barnacle settlement: day to day changes during the settlement season. *J. Mar. Biol. Ass. U.K.*, **64**, 687–98.

Wethey, D.S. (1984b) Sun and shade mediate competition in the barnacles *Chthamalus* and *Semibalanus*: a field experiment. *Biol. Bull. Mar. Biol. Lab., Woods Hole*, **167**, 176–85.

White, K.M. (1937) *Mytilus*, L.M.B.C. Memoir 31, Liverpool University Press, Liverpool.

Whittaker, R.H. and Feeny, P.P. (1971) Allelochemics: chemical interactions between species. *Science, N.Y.*, **171**, 757–70.

Whittingham, D.G. (1967) Light induction of shedding of gametes in *Ciona intestinalis* and *Mogula manhattensis*. *Biol. Bull. Mar. Biol. Lab., Woods Hole*, **132**, 292–8.

Wiese, L. (1974) Nature of sex specific glycoprotein agglutinins in *Chlamydomonas*. *Ann. N.Y. Acad. Sci.*, **234**, 283–95.

Wilbur, K.M. (1964) Shell formation and regeneration, in *Physiology of Mollusca*. Vol. 1 (eds K.M. Wilbur and C.M. Yonge), Academic Press, New York and London, pp. 243–82.

Wilbur, K.M. (1972) Shell formation in molluscs, in *Chemical Zoology*. Vol. VII, *Mollusca* (eds M. Florkin and B.T. Schoer), Academic Press, New York and London, pp. 103–45.

Willenz, P. (1980) Kinetic and morphological aspects of particle ingestion by the freshwater sponge *Ephidatia fluviatilis* L., in *Nutrition in the Lower Metazoa* (eds D.C. Smith and Y. Tiffon), Pergamon Press, Oxford and New York, pp. 163–78.

Wilson, S.R. (1982) Horizontal and vertical density distribution of polychaete and cirripede larvae over an inshore rock platform off Northumberland. *J. Mar. Biol. Ass. U.K.*, **62**, 907–17.

Wineera, J.S. (1972) The body wall of the hydroid *Syncoryne tenella* (Farquhar, 1895). Part I, historical and histochemical observations. *Zoology Publs. Vict. Univ. Wellington*, **61**, 1–29.

Wisley, B. (1960) Observations on the settling behaviour of larvae of the tubeworm *Spirorbis borealis* Daudin (Polychaeta). *Aust. J. Mar. Freshwat. Res.*, **11**, 55–72.

Wolfson, A., Van Blaricom, G., Davis, N. and Lewbel, G.S. (1979) The marine life of an offshore oil platform. *Mar. Ecol. Prog. Ser.*, **1**, 81–9.

Woollacott, R.M. and Zimmer, R.L. (1975) A simplified placenta-like system for the transport of extraembryonic nutrients during embryogenesis of *Bugula neritina* (Bryozoa). *J. Morph.*, **147**, 355–78.

Wright, R.K. (1981) Urochordates, in *Invertebrate Blood Cells*, Vol. 2 (eds N.A. Ratcliffe and A.F. Rowley), Academic Press, London, pp. 327–641 + i–XLV.

Wright, S.H. (1982) A nutritional role for amino acid transport in filter feeding marine invertebrates. *Am. Zool.*, **22**, 621–34.

Wyttenbach, C.R. (1968) The dynamics of stolon elongation in the hydroid, *Campanularia flexuosa*. *J. Exp. Zool.*, **167**, 333–52.

Wyttenbach, C.R. (1969) Genetic variations in the mode of stolon growth in the hydroid *Campanularia flexuosa*. *Biol. Bull. Mar. Biol. Lab., Woods Hole*, **137**, 547–56.

Yanagita, T.M. and Wada, T. (1954) Effects of trypsin and thioglycollate upon the nematocysts of the sea anemone. *Nature, Lond.*, **173**, 171.

Yonge, C.M. (1962) On the primitive significance of the byssus in the Bivalvia and its effects in evolution. *J. Mar. Biol. Ass. U.K.*, **42**, 113–25.

Yonge, C.M. (1963) The biology of coral reefs. *Adv. Mar. Biol.*, **1**, 209–60.

Yonge, C.M. (1979) Cementation in bivalves, in *Pathways in Malacology* (eds S. Van der Spoel, A.C. Bruggen and J. Lever), Bohn, Scheltema and Holkema, Utrecht, pp. 83–106.

Young, C.M. and Braithwaite, L.F. (1980) Larval behaviour and post-settling morphology in the ascidian *Chelyosoma productum* Simpson. *J. Exp. Mar. Biol. Ecol.*, **42**, 157–69.

Yule, A.B. and Walker, G. (1984a) The adhesion of the barnacle, *Balanus balanoides*, to slate surfaces. *J. Mar. Biol. Ass. U.K.*, **64**, 147–56.

Yule, A.B. and Walker, G. (1984b) The temporary adhesion of barnacle cyprids: effects of some differing surface characteristics. *J. Mar. Biol. Ass. U.K.*, **64**, 429–39.

Yule, A.B. and Walker, G. (1985) Settlement of *Balanus balanoides*: the effect of antennular secretion. *J. Mar. Biol. Ass. U.K.*, **65**, 707–12.

Zamer, W.E. (1986) Physiological energetics of the intertidal sea anemone *Anthopleura elegantissima*. 1. Prey capture, absorption efficiency and growth. *Mar. Biol.*, **92**, 299–314.

Zamer, W.E. and Shick, J.M. (1987) Physiological energetics of the intertidal sea anemone *Anthopleura elegantissima*. 2. Energy balance. *Mar. Biol.*, **93**, 481–91.

Zs-Nagy and Ermini, M. (1972) A.T.P. production in the tissues of the bivalve *Mytilus galloprovincialis* (Pelecypoda) under normal and anoxic conditions. *Comp. Biochem. Physiol.*, **B**, **43**, 593–600.

Zwaan: *see* De Zwaan

Systematic index

Page numbers in italics refer to illustrations

Subject index

Page numbers in italics refer to illustrations